Electrochemical Sensors in Immunological Analysis

Electrochemical Sensors in Immunological Analysis

Edited by
T. T. Ngo
University of California, Irvine
Irvine, California

Springer Science+Business Media, LLC

Library of Congress Cataloging in Publication Data

Electrochemical sensors in immunological analysis.

Includes bibliographies and index.
1. Immunoassay. 2. Electrochemical analysis. I. Ngo, T. T. (That Tjien), 1944–
[DNLM: 1. Electrochemistry. 2. Immunoassay—methods. QW 570 E38]
QP519.9.I42E44 1987 574′.028 87-14116
ISBN 978-1-4899-1976-2

ISBN 978-1-4899-1976-2 ISBN 978-1-4899-1974-8 (eBook)
DOI 10.1007/978-1-4899-1974-8

Originally published by Plenum Press, New York in 1987
Softcover reprint of the hardcover 1st edition 1987

Dedicated to the memory of my parents

Mr. Ngo Nam Seng 伍南生 (1913–1984)

and

Mrs. Thjin Hie Foeng 陳喜鳳 (1913–1986)

PREFACE

The development of radioimmunoassay (RIA) by R.S. Yalow and S.A. Berson in 1959 opens up a new avenue in ultra-sensitive analysis of trace substances in complex biological systems. In recognition of the enormous contributions of RIA to basic research in biology and to routine clinical tests in laboratory medicine, R.S. Yalow, the co-developer of RIA, was awarded, in 1977, the Nobel Prize for Medicine and Physiology. The basic principle of RIA is elegantly simple. It is based on a specific, competitive binding reaction between the analyte and the radio-labeled analog of the analyte for the specific antibody raised to the analyte. The combination of high specificity and affinity of an antibody molecule makes it a very versatile analytical reagent capable of reacting specifically with analytes at a very low concentration in a complex solution such as serum. The sensitivity of RIA is provided by using a radioactive tracer. RIA is a general technique applicable to the analysis of a wide range of substances provided that the antibodies to these substances can be obtained. The two essential reagents in a RIA are (1) the antibody raised to an analyte and (2) a radio-labeled analyte or its analog. RIA is typically performed by simply mixing isotopically labeled analytes and the specific antibody with the sample, and then separating the radio-labeled and unlabeled analytes (i.e. the analyte to be measured) that are unbound to the antibody from the antibody-bound ones and finally counting the radioactivity associated with either fraction. The radioactivity is related to the amount of the analyte by reference to a standard curve.

Recently, there are intense research efforts to simplify immunoassays and to expand the applications of this technology into the physician's office, field-use in environmental, agricultural and veterinary monitoring and control where the use of radioactive labels is impractical or economically infeasible. These concerted research efforts have led to the development of a number of ingenious non-isotopic labels capable of simplifying the assay by replacing radioisotopic labels. Some non-isotopic labels have advantages of allowing the development of a separation-free (homogeneous) assay format, i.e. a physical separation of the antibody-bound from the unbound analytes is not required. This is in contrast to the heterogeneous assay which requirs a separation step at some point in the assay. Potential health and environmental hazards associated with the use of radioisotopes and the limited shelf life of reagents labeled with isotopes have further stimulated wider applications of these non-isotopic labels in immunoassay systems. The most widely used non-isotopic labels are enzymes, fluorescent molecules and bio-and chemi-luminogenic reagents. These labels have been measured mostly by spectrophotometric methods. Many of these labels can also be conveniently and sensitively measured electrochemically. The advantages of using electrochemical sensors over spectrophotometers are: (1) the capability of making measurements in a highly turbid sample; (2) excellent detection limits and (3) the low cost and simplicity of electrochemical devices.

The purpose of this volume is to bring together works of scientists of diverse backgrounds and to focus attention on concepts, methodology, limitations and future potentials of electrochemical sensors in immunological analysis. Topics includes (1) direct reading immunosensor devices; (2) separation-free (homogeneous) enzyme mediated immunoassays with electrochemical detections; (3) liposome mediated electrochemical immunoassays; (4) applications of ion selective membrane electrodes in immunoassays; (5) the use electric pulse to accelerate antigen-antibody interactions in immunoassays and (6) the combined use of a flow system and an electrochemical detector to monitor the immunochemical reaction.

This volume should complement the recently published "Enzyme-Mediated Immunoassay", edited by T.T. Ngo and H.M. Lenhoff, 1985, Plenum Press. Both volumes should be useful to clinical chemists, clinicians, biochemists, analytical chemists, toxicologists and biologists.

I am grateful to my wife, Ping Ying, for her understanding, encouragements and patience while I spent seemingly endless evenings and week-ends working on this volume. To my brothers and sisters, in Indonesia and Canada, who suffered with me the loss of our parents within the last two years, I wish them well and share with them whatever recognition this work may bring.

December, 1986. T.T. Ngo

CONTENTS

ON THE DIRECT IMMUNOCHEMICAL POTENTIOMETRIC SIGNAL

Michael Thompson, Joseph S. Tauskela and Ulrich J. Krull
Chemical Sensors Group, Department of Chemistry
University of Toronto
Toronto, Ontario, M5S 1A1, Canada

INTRODUCTION

In recent years, there have been several attempts to apply the high selectivity of an immunochemical reaction towards the development of an electrode which could measure solution concentrations of antigen or antibody. Conventional ion selective electrode (ISE) technology has not been used to design an ISE immunoelectrode per se because the molecular size of the immunochemical species places special requirements on the electrochemical properties of a sensing membrane; i.e., it would be difficult to build a membrane in such a way that it would allow high exchange current density of the immunochemical species of interest and exclude the permeation of small inorganic ions. Thus, since the mechanism that makes ion selective electrodes selective can not be directly applied in the case of an immunochemical reaction, several researchers have resorted to modifying the conditions under which a conventional ISE works so that an indirect determination of antibody/antigen concentration can be made. As the example electrode designs given in Table 1 show, the way in which these indirect immunoelectrodes function is the local ion activity that is measured at the ISE surface becomes modified in a way that is directly proportional to the antibody/antigen concentration. For instance, in the enzyme immunoelectrode technique, the antibody-bound enzyme is only activated by the specific antibody-antigen reaction, thereby releasing a potentiometrically measureable ionic product. The antigen-ionophore electrodes represent another indirect approach

Table I

Examples of the use of ion selective electrodes (ISE) for the indirect determination of antibody/antigen concentrations

Method	ISE	Antibody/Antigen	References
Enzyme immunoassay	I^-	hepatitis B	Boitieux et al., 1979, 1981
	F^-	IgG	Alexander and Maitra, 1982
	$TMPA^+$	rabbit γ-globulin	Meyerhoff and Rechnitz, 1977
	NH_4^+	human serum albumin	Brontman and Meyerhoff, 1984
	NH_3 gas electrode	dinitrophenol	Gebauer and Rechnitz, 1981
	CO_2 gas electrode	digoxin	Keating and Rechnitz, 1985
Antigen-ionophore fixation	K^+	dinitrophenol, bovine serum albumin	Solsky and Rechnitz, 1979
	K^+	dinitrophenol	Solsky and Rechnitz, 1981
	K^+	cortisol	Keating and Rechnitz, 1983
	K^+	digoxin	Keating and Rechnitz, 1984
	H^+	quinidine	Bush and Rechnitz, 1986
Antigen-liposome fixation	$TMPA^+$	hemolysin	D'Orazio and Rechnitz, 1977, 1979
	TPA^+	serum	Shiba et al., 1980a
	TPA^+	ε-DNP aminocuproyl PE	Shiba et al., 1980b
	TPA^+	cardiolipin	Umezawa et al., 1981
	TPA^+	C-reactive protein (CRP)	Umezawa et al., 1983
	TPA^+	syphilis	Umezawa et al., 1984
Precipitin formation	Ag_2S	human serum albumin	Alexander and Rechnitz, 1974

since the membranes of the ISE's have been modified to incorporate an antigen-(K^+) carrier conjugate so that reaction with specific antibodies causes an apparent change in K^+ activity that is in turn proportional to the antibody concentration. Another approach to indirectly measure solution antibody/antigen concentrations is achieved by the use of electroactive marker-loaded sensitized liposomes. The liposomes respond to varying levels of antibody or complement through a process that involves lysis of the liposomes, allowing the marker to escape and subsequently be detected by the ISE. A number of these indirect systems are reviewed by other authors in this text.

Although these methods have been shown to be sensitive measures of antigen/antibody concentration under very controlled conditions, much research has proceeded on an altogether different design of immunoelectrode. Development of the "direct" electrode has been based upon the fact that since antibodies and some antigens have a net electrical charge, the charge of the resulting complex will be different from that of the antibody or antigen alone. Thus, by attaching the antibody or antigen to an electrode that is capable of measuring surface charge density, the resulting immunochemical reaction that takes place at the interface would be detected (Figure 1). Furthermore, since the reaction between antibody and antigen is an equilibrium process, the potential difference between the reference electrode and immunoelectrode should depend on the concentration of the corresponding free immunochemical counterpart in solution.

This idea has been tested by several researchers and, based on these studies, it has become apparent that several problems must first be overcome before a reliable direct-measuring immunoelectrode becomes a reality. A significant problem has been the fact that a number of "immunoelectrode" studies have included only a cursory explanation of the processes occurring at the electrode interface. The purpose of the present paper is to address this issue and outline the specific constraints involved in the development of this type of device.

DETERMINATION OF CHARGE AT AN IDEALLY POLARIZED INTERFACE - A BRIEF REVIEW

The type of interface that is desired for a direct-measuring immunoelectrode is one which will not allow charge to cross from the bulk of solution to the bulk of the immobilizing medium, (the nature of the latter is discussed in more detail later). Such an interface is said to be blocked, or polarized, and is electrically equivalent to a capacitor. Because ion and electron exchange do not occur, any potential developed at an ideally polarizable interface arises due to nonfaradaic processes of accumulation of charge (diffuse and specifically adsorbed or bound) and/or through alignment of dipoles.

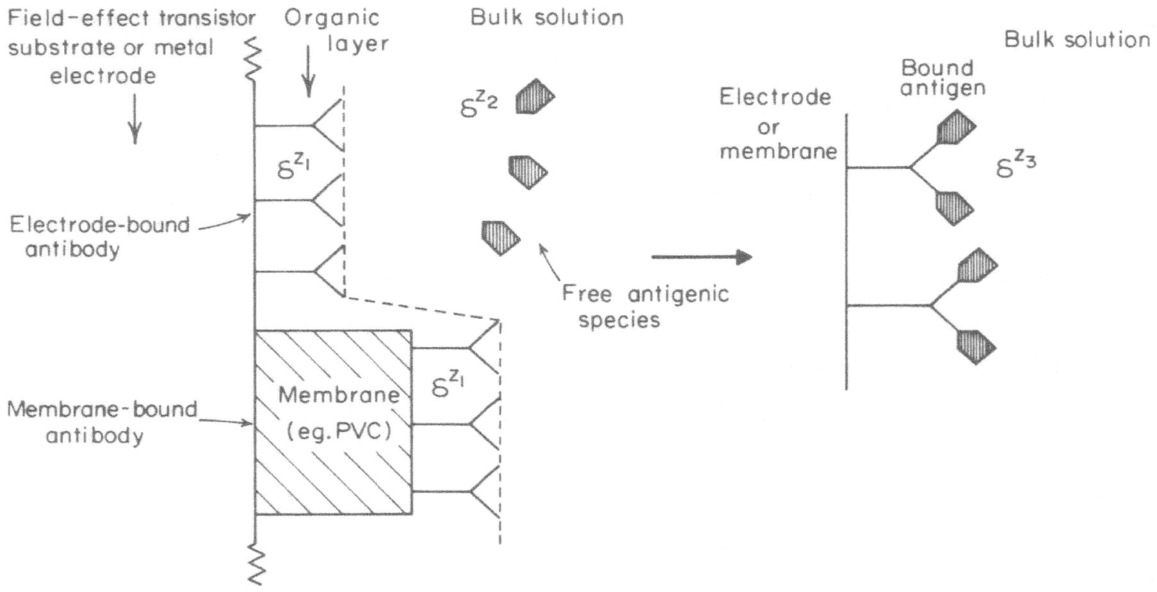

Figure 1. The concept of a "direct" potentiometric immunoelectrode in its simplest form. Immunochemical reaction between an immobilized antibody of charge $\delta^{Z}1$, and a free antigen of charge $\delta^{Z}2$, gives rise to a new charge of $\delta^{Z}3$.

Based upon this premise, immunoelectrode devices have been fashioned out of both conventional metal electrodes and field effect transistors (FET's). For molecular recognition, the biospecific ligand (antibody or antigen) is covalently bound to an electrode or to the surface of a thin layer of a hydrophobic polymer which, in turn, is deposited on an appropriate substrate. In a somewhat analogous fashion, such ligands have also been attached to Langmuir-Blodgett films. The following short critique will show that the potential measured by both device types will theoretically be equivalent, although the actual measuring mechanism will be quite different.

If we apply the rigorous condition that an interface is ideally polarized, it can be described by the Gibbs-Lippmann equation (Janata, 1978, 1983):

$$d\gamma = q^M dE + \sum_i \Gamma_i d\mu_i \tag{1}$$

where γ is surface energy, q^M is the charge on the (metal) electrode, E is the interfacial potential, Γ_i is the surface excess of adsorbed ion i, and μ_i is the chemical potential of adsorbed species i.

The change of the interfacial potential with the activity of the adsorbing ion in bulk solution, a_i, at constant charge on the metal, is the Esin-Markov coefficient and it can be derived from equation (1) giving:

$$\left(\frac{\partial E}{\partial \ln a_i}\right)_{q^M} = -RT \left(\frac{\partial \Gamma_i}{\partial q^M}\right)_{a_i} \tag{2}$$

With conventional metal electrodes, the Esin-Markov coefficient can be evaluated only indirectly by measuring the differential capacitance at different applied potentials and different concentrations of adsorbing species; the reasoning behind this is that since the Gibbs-Lippman equation is the equation of state, the following relationship holds for the state variables:

$$\left(\frac{\Delta E}{\delta \mu}\right)_{q^M} \left(\frac{\delta \mu_i}{\delta q^M}\right)_E \left(\frac{\delta q^M}{\delta E}\right)_{\mu_i} = -1 \tag{3}$$

where the first term is the Esin-Markov coefficient and the third is the differential capacitance. Rearranging gives:

$$\left(\frac{\delta E}{\delta \ln a_i}\right)_{q^M} = \frac{1}{C_d} \left(\frac{\delta q^M}{\delta \ln a_i}\right)_E \tag{4}$$

where C_d is the differential capacitance. Experimentally this means the measurement of interfacial potential at constant charge has been substituted for the measurement of interfacial charge at constant potential.

The ChemFET, on the other hand, can be used for direct determination of the Esin-Markov coefficient because it is basically a charge measuring device (Janata, 1980). The transistor gate potential, V_G, consists of several components and they are as follows:

$$V_G = \phi_R + \phi_{dl} + \phi_{ins} + \phi_s \qquad (5)$$

where ϕ_R, ϕ_{dl}, ϕ_{ins} and ϕ_s represent potential differences across the reference system, the double layer present at the insulator-solution interface, the insulator and the insulator-semiconductor interface, respectively. Clearly ϕ_{dl} is directly related to V_G if the other potentials can be considered as constants, (ϕ_{ins} can be assumed to be constant as long as there is constant interfacial charge q_i). Since the double layer potential is identical to the interfacial potential term (E) in the Gibbs-Lippman equation, the change of ϕ_{dl} with bulk activity of adsorbing species is actually the Esin-Markov coefficient:

$$\left(\frac{\delta E}{\delta \ln a_i}\right)_{P,T,q_i} = \left(\frac{\delta \phi_{dl}}{\delta \ln a_i}\right)_{P,T,q_i} = \left(\frac{\delta V_G}{\delta \ln a_i}\right)_{P,T,q_i} \qquad (6)$$

For measurement purposes involving a given V_G and drain current (I_D) the effect of adsorption of cations at the interface is to increase I_D. Equation (6) states that the change in ϕ_{dl} can only be measured at constant interfacial charge so the drain current is readjusted back to its original value by adjusting V_G. This new value of V_G thus gives a measurement of the interfacial potential drop.

This theory indicates that it might be possible to measure the change in interfacial charge density due to an immunochemical reaction of antigen bound to a device surface with free antibody in solution. It is important to recall however, that this treatment is only strictly valid for an ideally polarized interface. Such a condition implies that the "organic"-to-solution interface can be considered to be no more than a special case of the metal electrode-to-solution model.

ATTEMPTS AT IMPLEMENTATION OF THE POTENTIOMETRIC MODEL

The first direct immunoelectrode was described by Janata (1975). In this device the lectin concanavalin A was covalently attached to a film of 5 µm thick polyvinyl chloride which was deposited on a platinum wire. This particular protein is not an antibody, but the nature of its interaction with certain polysaccharides has been used extensively as a model for the immunochemical system. The results of this study did indeed confirm the development of change in electrode potential upon reaction of the bound lectin with yeast mannan.

Aizawa and co-workers (1977), showed that an immunochemical reaction between membrane-bound antigen with free antibody could also induce a potential shift. The antigen used was cardiolipin (immobilized to a triacetyl cellulose membrane) which, in the presence of cholesterol and phosphatidylcholine, reacted with the Wasserman (syphilis) antibody. A relationship between the immunochemically induced potential change and the concentration of the antibody solution was demonstrated. The proposition was made that the potential change resulted from a change in the charge density as a direct result of specific adsorption of antibody onto the membrane-bound antigen. Furthermore, the possible occurrence of non-specific adsorption was recognized (since the antigen-binding membrane was exposed to other proteins in contact with the serum), but this effect was minimized by extensively washing the reacted membrane with physiological saline and water and waiting (about 30-60 minutes) until the transmembrane potential reached a steady-state value. Finally, the important observation was made that the PVC/antigen membrane also exhibited sensitivity to concentration changes of inorganic ions.

Collins and Janata (1982) effectively repeated this work with antibody-containing serum and its interaction with a PVC/antigen membrane. This study concentrated on a comparison of the potentiometric response of the electrode to an antibody-free serum. They found the magnitude of the response was as large for the latter as for the genuine antibody-antigen case, and this was attributed to the effect of non-specific adsorption on an experimentally observed mixed potential. The significant conclusion of this study was that current could pass through the PVC/antigen combination (with an exchange current of about 10^{-6} A cm^{-2}) and, therefore, the membrane was non-polarized by definition. On a qualitative level this result implies that observed changes in potential can not be solely ascribed to the build-up of interfacial charge, but are also caused by perturbation of ion crossings at the interface. In view of this result, the next stage in the rationale concerning the development of the so-called immunoelectrode was to perform immunochemistry at the interface of a Langmuir-Blodgett film (multilayer) held at the gate of a FET (Janata, 1984). However, similar conclusions were drawn to those described above in that, apparently, there is sufficient partition of ions into LB films to classify them as non-polarized and to attribute experimental non-selectivity to an analogous generation of mixed potentials.

Finally, we note that Yamamoto and co-workers (1978, 1980, 1983) have carried out the direct attachment of protein to titanium electrode surfaces for the study of potentiometric detection of complementary biological species. However, in this work the difficulties alluded to by others are not described.

PROCESSES AT NON-POLARIZED INTERFACES

It is clear from the experiments outlined in the previous section that, in the case of the immunoelectrode, the Gibbs-Lippman expression can not be employed to describe changes in interfacial potential. In fact, a new model is required which encompasses both the immunochemistry and the ion crossings that occur at the interface. With respect to the physical chemistry of any such interplane, it is very important to recognize that the interface in question will be located between the organic layer and solution. Accordingly, we can consider ion transfer into a close-packed, directly immobilized biological species in the same fashion as we would deal with an auxiliary layer such as PVC or an LB film.

The quantitative aspects of interfacial electrochemistry at stake here have been excellently described by Buck (1982) and we reproduce his arguments here. For ion crossings or ion attachments, the individual rate constants (k_f and k_b) are described as:

$$k_f = k_f^{eq} \exp[-\frac{\alpha zF}{RT}(\Delta\phi - \Delta\phi^{eq}) \qquad (7)$$

$$k_b = k_b^{eq} \exp[\frac{(1-\alpha)zF}{RT}(\Delta\phi - \Delta\phi^{eq})] \qquad (8)$$

These equations are potential dependent, and invoke the usual electrochemical terminology, i.e. α is the symmetry coefficient, z is the charge, $\Delta\phi$ is the mixed potential at steady state and $\Delta\phi^{eq}$ is the overall membrane potential at equilibrium.

The partitioning of a single negative or positive ion corresponds to a hypothetical, unmeasurable single ion extraction equilibrium constant, $K_{\pm}$. This equilibrium state is described in terms of the above rate constants and of the intrinsic thermodynamic stabilities of the ions in each phase in their standard states ($\mu^{o}{}_{\pm}$):

$$K_{\pm} = \left(\frac{k_f}{k_b}\right)^{eq} = \frac{\overline{a}}{a}\left[\exp \frac{zF\Delta\phi^{eq}}{RT}\right] = \exp\left[(\mu_{\pm}{}^{o} - \overline{\mu_{\pm}{}^{o}})/RT\right] \qquad (9)$$

where the overbar indicates an organic layer parameter and, as usual, a is the ion activity, from which follows the Nernst equation:

$$\Delta\phi^{eq} = \frac{RT}{zF} \ln\left(K_{\pm}\frac{a}{\overline{a}}\right) \qquad (10)$$

When an "interfering" ion of the same charge is also present, the equilibrium interfacial potential difference is given by:

$$\Delta\phi^{eq} = \frac{RT}{zF} \ln(a_1 + k_{i,exch}a_2) + \text{constant} \qquad (11)$$

where a_1 and a_2 are the activities of the two ions now in question, and $k_{i,exch}$ is the ion exchange constant for the reaction (positive ions):

$$(\text{ion } 1)^{+} + \overline{(\text{ion } 2)^{+}} \rightleftharpoons (\text{ion } 2)^{+} + \overline{(\text{ion } 1)^{+}} \qquad (12)$$

$$\text{and } k_{i,exch} = \frac{K_{\pm}(\text{ion } 1)^{+}}{K_{\pm}(\text{ion } 2)^{+}} \qquad (13)$$

Thus, two or more ions of equal charge will exchange with a measurable exchange constant, which is a ratio of single ion partition constants. Essentially this means that ion exchange is merely a manifestation of the relative extraction coefficients, which in turn reflects the intrinsic relative stabilities of ions in the two phases.

Extending this to a situation which is close to real conditions, that is, the occurrence of two ions of equal but opposite charge being partitioned at equilibrium, the interfacial potential difference is now given by:

$$\Delta\phi^{eq} = \frac{RT}{zF} \ln \frac{K_+\gamma_+\overline{\gamma_-}}{K_-\gamma_-\overline{\gamma_+}} \tag{14}$$

This single-ion interfacial potential difference expression has the usual logarithmic Nernstian form, with γ_+ and γ_- being the activity coefficients of the positive and negative ions, respectively. As this equation shows, when two ions of equal but opposite charge can cross an interface, the equilibrium partitioning leads to a potential difference that is dependent on the relative magnitudes of the individual ionic currents. Figure 2(a) depicts this situation for some arbitrarily chosen values of K_+ and K_-, such that the sum of the ionic currents gives rise to a mixed potential (Janata and Blackburn, 1984). As shown in figure 2(b), the resulting state represents a condition in which the ionic fluxes in one direction are equal and opposite, while still maintaining a zero net current. Thus, in an analogous manner to equation (12), this state can be represented by:

$$(\text{ion } 1)^- + \overline{(\text{ion } 2)^+} \rightleftharpoons (\text{ion } 2)^+ + \overline{(\text{ion } 1)^-} \tag{15}$$

It should be emphasized that contrary to what Collins and Janata (1982) stated, there are <u>not</u> equal fluxes in opposite directions for each type of ion. If this were indeed the case, an interfacial potential difference (or mixed potential) would not exist since this implies that the relative stabilities of these ions are equal in all phases.

We now turn to some brief remarks regarding the concept of ion trapping or crossing to the bulk of an organic phase. It is our view that there has been a tendency to ascribe ideally polarizable behaviour to the organic layer, such as the PVC-to-solution interface. Perhaps this has resulted from the classical electrochemical treatments which incorporate the fundamental laws of electrostatics and electrodiffusion. Traditionally, these laws consider condensed organic layers (such as PVC or possibly even LB films) as uniform macroscopic phases, devoid of chemistry. In fact, as pointed out by Dorn (1985) in a thorough review of membrane ion transport, local intrinsic phenomena can be very important. Certain variations in measurements of ion currents in bilayer lipid membranes have been postulated to be associated with local perturbation of electrostatics and also steric factors in the membrane structure (Krull et al., 1985, 1986). Moreover, it has been shown that such membranes, completely free of protein, may exhibit

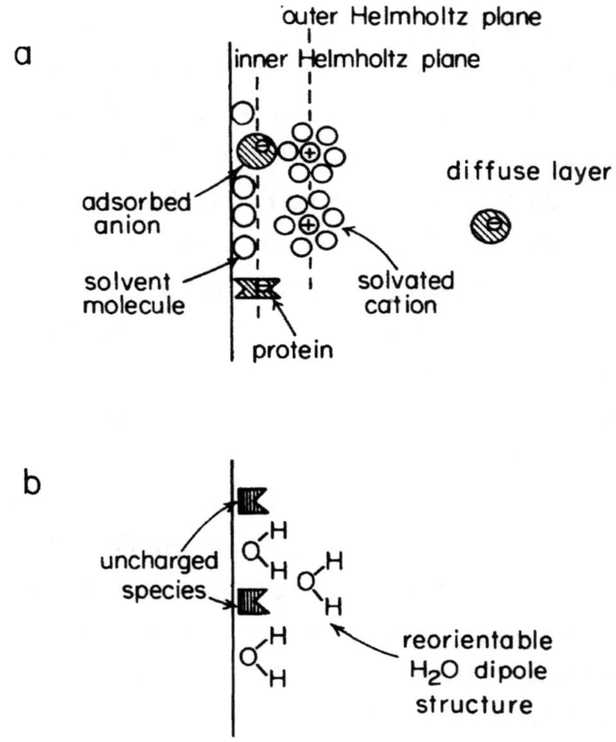

Figure 2. (a) Schematic showing the origin of the mixed potential (after Janata and Blackburn, 1984). Overbars represent ionic species within the organic layer. The intercepts of the solid current-voltage lines at the zero current axis show the position of $\Delta\phi^{eq}$ for (ion 1)$^-$, top, (ion 2)$^+$, centre, and (ion 1)$^-$ and (ion 2),$^+$ bottom.

(b) Description of the movement of two ions of equal but opposite charge across the membrane at the mixed potential.

interfacial selectivity to certain ions (Thompson et al., 1985). The end result of all this is that significant ion crossing to the bulk organic phase can occur, placing a question mark against the invocation of the concept of a polarized organic film-to-solution interface (at least for the conditions required for the operation of a true direct immunochemical potentiometric electrode.)

GENERATION OF MIXED POTENTIALS

From the above it can be seen that if the relative magnitudes of the individual ionic currents change, then $\Delta\phi^{eq}$ shifts. Since the organic layers envisaged for a direct immuno-electrode are not designed to exhibit selectivity with respect to inorganic ion trapping, many ions of both signs of charge can influence the interfacial potential difference and one is faced with a complex general salt response rather than selective ion responses. The various trends that these salt responses take can be understood to some degree in terms of the Frumkin effect, which describes the influences that both the structure of the double layer and specific adsorption of ions have on the kinetics of electrode reactions. These effects can be interpreted in terms of the variation of potential at the double layer; i.e., the potential at the outer Helmholtz plane is not equal to the potential in bulk solution due to the potential drop through the diffuse layer and also due to ions that are specifically adsorbed. These potential differences affect the electrode reaction kinetics in two ways:

(i) examination of equations (9) and (14) together indicate that the actual concentrations of the transferring species at the outer and inner Helmholtz planes may affect the rates of slow ion transfers,

(ii) as equation (9) demonstrates, the ion transfer rate constant will be potential dependent.

Consideration of both these points allows one to understand how adsorption of a protein at an interface causes a change in the ionic exchange currents. What is not so clear, though, is how the individual ionic exchange currents are affected in a particular situation.

Although only those immunochemical reactions which result in a change of charge have been considered up to this point, it is seen that specific adsorption of uncharged molecules will also have similar effects. The explanation for these processes is based on the fact that the organic molecules can physically displace water molecules and ions at the interface and therefore can affect the potential that is due to the layer of oriented dipoles and adsorbed ions.

The complex nature of ion exchange currents through an interface are likely to remain one of the most significant problems in the field of immunoelectrode design since several researchers have grave doubts as to whether it is possible to achieve an appropriately non-polarizable interface. This being the case, the only situation in which an immunoelectrode of this type will work is under carefully controlled conditions; otherwise, as equation (14) indicates, even a very small change in activity of one salt will dramatically affect the mixed potential. Thus, it is not surprising that an integral portion of the experimental protocol for measuring antibody/antigen concentrations involves extensive washings. This step does not appear to cause significant desorption of the complexed immunochemical species from the surface, since a potentiometric response is still generated.

THE ROLE OF NON-SPECIFIC ADSORPTION OF PROTEINACEOUS MATERIAL

Another major factor that must be compensated for in order to develop a functioning immunoelectrode is the effect of non-specific adsorption. Non-specific adsorptive effects may be defined to include both close- and long-range interactions. The former will involve adsorption of proteins and other interferents to the electrode (for which the electrode is not designed) while the latter will be those long-range electrostatic forces present which could perturb the distribution of ions near the electrode surface. Note that when describing long-range effects, species are not really adsorbed at all in the context usually attributed to the term adsorption. Within this definition, it should be pointed out that non-specific adsorption will not include the adsorption of various inorganic ions to the interface, but rather will describe those events that cause a change in the apparent activities or partition coefficients of those inorganic ions whose fluxes make up the mixed potential.

It appears that non-specific adsorption will have similar effects as compared to specific adsorption since the concentrations of the ion-exchanging species, as well as the potential distribution, near the electrode will be affected by these events (Figure 3). Of course, it is hoped that the desired specific adsorptive processes will dominate, but it is not at all apparent that this is indeed the case. In fact, all the papers that have so far been published to date have indicated that immunoelectrodes do respond, to varying degrees, to species for which they are not designed.

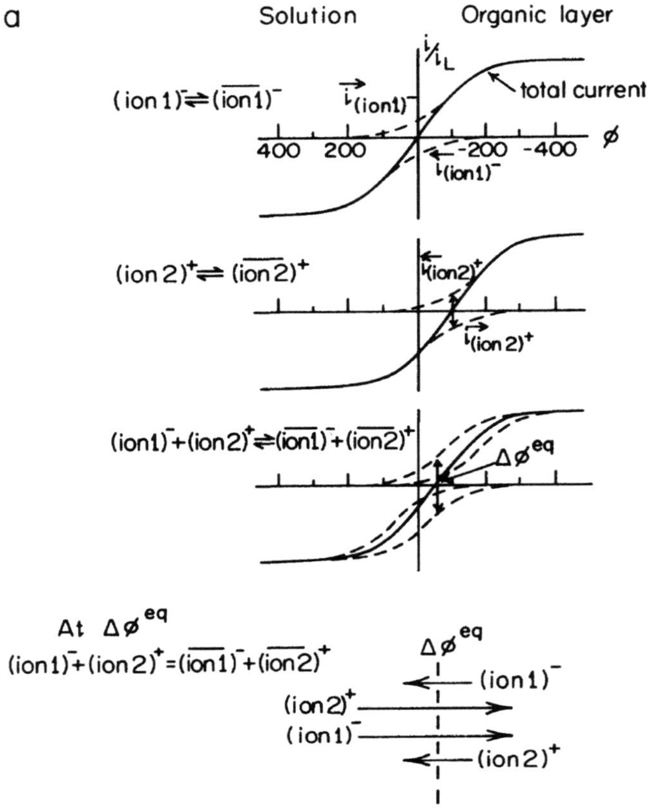

Figure 3. The effect of non-specific adsorption of various species on the potential at the interface. a) - double layer effects - surface charge originates from the reversible double layer (this includes not only adsorption of simple electrolytes, but also of other charged species. b) - adsorption of uncharged species causes changes in H_2O orientation, thereby perturbing the local dielectric. At a non-polarized electrode these effects will alter the relative ionic currents, which make up the mixed potential.

Several courses of action have been detailed in the literature which could help minimize this problem, thereby enhancing both the selectivity and sensitivity of the immunoelectrode. One constraint that should be used to reduce the effects of nonspecific adsorption, although not so much from a selectivity point of view but rather from a sensitivity one, is to use as low a salt concentration as is practical (Schenk, 1978). This is based on Debye length restrictions; the electric field produced by an ion in an electrolyte is exponentially damped by a cloud of oppositely charged ions - the characteristic distance for this effect is called the Debye length. It is strongly dependent on the total ionic concentration in the electrolyte; for instance, in a 0.001 N NaCl solution the Debye length is 96.5 Å, while for a 1.0 N solution, it is reduced to 3.0 Å. Thus, the Debye shielding effect can be expected to markedly reduce the sensitivity of an immunoelectrode to charge dependent reactions if the reaction occurs at a distance of more than a Debye length from the surface. Since adsorbed protein monolayers can be as thick as 50 Å or more, it is clear that at high salt concentrations the shielding effect could greatly reduce the potentiometric signal observed at the electrode surface.

It is becoming increasingly apparent that the problem of non-specific adsorption is linked with the lack of a suitable immobilization technique to adhere the antibody (or antigen) on to the surface of an inert hydrophobic substrate. Since a hydrophobic substrate does not generally have polar groups which could be used for covalent attachment, such groups must be introduced to the surface of the substrate without substantially altering its hydrophobic character. The various techniques developed so far to achieve this aim have not proved entirely satisfactory because the presence of extensive non-specific adsorption onto an immunoelectrode indicates that particular surfaces are not saturated with monolayer coverage of the desired receiving site. The electrochemical theory clearly indicates that non-specific adsorption effects will influence the extraction coefficients and exchange currents of ions transferring into the organic layer (membrane or directly immobilized protein), which in turn yields shifts in mixed potential. In lieu of the "ideal" interface where no non-specific adsorption occurs, one technique is the measurement of the potential of an active immunoelectrode for comparison with that of an identical electrode in which specific binding sites are blocked.

CONCLUDING REMARKS

There is no doubt that an equilibrium potentiometric immunosensor (direct) would be an attractive proposition for the clinical biochemistry laboratory and for *in vivo* measurements. In our view, however, there are two major difficulties associated with the concept. First, the immunochemical reaction is not designed for the reversible behaviour required in a true sensor configuration, although it is possible that the system can be manipulated to yield the desirable properties (Liu and Schultz, 1986). Second, it is clear from the work reviewed in this paper that severe constraints must be applied in order to contemplate the understanding of the potential that is generated at a direct immunochemical interface. The development of such a system requires an interface with a very high charge resistance. In fact Janata and Blackburn (1984) place this at around 6×10^7 ohm cm^{-2}. There is always the possibility that "calibration" could take care of some of the problems, but from the evidence presented in this article it is apparent that even slight changes in the sample matrix will cause severe difficulties with respect to analytical control. In light of these remarks it is interesting to note that in natural biological electrochemical sensing the strategy invokes the use of reversible receptor sites, with a reduced degree of selectivity, and operation via digital threshold switching processes. Perhaps a future direction for electrochemical sensing lies in this type of physical chemistry (Thompson and Krull, 1986).

ACKNOWLEDGEMENTS

We are grateful to the Natural Sciences and Engineering Council of Canada and DARPA for support of our work in the field of Chemical Sensor technology.

REFERENCES

Aizawa, M., Kato, S. and Suzuki, S., 1977, Immunoresponsive membrane I. Membrane potential change associated with an immunochemical reaction between membrane-bound antigen and free antibody, J. Membr. Sci., 2:125-132.

Alexander, P.W. and Rechnitz, G.A., 1974, Ion-electrode based immunoassay and antibody-antigen precipitin reaction monitoring, Anal. Chem., 46:1253-1257.

Alexander, P.W., Maitra, C., 1982, Enzyme-linked immunoassay of human immunoglobulin G with the fluoride ion selective electrode, Anal. Chem., 54(1): 68-71.

Boitieux, J.-L., Desmet, G. and Thomas, D. 1979, An "antibody electrode": Preliminary report on a new approach in enzyme immunoassay, Clin. Chem., 25(2):318-321.

Boitieux, J.-L. Lemzy, C., Desmet, G. and Thomas, D., 1981, Use of solid phase biochemistry for potentiometric enzyme immunoassay of 17 β-estradiol - preliminary report, Clin. Chim. Acta, 113(2):175-182.
Brontman, S.B. and Meyerhoff, M., 1984, Homogeneous enzyme-linked assays mediated by enzyme antibodies; a new approach to electrode-based immunoassays, Anal. Chim. Acta, 162:363-367.
Buck, R.P., 1982, Kinetics and drift of gate voltages for electrolyte bathed chemically sensitive semiconductor devices, IEEE Trans. Electron Dev., ED-29:108-115.
Bush, D.L., Rechnitz, G.A., 1986, Antibody-sensing polymer membrane electrode using a proton carrier, Fresenius' Z. Anal. Chem., 323(5):491.
Collins, S. and Janata, J., 1982, A critical evaluation of the mechanism of potential response of antigen polymer membranes to the corresponding antiserum, Anal. Chim. Acta, 136:93-99.
D'Orazio, P. and Rechnitz, G.A., 1977, Ion electrode measurements of complement and antibody levels using marker-loaded sheep red blood cell ghosts, Anal. Chem., 49:2083-2086.
D'Orazio, P. and Rechnitz, G.A., 1979, Potentiometric electrode measurement of serum antibodies used on the complement fixation test, Anal. Chim. Acta, 109(1):25-31.
Dorn, W.H., 1985, "The Dipole Potential in Monolayers, Bilayers and Biological Membranes," University of Toronto.
Gebauer, C.R. and Rechnitz, G.A., 1981, Immunoassay studies using adenosine deaminase enzyme with potentiometric rate measurement, Anal. Lett., 14(B2):97-109.
Janata, J., 1975, An immunoelectrode, J. Am. Chem. Soc., 97:2914-2916.
Janata, J., 1978, Thermodynamics of chemically sensitive field effect transistors, in: "Theory, Design and Biomedical Applications of Solid State Chemical Sensors," P. Cheung, D. Fleming, W. Ko and M. Neuman, eds., CRC Press, West Palm Beach, Florida.
Janata, J. and Huber, R.J., 1980, Chemically sensitive field effect transistors, in: "Ion Selective Electrodes," M. Freiser, ed., Plenum Press, New York.
Janata, J., 1983, Electrochemistry of chemically sensitive field effect transistors, Sensors and Actuators, 4:255-265.
Janata, J., and Blackburn, G.F., 1984, Immunochemical potentiometric sensors, Ann. N.Y. Acad. Sci., 428:286-292.
Keating, M.Y., and Rechnitz, G.A., 1983, Cortisol antibody electrode, Analyst, 108:766-768.
Keating, M.Y., and Rechnitz, G.A., 1984, Potentiometric digoxin antibody measurements with antigen-ionophore based membrane electrodes, Anal. Chem., 56:801-806.
Keating, M.Y. and Rechnitz, G.A., 1985, Potentiometric enzyme immunoassay for digoxin using polystyrene beads, Anal. Lett., 18(B1):1-10.
Krull, U.J., Thompson, M., Vandenberg, E.T. and Wong, H.E., 1985, Langmuir-Blodgett film characteristics and phospholipid membrane ion conduction. I. Modification by cholesterol and oxidized derivatives, Anal. Chim. Acta, 174:83-94.
Krull, U.J., Thompson, M. and Wong, H.E., 1986, Chemical modification of the bilayer lipid membrane biosensor dipole potential, Bioelectrochem. Bioenerg., 15:371-382.
Liu, B.L. and Schultz, J.S., 1986, Equilibrium binding in immunosensors, IEEE Trans. Biomed. Eng., BME-33(2):133-138.

Meyerhoff, M. and Rechnitz, G.A., 1977, Antibody binding measurements with hapten-selective membrane electrodes, Science, 195:494-495.

Schenk, J.F., 1978, Technical difficulties remaining to the application of ISFET devices, in: "Theory, Design and Biomedical Applications of Solid State Chemical Sensors," P. Cheung, D. Fleming, W. Ko, M. Neuman, eds., CRC Press, West Palm Beach, Florida.

Shiba, K., Watanabe, T., Umezawa, Y., Fujiwara, S. and Momoi, H., 1980a, Liposome immunoelectrode, Chem. Lett, 2:155-158.

Shiba, K., Umezawa, Y., Watanabe, T., Ogewa, S. and Fujiwara, S., 1980b, Thin-layer potentiometric analysis of lipid antigen-antibody reaction by tetrapentylammonium (TPA^+) ion loaded liposomes and TPA^+ ion selective electrode, Anal. Chem., 52(11):1610-1613.

Solsky, R.L.and Rechnitz, G.A., 1979, Antibody-selective membrane electrodes, Science, 204:1308-1309.

Solsky, R.L. and Rechnitz, G.A., 1981, Preparation and properties of an antibody-selective membrane electrode, Anal. Chim. Acta, 123:135-141.

Thompson, M., Krull, U.J., Bendell-Young, L.I., Lundstrom, I. and Nylander, C., 1985, Local surface dipolar perturbation of lipid membranes by phloretin and its analogues, Anal. Chim. Acta, 173:129-140.

Thompson, M. and Krull U.J., 1986, The chemoreceptor-transducer interface in the development of biosensors, in: "Electrochemistry, Sensors and Analysis," M.R. Smyth, J.G. Vos, eds., Proceedings of the International Conference Electroanalysis na h'Eireann, Ireland, Elsevier, Amsterdam.

Umezawa, Y., Shiba, K., Watanabe, T., Ogawa, S. and Fujiwara, S., 1981, A microelectrode, in: "3rd Symp. on Ion Selective Electrodes," E. Pungor, ed., Elsevier Co. & Akademiai Kiado, Mitrafured.

Umezawa, Y., 1983, Ion-selective immunoelectrode, in: "Anal. Chem. Symp Ser.," 17 (Chem. Sens.), Tokyo & Elsevier Science Publishers B.V., Amsterdam.

Umezawa, Y., Sofue, S. and Takamoto, Y., 1984, Thin-layer ion-selective electrode detection of anticardiolipid antibodies in syphilis serology, Talanta, 31:375-378.

Yamamoto, N., Nagasawa, Y., Sawai, M., Sudo, T. and Tsubomura, H., 1978, Potentiometric investigations of antigen-antibody and enzyme-enzyme inhibitor reactions using chemically modified metal electrodes, J. Immunol. Meth., 22(3-4):309-317.

Yamamoto, N., Nagasawa, Y., Shuto, S., Tsubomura, H., Sawai, M. and Okumura, H., 1980, Antigen-antibody reaction investigated with use of a chemically modified electrode, Clin. Chem., 26(11):1569-1572.

Yamamoto, N., Nagasawa, Y., Shuto, S. and Tsubomura, H., 1983, Potentiometric detection of biological substances by using chemically modified electrodes, in: "Anal. Chem. Symp. Ser., 17 (Chem. Sens.), Tokyo & Elsevier Science Publishers, Amsterdam.

SELECTIVE ANTIBODY RESPONSIVE MEMBRANE ELECTRODES

M.Y. Keating*

Department of Chemistry
University of Delaware
Newark, DE 19716

INTRODUCTION

Demands of sensitivity, speed and accuracy of immunoassay methods in the health care industry have stimulated the interest of scientists in many disciplines. Although there exist well-established assays, alternative methods are still being sought. A considerable advance has been made in potentiometric immuno-sensors alone. The method described in this chapter is different from straight forward enzyme labeled immuno techniques (e.g., Meyerhoff and Rechnitz, 1979; Boitieux, Desmet and Thomas, 1979; Alexander and Maitra, 1982) where an electrometrically measurable product is liberated by the enzyme label. This class of membrane electrodes respond to specific antibodies through modulation of a background potential fixed by a marker ion. Solsky and Rechnitz (1981) showed that a PVC (polyvinyl chloride) matrix containing DNP-ionophore conjugate responded to anti-DNP antibodies specifically. Keating and Rechnitz (1983, 1984) found that cortisol-ionophore and digoxin-ionophore membrane electrodes are specific to their corresponding antibodies. The term "ionophore" refers to a class of macro-organic molecules which are capable to solubilize specifically an inorganic ion of the suitable size in organic mediums. The ionophore used were benzo-crown ethers and the marker ion was potassium ion.

The origin of potential changes across the membrane before and after the specific binding of antibodies is not yet fully understood. Attempts of rationalization have been reported by Solsky and Rechnitz (1981) and later by Collins and Jañata (1982). This chapter will present the facts and observations and will not deal with the mechanism.

The method is simple. A hapten, its corresponding antibody is to be measured, is chemically coupled to an ionophore to form a hapten-ionophore conjugate. The conjugate is then incorporated into a polymeric membrane support and that membrane is in turn mounted in the sensing tip of a conventional potentiometric membrane electrode (Figure 1).

---

*Present address: Central Research and Development Department, Du Pont Company, Wilmington, DE 19898.

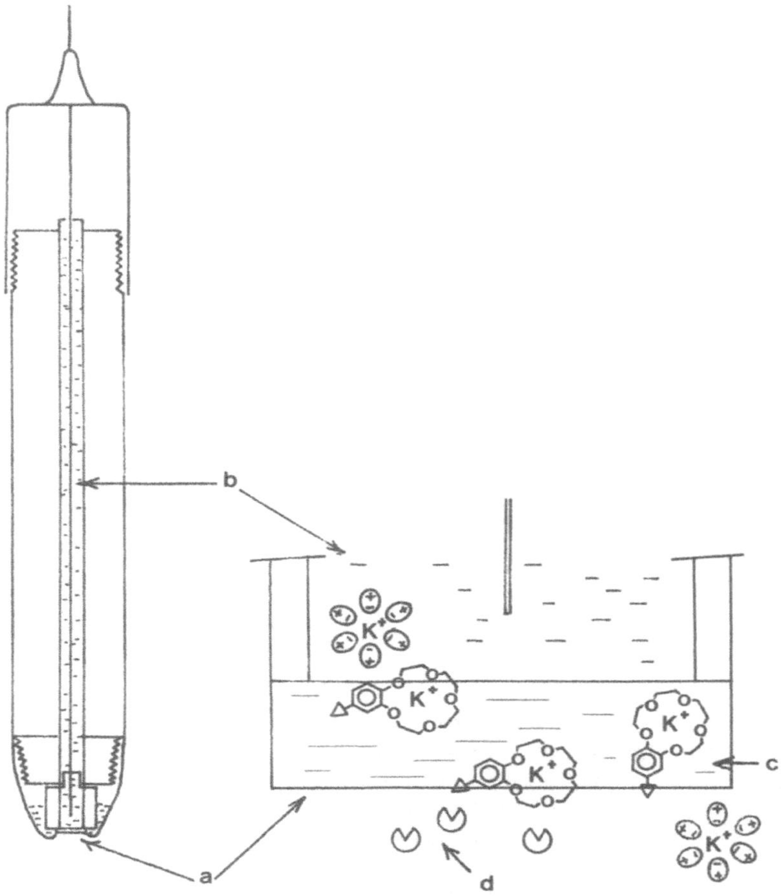

Figure 1. Pictorial diagram of antibody sensing electrode: (a) PVC membrane containing hapten-ionophore; (b) inner filling solution, 0.01 M KCl; (c) plasticizer, dibutylsebacate; (d) antibodies.

The resulting electrode is exposed to a constant activity of a marker ion chosen for its compatibility with the ionophore portion of the conjugate, under conditions which produce a stable and reproducible background potential. When an antibody capable of binding the hapten (or antigen) portion of the conjugate is added to the background electrolyte, a potential change (ΔE) proportional to the antibody concentration is produced. If the electrode is indeed functioning as an antibody sensor, a calibration curve can be constructed by measuring ΔE as a function of antibody concentration. The selectivity of the sensor can be checked by similarly measuring the effect of possible interferences, and other experimental variables can be controlled.

Also, through a competitive binding approach as illustrated in Figure 2, the hapten or antigen concentration can be determined from a calibration curve, established by measuring the percentage of membrane potential changes (%ΔE) as a function of hapten or antigen concentration. This approach is based on measuring the concentration of antibody binding sites which are still uncomplexed after various amounts of ligand analyte have been added.

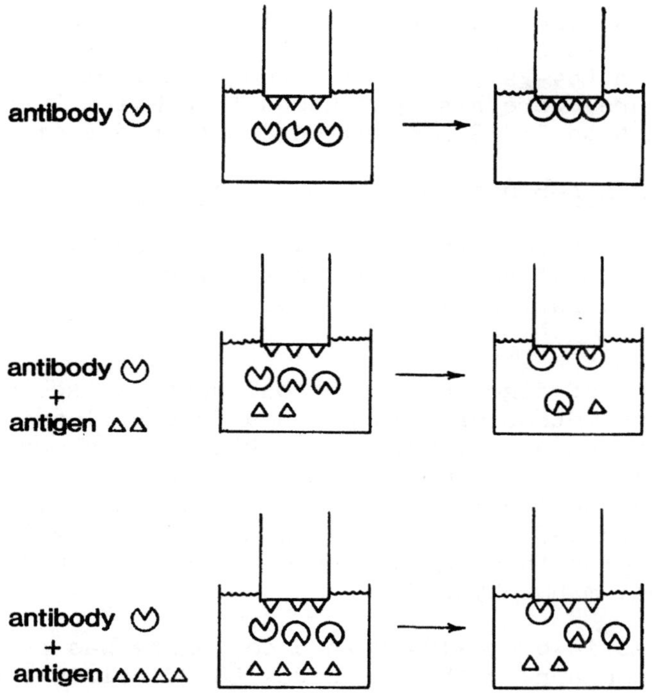

Figure 2. Hapten or antigen measurements with competitive binding approach.

MATERIALS AND METHODS

To illustrate the antibody responsive electrodes in details, the materials and methods for development of a selective electrode for antibodies to digoxin, a steroidal cardiac drug, are described. The selectivity and high affinity of digoxin antibodies raised in rabbits for the drug are especially favorable (the intrinsic affinity constant has been reported to be 1.7 x 10^{10} $\underline{M}^{-1}$ by Smith, Butler, and Haber, 1970). The selected ionophores, benzo-15-crown-5 and cis-dibenzo-18-crown-6, have good solubility in the polyvinyl chloride support membranes employed.

Reagents and Materials

All chemicals used were of analytical reagent grade. Deionized water was used throughout.

Digoxin antibodies were obtained from Miles Laboratories as rabbit anti-digoxin-bovine serum albumin antisera (65-866, lot 5552, titer 1:25000, K_a = 2.0 x 10^{10} $\underline{M}^{-1}$; lot 5553, titer 1:20000, K_a = 2.7 x 10^{10} $\underline{M}^{-1}$). With the information provided by the manufacturer, the respective digoxin antibodies concentration were estimated by the use Scatchard's equation, $r/c = nK_a - rK_a$, at 50% binding of antibody sites (n = 2, r = 1), to be 0.22 mg/ml and 0.26 mg/ml. (The term r corresponds to the moles of hapten bound per mole of antibody at a free hapten concentration c, n equals moles of hapten bound at binding site saturation, and K_a is the average intrinsic association constant.)

BSA antibodies were supplied (Miles) as rabbit anti-bovine serum albumin in lyophilized form (65-111, lot R725, titer 2.6 mg/ml). Normal pooled rabbit serum (64-291, lot 0015) and rabbit albumin (82-451, lot 42, fraction V, 99% pure) were also purchased from Miles. Rabbit γ-globulins (G 0261, Cohn fraction II, 99% pure) were obtained from Sigma.

Preparation of Digoxin-Benzo-15-Crown-5 Conjugate

The ionophore, benzo-15-crown-5, was modified by the introduction of an amino group at the 4' position via nitration (Ungaro and El Haj, 1976) and catalytic hydrogenation with Raney Ni as the catalyst (Feigenbaum and Michel, 1971).

The attachment of digoxin to the ionophore was accomplished by the periodate oxidation method of Erlanger and Beiser (1964) and the modified procedure of Butler et al. (1967, 1970, 1982). As shown in Figure 3, the digoxin was oxidized by sodium periodate at the terminal digitoxose to form the dialdehyde derivative. The dialdehyde was allowed to react with 4'-aminobenzo-15-crown-5 to form a seven-membered-ring Schiff base compound, which yields the digoxin-benzo-15-crown-5 conjugate after reduction with sodium borohydride.

The crude digoxin-benzo-15-crown-5 conjugate was purified by gel filtration chromatography with a Sephadex LH 20 column (bed dimensions 2.2 x 42 cm). The fractions were pooled according to a plot of UV absorbance at 193 nm against eluent volume in milliliters. The molecular weight was first estimated by comparing the eluent volume of a MW marker

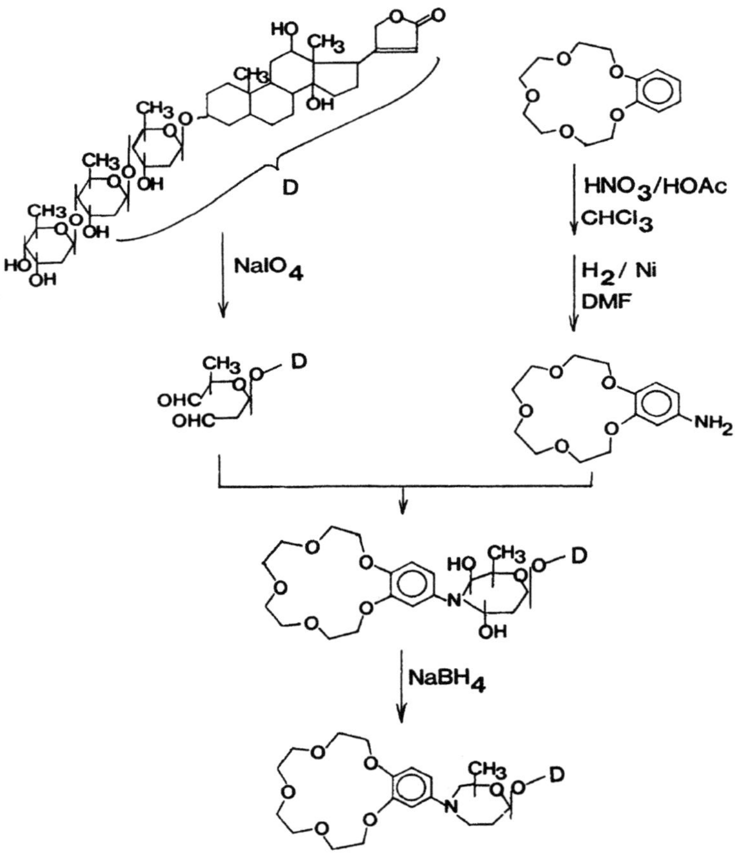

Figure 3. Scheme for preparation of digoxin-benzo-15-crown-5 conjugate. (With permission)

and then confirmed on a Kratos MS-50 mass spectrometer using fast atom bombardment ionization. The digoxin-benzo-15-crown-5 is a potassium ion scavenger. A peak corresponding to the (M + K)+ ion was found in the spectrum (Figure 4) with the most abundant monoisotopic mass of 1070.

The detail preparation procedures were published elsewhere (Keating and Rechnitz, 1984).

Preparation of Digoxin-Dibenzo-18-Crown-6 Conjugate

The ionophore, cis-diaminobenzo-18-crown-6, was prepared from dibenzo-18-crown-6 following Feigenbaum and Michel's procedure (1971), but the purification was accomplished by recrystallization in hot absolute ethanol.

The attachment procedure of digoxin to cis-diaminobenzo-18-crown-6 and purification of the crude conjugate were similar to that of digoxin-benzo-15-crown-5. The molecular weight of the monosubstituted digoxin-dibenzo-18-crown-6 was confirmed on a Kratos MS-50 mass spectrometer. A peak corresponding to (M + K)+ was found in the spectrum (Figure 5) with the monoisotopic mass at 1177.

Preparation of Digoxin-BSA Conjugate for Inhibition Study

The conjugate was prepared and characterized by closely following the procedure of Butler and Tse-Eng (1982). The Digoxin to BSA ratio was 20 to 1.

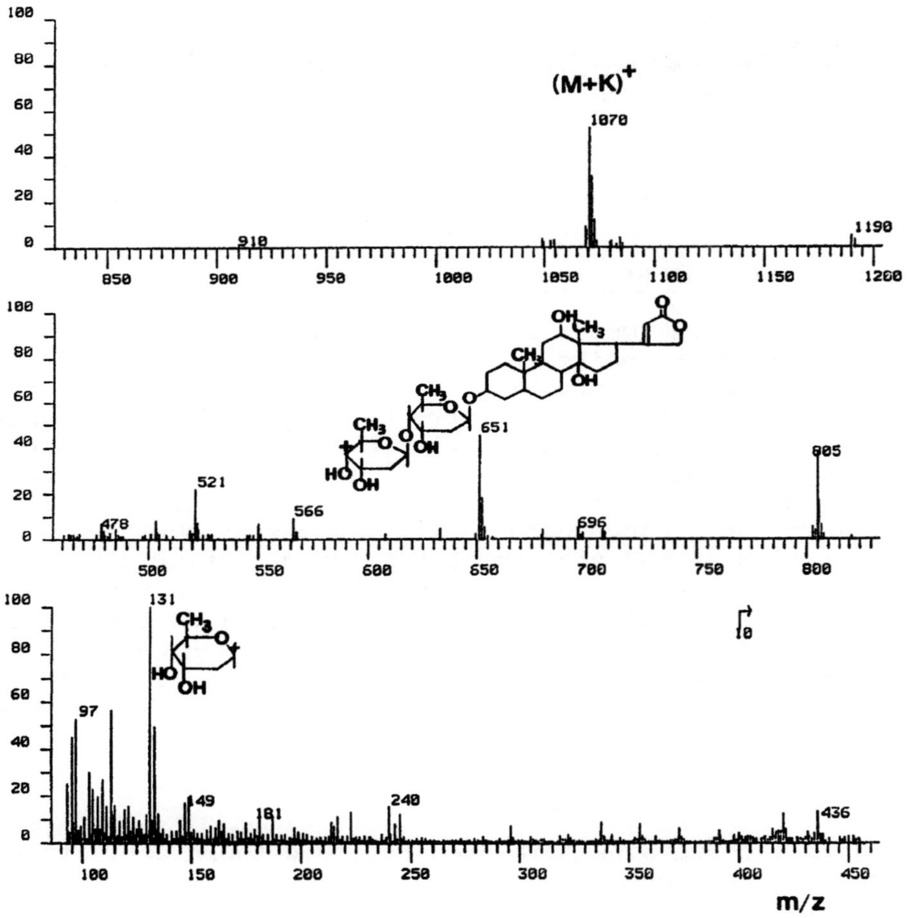

Figure 4. Positive-ion fast-atom-bombardment mass spectrum of digoxin-benzo-15-crown-5 plus K+. (With permission)

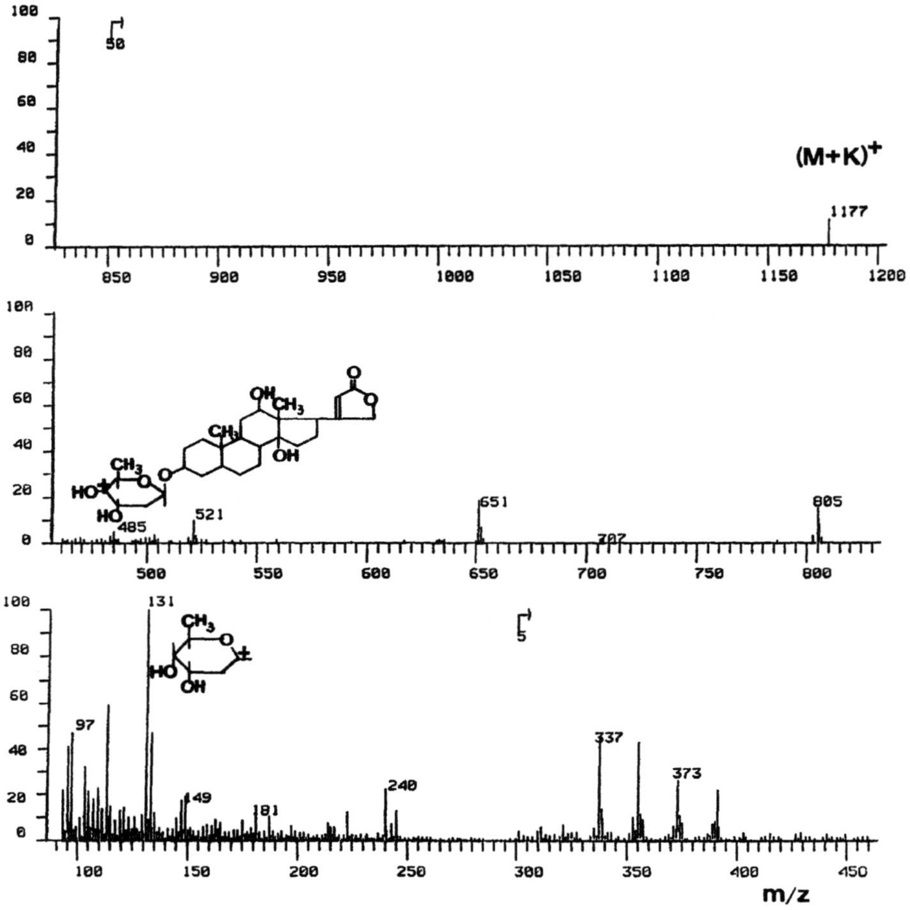

Figure 5. Positive-ion fast-atom-bombardment mass spectrum of digoxin-dibenzo-18-crown-6 plus K^+. (With permission)

Construction of Membrane Electrode

The conjugates were immobilized in 0.2 mm thick poly(vinyl chloride) film with plasticizer (Rechnitz and Solsky, 1981). The membrane films were cast from a solution containing PVC powder (0.25 g), tetrahydrofuran (6.0 ml), dibutyl sebacate (plasticizer, 0.25ml), and various amounts of conjugates (0.3 to 3.0 mg) in 48-mm culture dishes. In the case of digoxin-dibenzo-18-crown-6 which is poorly soluble in THF, 2 drops of methanol were needed to dissolve the conjugate initially. Further reduction of membrane conjugate concentration was accomplished by dissolving a portion of the stock membrane and recasting with additional PVC and plasticizer. Disks of 3 mm diameter were cut out for placement into the Orion 92 series electrode bodies (Figure 1) employed with 0.01 M KCl internal filling solution. Before use, electrodes were conditioned in 0.01 M KCl overnight, because the crown moiety of the conjugate is a potassium ion carrier. Subsequently, a 0.1 M Tris-HCl (pH 7.4) working buffer, containing 3 x 10^{-3} M KCl, was used.

Stability of Membrane Electrodes

The conjugate containing membranes retained their activity for at least six months when stored at 4°C prior to mounting. Fully assembled electrodes, stored in buffer between measurements, were routinely used for 1 to 2 weeks before membrane replacement.

Apparatus

Potentiometric measurements were made with a Corning Model 12 pH/mV meter and recorded on a Heath-Schlumberger SR 204 strip chart recorder (Figure 6). All measurements were made in twin cells thermostated at 25°C with a Haake Model FS temperature controller. Membranes were assembled in Orion Research 92 series electrode bodies. Orion 90-01 single junction reference electrode with potassium chloride - agar salt bridges were employed and potential readings were taken with the antibody electrode inside a Faraday cage for optimum potentiometric stability.

Assay Procedures

Antibody Measurements. In order to eliminate any potentiometric contribution from changes in pH or ion activities, all antibody measurements were carried out in 0.1 M Tris-HCl buffer (pH 7.4) with fixed potassium ion concentration at millimolar levels. All the sera and protein samples (1 to 3 ml) were extensively dialyzed (Spectropor 2 dialysis tubing, molecular weight cutoff (MWCO), 12000 to 14000) against 4 x 1 liter of the background solution to remove any interfering ions. The calibration curves for antibodies, normal pooled serum samples, and pure protein samples were constructed by adding the appropriate amounts of stock solutions to the background buffer to a final volume of 2 ml. The ΔE values plotted represent the difference between the base line potential and the steady-state potentials established in the test solution. Between the digoxin antibody measurements, it was found necessary to place

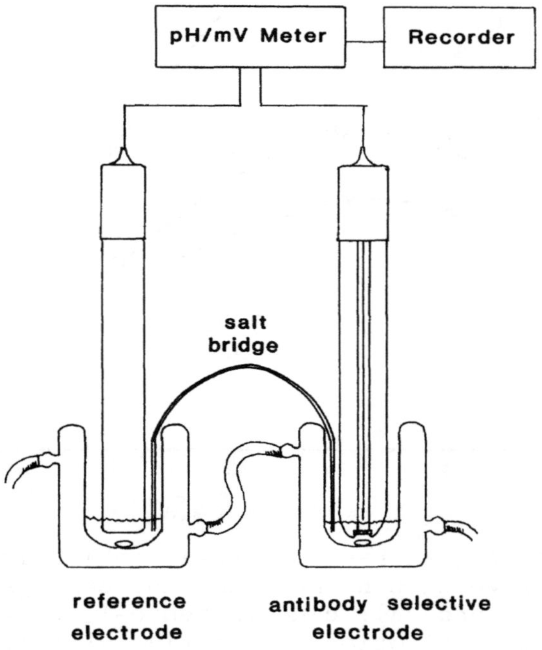

Figure 6. Potentiometric set-up for antibody measurement.

the electrode in 0.1 M glycine-HCl (pH2.8) buffer briefly (less than 60 seconds), rinse, and soak in background buffer or simply to soak the electrode in large amount (50 ml) of background buffer to reestablish the base line potential. Generally, it took 5 to 10 minutes to return to the base line potential.

Antigen or Hapten Measurements. Sets of four 100 μl antidigoxin stock solutions were incubated (about 30 minutes) with various concentrations of digoxin or structure related or nonrelated steroids in a total of 2 ml potassium ion activity and pH controlled background buffer. The final concentrations of antigen or hapten in the solutions were calculated at 1, 2, 10, 20, 100, 200, 1000 nM. Calibration curves of these haptens were established by plotting the %ΔE as a function of hapten concentrations. The 100 %ΔE is the net potentiometric response of antidigoxin alone. The degree of blockage of antidigoxin active sites with these steroids (same as the reduction in %ΔE) is directly related to the haptens (steroids) concentrations.

TYPICAL RESULTS AND DISCUSSION

It will be shown in this section that the magnitude of the ΔE values, the potentiometric response to antibody, can be controlled by adjustment of the marker ion activity, and the membrane composition and that the sensor has good selectivity for the desired antibody over other antibodies or proteins in general. The key parameters in the overall system are studied by separating the experimental variables.

The useful dynamic range of the method is obtained by determining the maximum ΔE generated at the simulated physiological buffer solution (0.1 M Tris-HCl). Figure 7 shows that the potentiometric responses of the

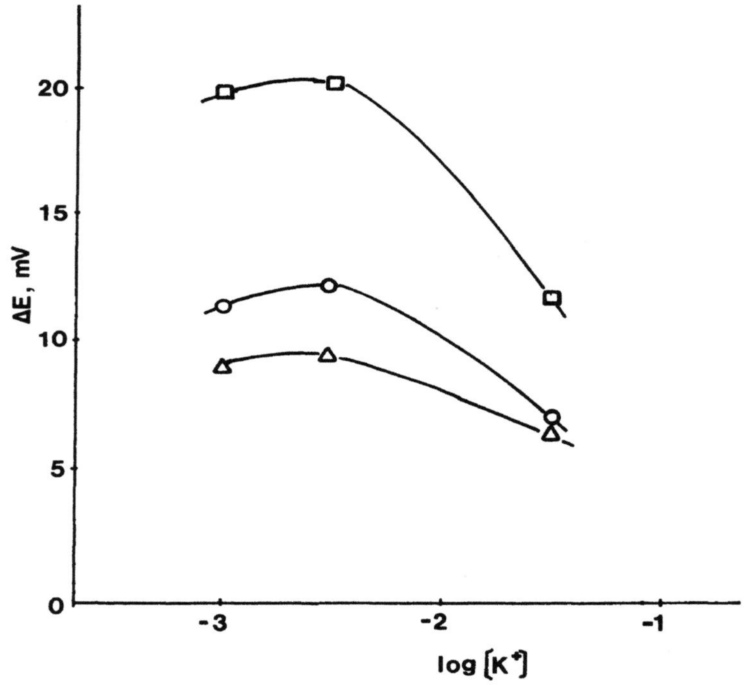

Figure 7. Potentiometric response to digoxin antibodies at varying K^+ marker ion levels (constant membrane conjugate level at 0.1 μg/disk): (□) and (Δ), digoxin-benzo-15-crown-5 with 19 μg/ml and 7.9 μg/ml antibodies, respectively; (o), digoxin-dibenzo-18-crown-6 with 7.9 μg/ml antibodies. (With permission)

electrode to digoxin antibodies peak at millimolar potassium ion activities irrespective of ionophore type and antidigoxin concentration. This coincides with the optimum range of anti-DNP electrode (Solsky and Rechnitz, 1981). The buffer solution containing millimolar of potassium ion activity is thus chosen for the background solution.

A series of electrodes were prepared for each of the conjugates, digoxin-benzo-15-crown-5 and digoxin-dibenzo-18-crown-6, in which the concentration of the conjugate in the sensing membrane varied over the range of 0.1 to 11.7 µg per membrane disk. From Figures 8 and 9 respectively for the above mentioned digoxin-ionophores, it is clear that the potentiometric response to digoxin antibodies increases dramatically as the conjugate level in the sensing membrane is lowered. The trend is similar for both conjugates, but the larger ΔE values are obtained for the digoxin-dibenzo-18-crown-6 case. This crown ionophore is known to have a larger and more suitable cavity for complexing potassium ion. Attempts to lower the conjugate level below 0.1 µg per disk yielded even greater potential changes but with very poor reproducibility and sluggish responses. The relationship between membrane conjugate level and response time is evident from Figure 10. We have noted that these response times become shortened when electrodes have been used repeatedly (exposed to antibody solution and subsequently to glycine-HCl antibody dissociation buffer).

These observations should be viewed in conjunction with Figures 11 and 12 where the response of the same series of electrodes to the marker ion, K+ ion, is presented. All of the electrodes function as potassium ion sensors, but the linear range declines at the lowest membrane conjugate levels exactly where the antibody response is greatest. It is interesting

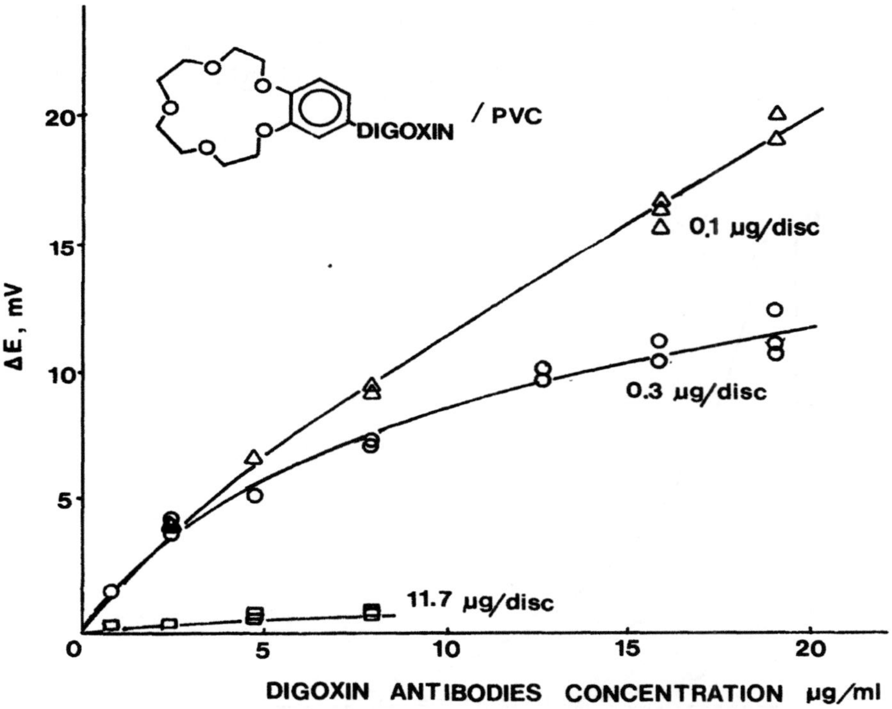

Figure 8. Effect of digoxin-benzo-15-crown-5 concentration in PVC membrane on net potential responses to rabbit anti-digoxin antibodies in Tris-HCl buffer (0.1 M), KCl 3 x 10^{-3} M. (With permission)

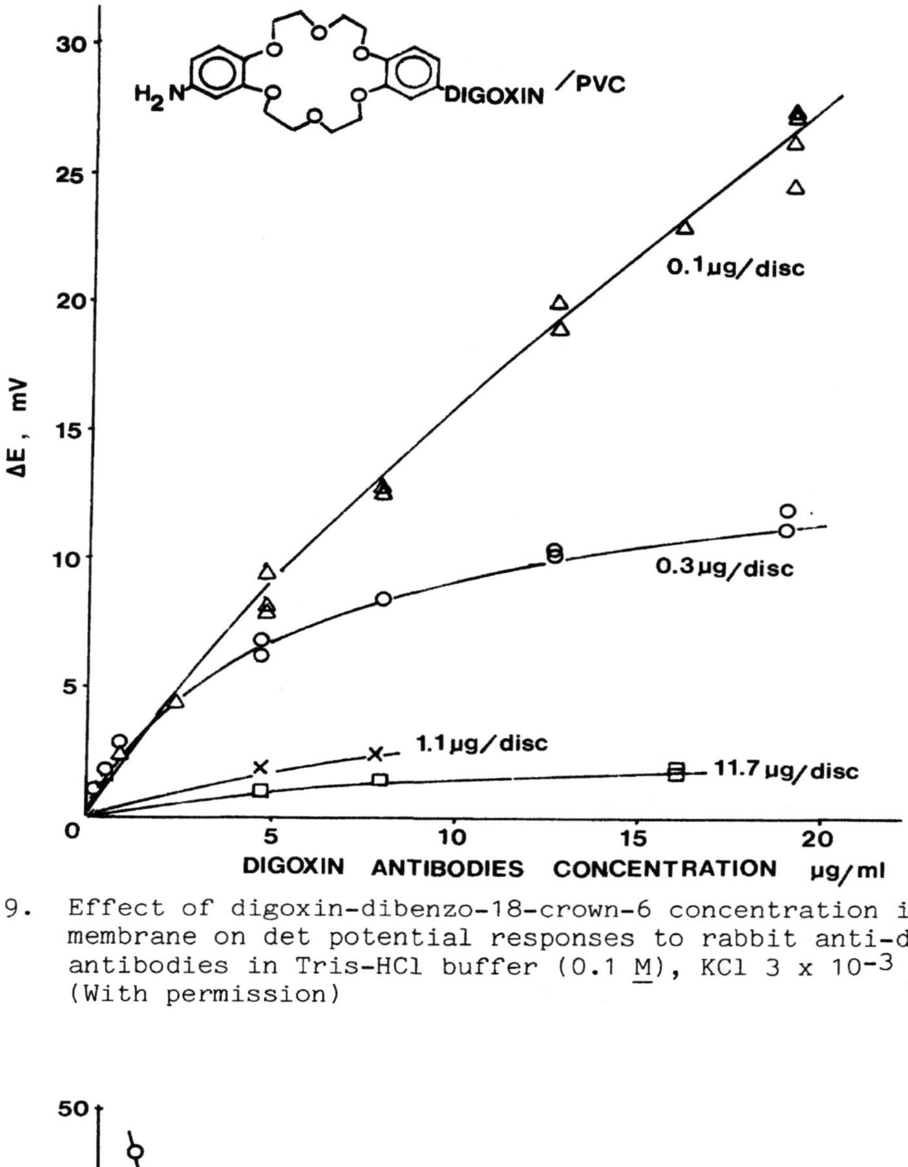

Figure 9. Effect of digoxin-dibenzo-18-crown-6 concentration in PVC membrane on det potential responses to rabbit anti-digoxin antibodies in Tris-HCl buffer (0.1 M), KCl 3 x 10^{-3} M. (With permission)

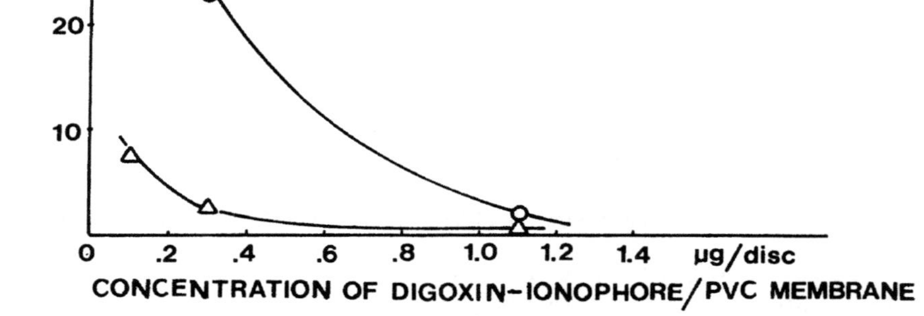

Figure 10. Effect of membrane conjugate concentration on electrode response time (time to reach steady-state) to 15.8 μg/ml antidigoxin antibodies (o) and 3×10^{-3} M K^+ ion (Δ).

to note that the crown ethers (ionophore) remain good potassium ion carriers even after being conjugated to the large digoxin hapten.

Separate experiments carried out to evaluate the absence of digoxin (hapten) moiety in the ionophore that membrane containing only the ionophore does not respond to antidigoxin antibodies. All of the above observations suggest that the experimentally obtained antibody responses may arise from the reversible binding of the primary antibody to the antigen (or hapten) portion of the conjugate molecules that are available at the membrane/solution interface. Such an effect would be expected to be greatest when the concentration of conjugate molecules in the membrane phase is low, because a larger fraction of the conjugate are exposed and can be bound by antibodies.

Two critical experiments supporting this hypothesis are presented in Figure 13 in which a comparison of the electrode response to digoxin antibodies alone and to antibodies which have been preincubated with an antigen, digoxin-BSA, were made using a membrane containing 0.1μg conjugate/disk. In the latter case, no significant potential changes are observed over the entire range of antibody levels studied, as would be expected if the antibodies are no longer available to bind to the digoxin portion of the conjugate in the electrode membrane due to the presence of digoxin-BSA. This finding and the selectivity results detailed below (Figure 14) rule out the possibility of a mechanism involving simple protein adsorption onto the membrane.

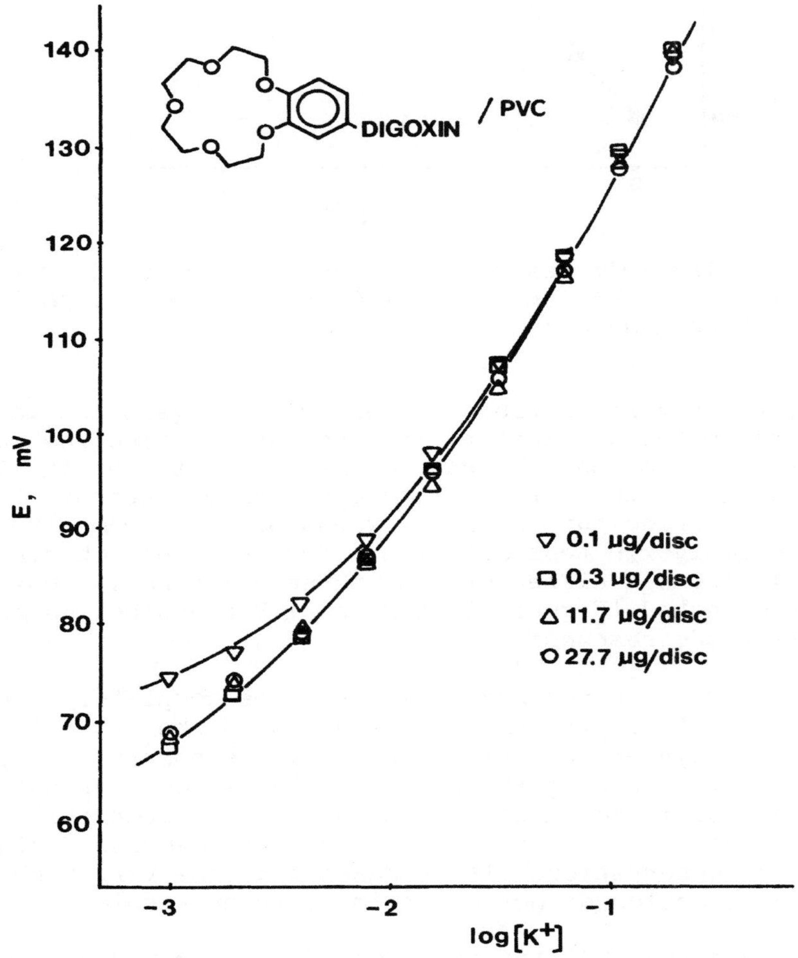

Figuare 11. Potentiometric response of Digoxin-benzo-15-crown-5 / PVC membrane to K^+ ion in Tris-HCl buffer (0.1 M, pH 7.4). (With permission)

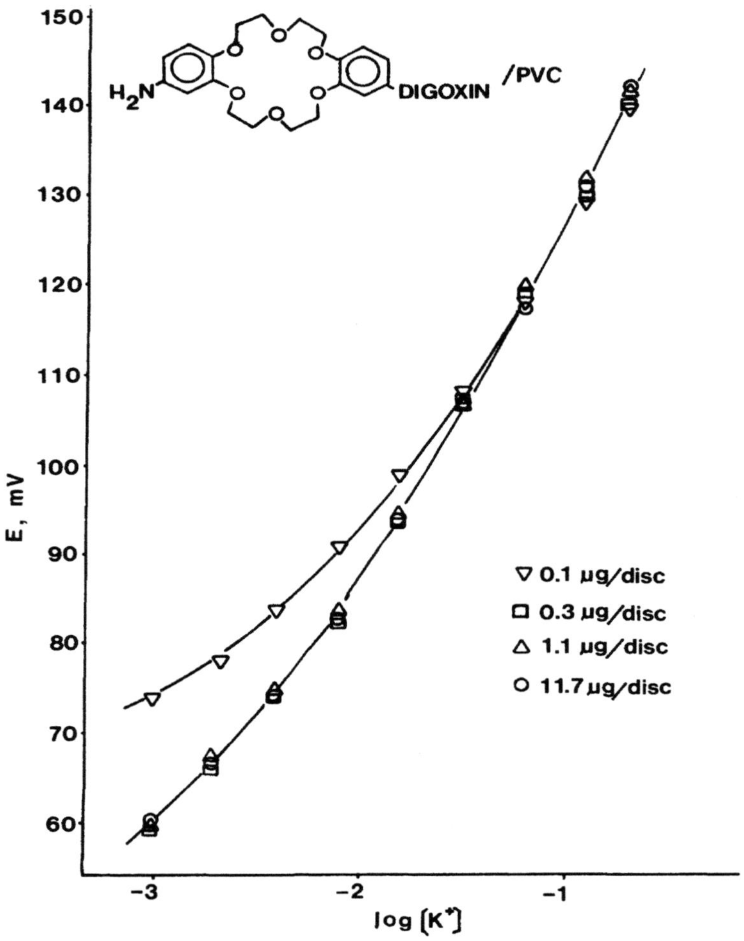

Figure 12. Potentiometric response of digoxin-dibenzo-18-crown-6 / PVC membrane to K+ ion in Tris-HCl buffer (0.1 M, pH 7.4). (With permission)

Since the antidigoxin antibodies were raised against digoxin-BSA immunogen, there is the possibility that anti-BSA antibodies may coexist with the primary antibodies. To ensure that coexistence of anti-BSA is not contributing to the potentiometric response, separate "control" experiments were carried out in which excess bovine serum albumin was added to the anti-digoxin test solution. The ΔE values obtained in the presence of BSA were not different from those using anti-digoxin alone. It should also be noted (Figure 14) that anti-BSA antibodies produce no appreciable electrode response either.

The observed digoxin antibody responses are completely reversible and the base line potentials can be reestablished when the electrodes are returned to the background buffer after a brief exposure to a pH 2.8 glycine-HCl buffer. Lowering the pH evidently dissociates antibody - antigen binding and, in conjunction with electrode reconditioning in the original background buffer, serves to restore the membrane/solution interface. These observations also suggest that the very high molecular weight antibody molecules do not enter the membrane phase.

It should be noted that the addition of antidigoxin antibodies always produces a potential change in the direction of increasing apparent potassium ion activity when membrane containing digoxin-ionophore conjugate is used as the sensing element. However, experiments carried

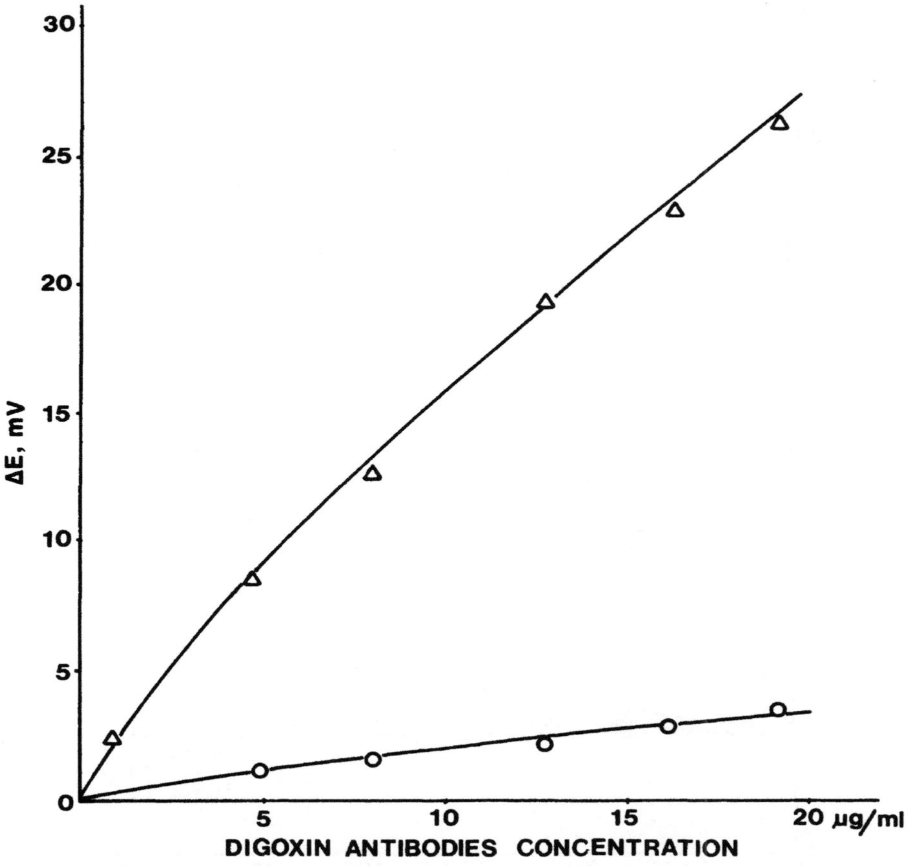

Figure 13. Electrode responses to antidigoxin antibodies alone (Δ) and to digoxin antibodies preincubated with 20 µg of digoxin-BSA antigen (o). Digoxin-dibenzo-18-crown-6 membrane concentration at 0.1 µg/disk.

out with conventional K^+ selective membrane electrode, e.g., Orien's 93-19 potassium ion electrode paired with 90-01 single junction reference electrode, showed no detectable changes in K^+ activity before and after the addition of antibodies. Taken together, these findings apparently indicate that neither a "titration" of K^+ by the antibodies (which would result in potential changes in the direction of decreasing K^+ ion activity) nor a release of K^+ from the membrane is part of the factors that give rise to the potentiometric response to the specific antibody.

Finally, the electrode is shown (Figure 14) to have excellent selectivity for antidigoxin antibodies over other antibodies and proteins. An additional experiment (not shown) with anticortisol antibodies also gave negligible response. This is particularly significant since the antidigoxin antibodies used in this study were raised, according to the supplier, in rabbits immunized with digoxin-bovine serum albumin antigen. It is known (Lukas, 1969; Evered, 1972; Voshall, 1975; Shaw, 1975; Holtzman, 1975) that human serum albumin binds to digoxin with an association constant of 10^5 $\underline{M}^{-1}$ and it has been shown that serum albumin from rabbit and other species bind to several cardiac glycosides (Farah, 1945). However, we found negligible interference of purified rabbit serum albumin on electrode response (Figure 11). The ΔE values obtained for the antidigoxin antibodies are routinely reproducible to +/-1.5 mV or better.

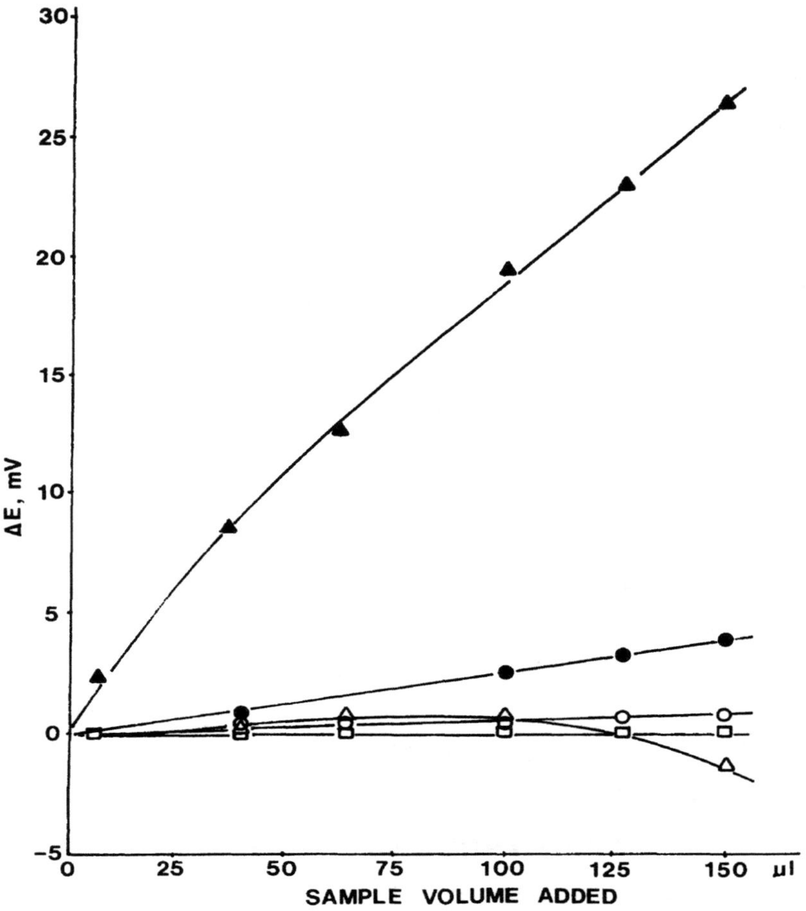

Figure 14. Comparison of responses of digoxin-dibenzo-18-crown-6 membrane (0.1 µg/disk) to antidigoxin antibodies (▲) and to normal rabbit serum (o); rabbit serum albumin stock solution (45 mg/ml) (Δ); rabbit γ-globulins stock solution (30 mg/ml) (□); BSA antibodies (●). (With permission)

All of these experimental results are consistent with a model (Figure 1) in which the modulation of the potentiometric responses results from selective and reversible binding of the antidigoxin antibodies to the digoxin portion of the conjugate molecules at the membrane interface. By use of the competitive binding approach, digoxin can be selectively measured with moderate sensitivity (Figure 15). The nonrelated steroid cortisol gives no reactivity even in large excess, while deslanoside, digitoxin show the modest immunological cross reactivity, which is expected for closely related compounds. Ouabain, however, shows little cross reactivity.

COMMENTS

Although the origin of potential difference ΔE across the membrane in the antibody electrode is not yet known. The experimental results are consistent with the model. The sensitivity for antibody measurements is in µg/ml range and the usable range for hapten measurements is 10 n<u>M</u> to 1 µ<u>M</u>.

This method is unique because it is economical and simple to use. The preparation of hapten-ionophore conjugates, a purely organic coupling

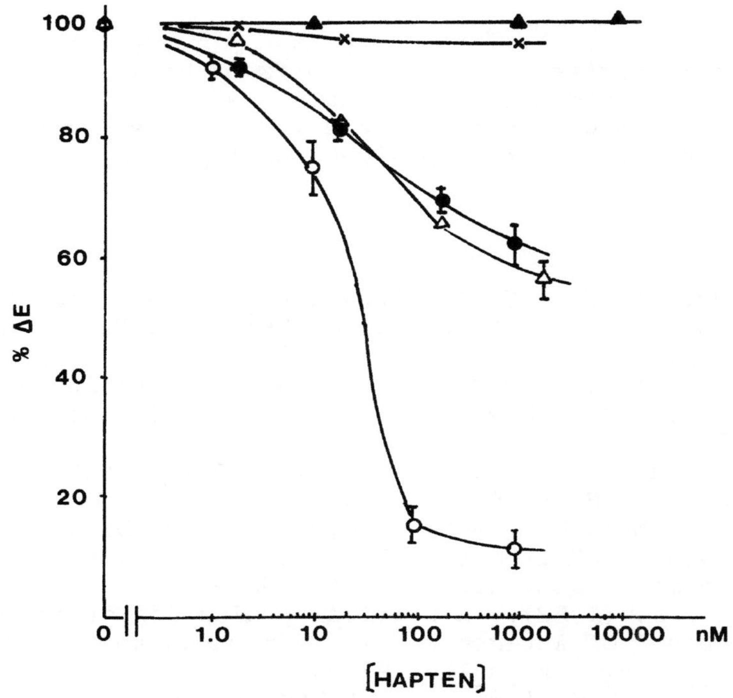

Figure 15. Assays for digoxin (o) using the digoxin-benzo-15-crown-5 / PVC electrode (0.1 µg/disk) at constant (19.0 µg/ml) anti-digoxin antibodies concentration; cortisol (▲); ouabain (x); digitoxin (Δ); deslanoside (●).

reaction, is much more robust and easier to handle than the delicate drug-enzyme conjugates required in enzyme immunoassays. Furthermore, the conjugates can be sold and stored as shelf chemicals. A single membrane casting is sufficient for the preparation of literally dozens of electrodes. The cost of the membrane is minimal.

The ionophores need not be limited to crown ethers. Today, new ionophores are still being discovered for ion-selective electrodes (ISE). The technique described in this chapter can most likely be used with other ion-ionophore pairs (e.g., the ionophores that are specific for ions that are not normally found in blood or other biological fluids). A background potential can be easily established by introducing a controlled amount of marker ion into the biological test medium.

SUMMARY

A method for the measurement of antibodies with hapten-ionophore conjugate based membrane is illustrated in this chapter using a steroidal cardiac drug as a model. The techniques for preparing the membrane, constructing the electrodes as well as the test procedure are described in detail. The membrane electrodes are capable of detecting antibodies at µg/ml levels with high degree of selectivity over other antibodies. By using a competitive binding approach, selective measurements of haptens can be made with the membrane electrodes.

ACKNOWLEDGEMENTS

I am indebted to Professor Garry Rechnitz of University of Delaware for his encouragement, guidance and valuable discussion during my postdoctoral appointment with him at the university, and grateful for the financial support of NIH Grant GM-25308 during 1982-84.

REFERENCES

Alexander, P.W.; Maitra, C. (1982). Enzyme-Linked Immunoassay of Human Immunoglobulin G with the Fluoride Ion Selective Electrode. Anal. Chem. 54: 68-71.

Boitieux, J.-L.; Desmet, G.; Thomas, D. (1979). An "Antibody Electrode," Preliminary Report on a New Approach in Enzyme Immunoassay. Clin. Chem. 25: 318-321.

Butler Jr., V.P.; Chen, J.P. (1967). Digocin Specific Antibodies. Proc. Natl. Acad. Sci. U. S. A. 57: 71-78.

Butler Jr., V.P.; Tse-Eng, D. (1982). Immunoassay of Digoxin and Other Cardiac Glycosides. "Methods in Enzymology", Langone, J.J.; Van Vunakis, H., Eds.; Academic Press: New York 84: 558-577.

Collins, S.; Janata, J. (1982). A Critical Evaluation of the Mechanism of Potential Response of Antigen Polymer Membranes to the Corresponding Antiserum. Anal. Chim. Acta 136: 93-99.

Erlanger, B.F.; Beiser, S.M. (1964). Antibodies Specific for Ribonucleosides and Ribonucleotides and their Reaction with Deoxyribonucleic Acid (DNA). Proc. Natl. Acad. Sci. U. S. A. 52: 68-74.

Evered, D.C. (1972). The Binding of Digoxin by the Serum Proteins. Eur. J. Pharmacol. 18:236-244.

Farah, A. (1945). On the Combination of Some Cardio-Active Glycosides with Serum Proteins. J. Pharmacol. Exp. Ther. 83:143-157.

Feigenbaum, W.M.; Michel, R.H. (1971). Novel Polyamides from Macrocyclic Ethers. J. Polym. Sci. 9: 817-820.

Holtzman, J.L.; Shafer, R.B. (1975). Radioimmunoassay of Digoxin: Effect of Albumin. Clin. Chem. 21: 636-637.

Keating, M.Y.; Rechnitz, G.A. (1983). Cortisol Electrode. Analyst (London) 108: 766-768.

Keating, M.Y.; Rechnitz, G.A. (1984). Potentiometric Digoxin Antibody Measurements with Antigen-Ionophore Based Membrane Electrodes. Anal. Chem. 56: 801-806.

Lukas, D.S.; De Martino, A.G. (1969). Binding of Digoxin and Some Related Cardenolides to Human Plasma Proteins. J. Clin. Invest. 48: 1041-1053.

Meyerhoff, M.E.; Rechnitz, G.A. (1979). Electrode-Based Enzyme Immunoassays Using Urease Conjugates. Anal. Biochem. 95: 483-493.

Shaw, W. (1975). Radioimmunoassay of Digoxin: Effect of Albumin. Clin. Chem. 21: 636.

Smith, T.W.; Butler Jr., V.P.; Haber, E. (1970). Characterization of Antibodies of High Affinity andSpecificity for the Digitalis Glycoside Digoxin. Biochem. 9: 331-337.

Solsky, R.L.; Rechnitz, G.A. (1981). Preparation and Properties of an Antibody-Selective Membrane Electrode. Anal. Chim. Acta 123: 135-141.

Ungaro, R.; El Haj, B.; Smid, J. (1976). Substituent Effects on the Stability of Cation Complexes of 4'-Substituted Monobenzo Crown Ethers. J. Am. Chem. Soc. 98: 5198-5202.

Voshall, D.L.; Hunter, L.; Grady, H.J. (1975). Effect of Albumin on Serum Digoxin Radioimmunoassays. Clin. Chem. 21: 402-406.

ELECTROIMMUNOASSAY OF PGE_2: AN ANTIBODY-SENSITIVE ELECTRODE BASED COMPETITIVE PROTEIN-BINDING ASSAY

George R. Connell and Kenton M. Sanders

Department of Physiology
University of Nevada
School of Medicine
Reno, Nevada 89557

INTRODUCTION

Radioimmunoassay revolutionized the capability of endocrinologists and clinical chemists to quickly and inexpensively analyze biological chemicals that occur naturally in small quantities. Only a few problems exist with RIAs: i) RIAs require the use of radioisotopes. These compounds are expensive and their use requires special training and handling. ii) The turn-around time for RIAs can range from several hours to several days. RIA is not very practical for monitoring biological compounds or changes in concentration as a function of time, because of the slow turn-around time of these assays. A technique that would allow on-line measurements would be extremely useful.

Recently Solsky and Rechnitz (1979) developed an electrode capable of measuring specific antibodies for dinitrophenol. These authors synthesized an ionophore-hapten conjugate and incorporated this novel molecule into polyvinyl chloride membranes. Antibodies specific for the hapten conjugated to the ionophore caused a shift in membrane potential (Solsky and Rechnitz, 1979; Solsky and Rechnitz, 1981). These authors speculated that it would be possible to construct antibody-sensitive electrodes for other antibody-hapten systems. We reasoned (Connell et al, 1983) that if the antibody-hapten interaction at the surface of antibody-sensitive membranes was reversible, then it would be possible to produce competition between the membrane-conjugated hapten and hapten molecules free in solution (Fig. 1). At high concentrations of solution phase hapten, the antibody molecules would be saturated and consequently unavailable to bind to the hapten-ionophore conjugates at the surface of the membrane, thus reducing the antibody effect on membrane potential. This should produce potential changes across the membrane equivalent to reduced antibody concentration. These are the conditions of a competitive, protein-binding assay where the dependent variable is a potentiometric measurement. The relative speed, freedom from having to use isotopes and hazardous chemicals, and the electrical output of the assay, which is easily processed, make this technique an attractive alternative to radioimmunoassays.

We tested the feasibility of "electroimmunoassay" by constructing plastic membranes containing covalent conjugates of dibenzo-18-crown-6 (a cation selective ionophore) and PGE_2 (Connell et al, 1983; Connell and Sanders, 1984). The membranes were installed in electrodes that responded with potential changes to antibodies raised against PGE_2-protein conjugates. This paper summarizes the steps necessary for the development of an electroimmunoassay, a competitive protein-binding assay that quantitates solution-phase PGE_2 concentration by measuring transmembrane potential.

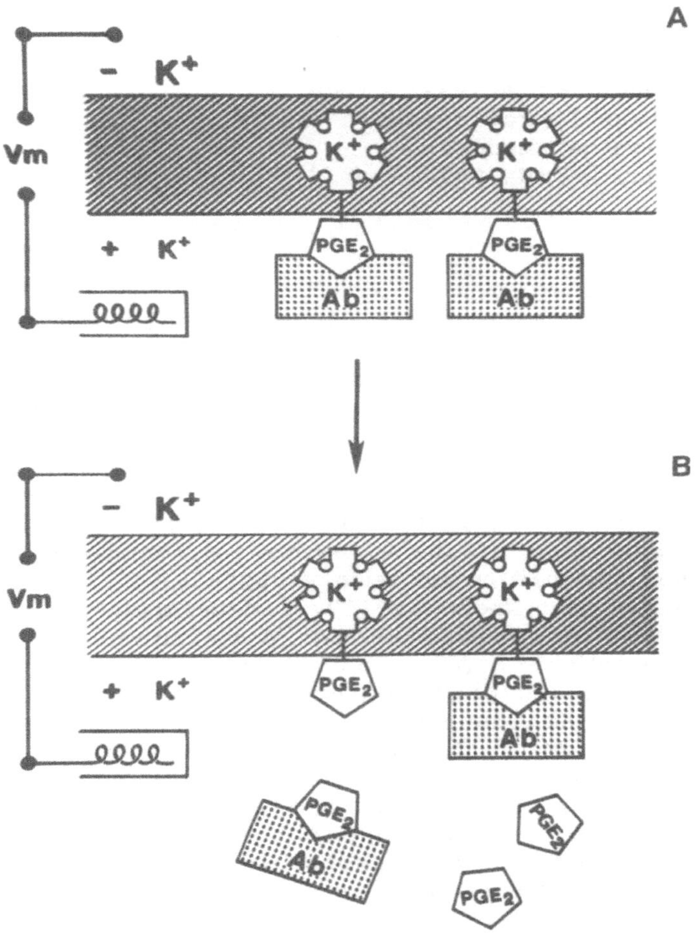

Fig. 1. Schematic representation of electroimmunoassay. Panel A shows a hypothetical interaction between "fixed" PGE_2 molecules conjugated to the ionophore and anti-PGE_2 antibodies. This interaction changes the transmembrane potential (V_m). The addition of "free" PGE_2 to the system results in competition between solution-phase and membrane-phase PGE_2 for antibody binding sites (panel B). As the concentration of "free" PGE_2 is increased, a greater percentage of antibody molecules are displaced from the membrane, resulting in a reduction in the voltage response caused by the antibody.

METHODS

Synthesis of ionophore hapten conjugate

To produce an antibody selective membrane the first step was to conjugate PGE_2 to an ionophore. Several antigenic compounds have been covalently linked to dibenzo-18-crown-6 (see Refs. Solsky and Rechnitz, 1979; Solsky and Rechnitz, 1981; Connell et al 1983, Connell and Sanders, 1984), which is a cation selective ionophore. This compound was useful because of its ion selectivity characteristics and its aromatic ring structure to which functional groups could be easily attached (Fig. 2). An amide bond utilizing the carboxylic group of PGE_2 was chosen for conjugation to the ionophore, because this was the site where PGE_2 was coupled to bovine thyroglobulin to produce antibodies directed against

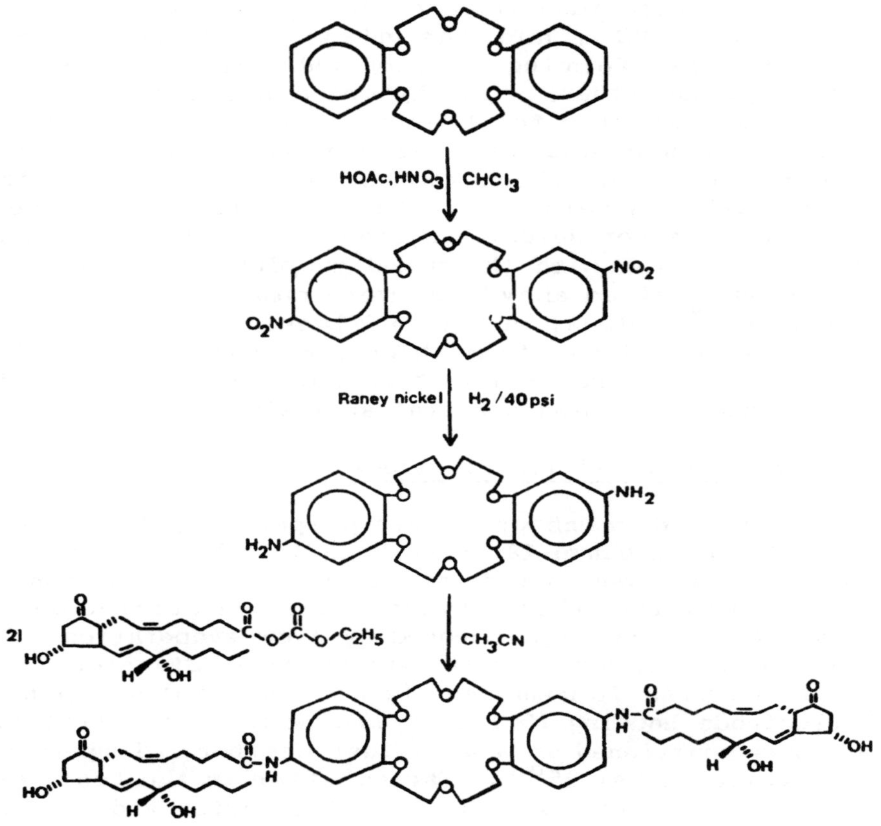

Fig. 2. Synthesis of the ionophore-hapten conjugate for PGE_2 antibody sensitive electrodes. The first step was to nitrate the benzene rings of the crown ether. The nitro groups were subsequently reduced to amines to facilitate the formation of amide linkages with the carboxyl groups of PGE_2.

PGE_2 (Sanders, 1976). Conjugation of PGE_2 and dibenzo-18-crown-6 was achieved in 4 steps: i) nitration of dibenzo-18-crown-6; ii) catalytic reduction of the trans dinitro compound to the corresponding diamine; iii) preparation of a mixed anhydride of PGE_2; iv) condensation of the mixed anhydride with the diamine of dibenzo-18-crown-6.

Dibenzo-18-crown-6 was dissolved in chloroform and acidified with glacial acidic acid. To this mixture a nitrating solution consisting of nitric and acetic acids was added. The reaction was stirred for 1 hour without heat and then refluxed for 3 hours. The trans dinitro product formed as yellow crystals and were harvested by filtration. The crude product was purified by recrystalization from N,N-dimethylformamide (DMF). The melting point of the product was 239-245° C (literature MP 237-246° C; Feigenbaum and Michel, 1971). The product was further verified by IR spectroscopy. The _cis_ dinitro product precipitated from the mother liquor after 1 hour of cooling. The _cis_ accounted for 43% of the yield. Only the _trans_ product was carried through the rest of the synthetic procedures.

The dinitro, dibenzo crown ether was reduced by catalytic hydrogenation to produce a diamine. The _trans_ dinitro-dibenzo-18-crown-6 was dissolved in DMF and a Raney nickel catalyst (Grace Raney nickel #28) was added. The mixture was shaken under hydrogen for 1 hour. At the completicn of the reaction the catalyst was removed by filtration, and the crude product was recovered from the DMF. The crude product was purified by sublimation and formed a white crystalline solid (MP 198-203° C). The product was also identified by IR spectroscopy.

Conjugation of PGE_2 and the diamino-dibenzo-18-crown-6 was accomplished by first forming a mixed anhydride of PGE_2. PGE_2 was dissolved in anhydrous ether. Triethylamine was added to the solution and the mixture was stirred for 1 hour. Then ethyl chloroformate was added and the mixture was stirred for an additional hour. The triethylamine hydrochloride which formed was removed by filtration and the mixed anhydride was retained in the ether solution. The diamino-dibenzo-18-crown-6 was not soluble in ether, so acetonitrile was used in the condensation reaction with the mixed anhydride. Diamino-dibenzo-18-crown-6 and the PGE_2 mixed anhydride were combined and stirred for 12 hours. The crude product, a white flocculent, was recovered by filtration and recrystallized from chloroform and ethanol. The melting point of the purified product was 134° C. The product was verified by NMR and fast atom bombardment mass spectral analysis.

Production of antibody-sensitive electrode

Polyvinyl chloride membranes were prepared by dissolving 1 mg of PGE_2-trans-diamide of dibenzo-18-crown-6 in tetrahydrofurane and dibutyl sebacate. This mixture was poured into a 50 mm petri dish containing 250 mg of polyvinyl chloride (PVC). The mixture was stirred until the PVC was dissolved. The solvent was removed by slow evaporation. The result was a flexible membrane, 0.2 mm thick and 50 mm in diameter.

A disk of the plastic membrane (4 mm x 0.2 mm) was placed into a commercial electrode housing (Orion 92-00), which was filled with 0.01 M KCl. The tip was positioned in a 4 ml test chamber. Temperature in the chamber was maintained at 25° C. The solution in the test chamber was stirred with an air driven magnetic stirring device and a small teflon stir bar. The solution could be exchanged via input and output ports in the test chamber. Transmembrane potential was measured in reference to a double junction reference electrode (Orion 90-02) by a standard electrometer (Orion 701 A). The External Buffer Solution (EBS) used in all tests consisted of 0.001 M KCl, 0.001 M TRIS, and 0.051 M $CaCl_2$. Ionic strength and pH were maintained at 0.154 M and pH 7.2 respectively. The conjugate containing membranes were shown to have low selectivity for calcium and TRIS ions. Data were recorded either manually from the digital readout of the electrometer, or by a strip chart recorder (Gould 2200).

RESULTS

Effects of anti-PGE_2 antiserum on transmembrane potential

Antibodies against PGE_2 were obtained by immunizing New Zealand white rabbits conjugates of bovine thyroglobulin and PGE_2 (Sanders,

1976). The sera of rabbits were collected before and after immunization. The sera were lyophilized at the time of harvest and later reconstituted in EBS. The protein concentrations of the reconstituted sera were analysed (Lowrey, 1951) and found to be similar. The effects of the sera on transmembrane potential were tested on membranes containing unconjugated dibenzo-18-crown-6 and on membranes containing PGE_2-dibenzo-18-crown-6 conjugates. Twenty ul of reconstituted sera were added to the test chamber resulting in a final protein concentration of 160 ug/ml. The solution was stirred and the steady-state transmembrane potential was recorded. The protein concentration was increased in 20 ul aliquots until a total of 100 ul had been added. After each 20 ul aliquot membrane potential was recorded. By this technique membrane potential as a function of protein concentration was evaluated for the immune and nonimmune sera. At the completion of each study the membranes were washed with 50 ml of fresh buffer.

Transmembrane potentials of membranes containing dibenzo-18-crown-6 were not affected by either immune or nonimmune sera (Fig. 3a). The sera of nonimmunized animals had little effect on transmembrane potential of membranes containing PGE_2-dibenzo-18-crown-6 conjugates. However, antisera from immunized animals produced a concentration-dependent shift in potential across these membranes (Fig. 3b). This effect was blocked if the antisera were pre-absorbed to saturation with PGE_2.

The shift in membrane potential in response to the presence of antibody was reversible. The average time constant of the potential response to antibody was 7.2 $\pm$ 0.3 min. When the antibodies were washed off the membrane with fresh buffer, the pre-antibody membrane potential was restored with an average time constant of 3.6 $\pm$ 0.2 min. An example of the time courses of the "on" and "off" responses are shown in Fig. 4.

The effect of "free" PGE_2 on the antibody responses

The fact that the interaction between antibodies and membrane bound PGE_2-dibenzo-18-crown-6 conjugates was reversible suggested that the shift in membrane potential due to antibody could be reversed by competition with "free" PGE_2. Competition for antibody binding between membrane and solution-phase PGE_2 molecules could thus provide the conditions necessary for a competitive protein-binding assay (see Fig. 1). In such an assay standard curves could be generated in which membrane potential would be a function of "free" PGE_2 concentration. This theory was tested by determining the ability of various concentrations of PGE_2 to reverse the effects of anti-PGE_2 serum on membrane potential.

Antibody was reconstituted at a concentration from the linear portion of the antibody-response curve (Figure 3b). Aliquots of the diluted serum were added to a series of tubes. PGE_2 was added to each of the tubes to achieve concentrations ranging from 3 to 860 nM. The electrode was first immersed in an antiserum solution containing no PGE_2, and the maximal voltage response (V_o) was recorded. Then the membrane was washed with buffer and the electrode was immersed in one of the antiserum solutions containing PGE_2. This cycle was repeated until potential measurements had been made on each of the PGE_2-containing antiserum samples. The V_o response of the membrane was reduced in a concentration-dependent manner by solution-phase PGE_2. This data were used to construct a standard curve, a plot of membrane potential as a function of solution-phase PGE_2 concentration. An example of a standard curve is shown in Fig. 5. Such a curve can be used as the basis for an "electroimmunoassay" (EIA). The response time constants for the standard concentrations of PGE_2 were similar to the time constants of samples without PGE_2, so an entire standard curve could be generated within an hour.

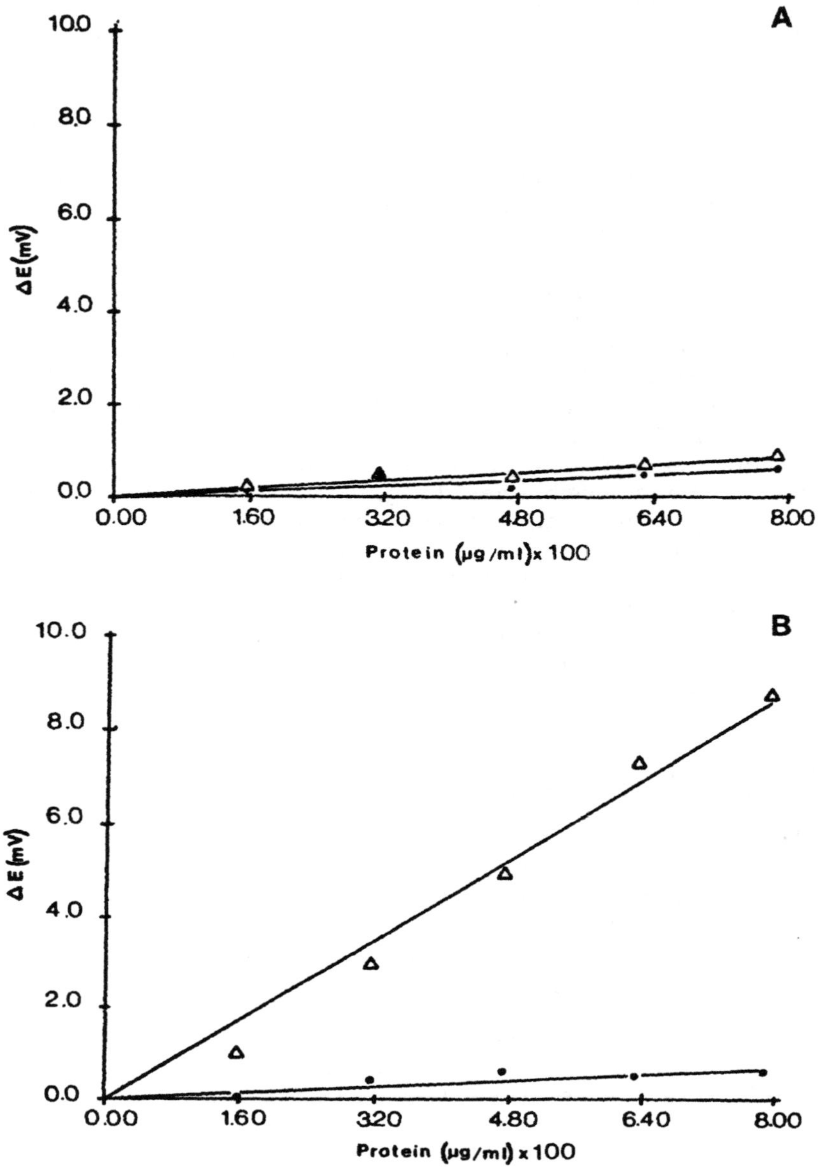

Fig. 3. Membrane potential as a function of serum protein concentration. Panel A shows the responses of a membrane containing unconjugated dibenzo-18-crown-6. Sera of either immunized (open triangles) or non-immunized (closed circles) animals had no effect on transmembrane potential. Panel B shows the responses of antibody-sensitive membranes in which PGE_2-crown ether conjugates were incorporated. Potential was changed in a concentration-dependent manner by immune serum added to the external solution. Concentration of antibody is expressed as total protein (Lowrey et al, 1951).

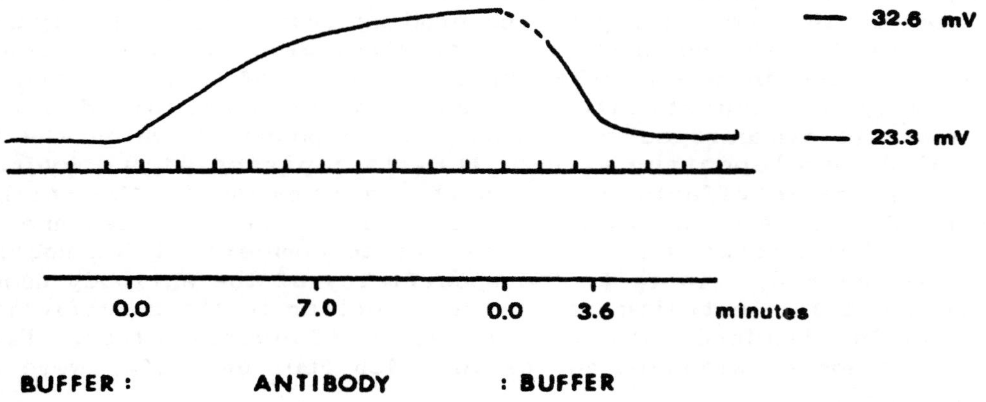

Fig. 4. Time course for antibody effect on membrane potential. The "on-time" constant was calculated to 67% of the maximal voltage response. The "off-time" constant was calculated as the time at which the maximal voltage response had decayed to 67%.

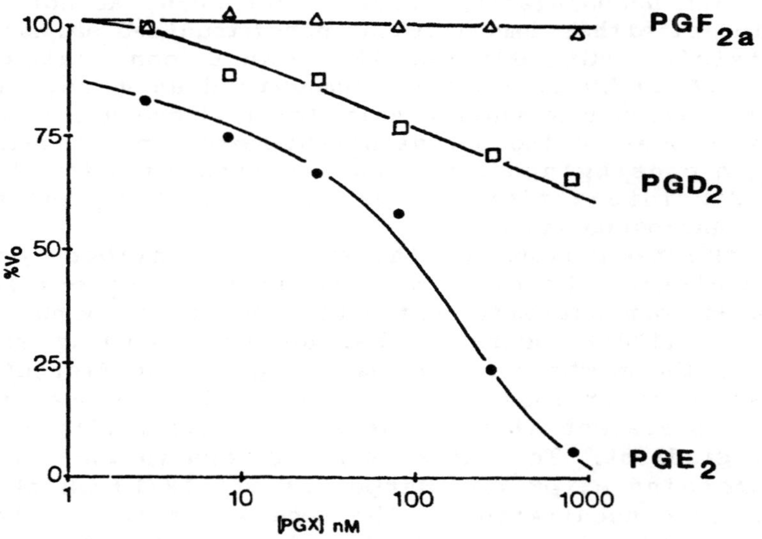

Fig. 5. Standard curves for electroimmunoassay. A concentration of antibody in the linear portion of the antibody-response curve (Fig. 3B.) was selected. The voltage response at this antibody concentration (V_o) was decreased as a function of the solution-phase concentrations of PGE_2 and PGD_2, but not by PGF_{2a} .

Assay specificity

A major determinant of the usefulness of a competitive protein-binding assay is the specificity of the binding protein for the ligand to be assayed. We tested the specificity of electroimmunoassay by conducting several trials in which other prostaglandins of similar structure (PGD_2 or PGF_{2a}) were substituted for PGE_2 in the procedure described above to generate standard curves. PGF_{2a}, a compound in which the keto function on the cyclopentane ring of PGE_2 is replaced with a hydroxyl group, had negligible effects on the membrane response to the antibody. PGD_2, in which the keto and hydroxyl functions of the cyclopentane ring are reversed, demonstrated limited ability to compete with membrane-bound, conjugated PGE_2. In fact, the specificity of the antibody used in these evaluations of electroimmunoassay was similar to the specificity of the antibody in standard radiometric assays (Sanders, 1976). Fig. 5 shows a comparison of standard curves in which PGD_2 and PGF_{2a} were substituted for PGE_2.

DISCUSSION

We have shown that the antibody-sensitive electrode technology introduced by Solsky and Rechnitz (1979) can be used in a competitive, protein-binding assay to measure solution-phase hapten concentration. The specificity of this "electroimmunoassay" was similar to that demonstrated by the antibody in standard radiometric assays, but the sensitivity was lower (1-1000 nM as compared to 0.1-100 nM in a radiometric assay; Sanders, 1976).

Our tests suggest that the effects of anti-PGE_2 antiserum on membrane potential was a specific antibody-hapten interaction. This reasoning is based on the observations that: i) membrane potential of membranes containing unconjugated dibenzo-18-crown-6 was not affected by the serum proteins of either immunized or non-immunized animals; ii) with membranes containing PGE_2-dibenzo-18-crown-6 conjugates membrane potential was shifted by serum from immunized animals, but was not affected by serum from non-immunized animals; iii) pre-absorbed anti-PGE_2 antibodies had no effect on the potentials of membranes containing conjugate; iv) PGF_{2a}, a prostaglandin of similar structure to PGE_2, did not alter the specific interaction between the anti-PGE_2 antibodies and conjugated, membrane-bound PGE_2.

At present the mechanism for the effect of antibody on membrane potential is not clear. The fact that the antibody effect is specific suggests it depends upon antibody-hapten binding at the membrane surface. Solsky and Rechnitz (1981) suggested that antibody binding may increase the selectivity of the membrane. Our data suggest a different mechanism because the presence of antibody caused a shift in transmembrane potential that is more consistent with a **decrease** in selectivity or a decrease in the potassium gradient. The latter might be due to the acidic properties of serum proteins at pH 7.2 (Harper et al, 1979) which might increase the potassium concentration at the membrane surface. It is easier to explain the mechanism of the solution-phase hapten on the antibody-membrane effect, which is likely to be consistent with the concept shown in Fig. 1. Its likely that the "free" PGE_2 competes with the membrane-bound PGE_2 for antibody binding sites. At high concentrations of solution-phase PGE_2 most of the antibody binding sites are occupied and unavailable to interact with the membrane.

Electroimmunoassay, as present here, is highly dependent on the ionic gradient, ionic strength, and pH of the solutions in contact with the membrane. This limits the usefulness of these assays somewhat because they require preconditioning of the ionic composition of samples. This may be achieved by: i) deionization of test solutions and reconstitution, ii) dilution of test solutions to decrease the concentration of contaminant ions. Another possible method to decrease the contaminant ion problem is to use more selective ionophores. Dibenzo-18-crown-6 is cation specific, but shows significant selectivity to several monovalent cations.

Electroimmunoassays, like other competitive, protein-binding assays, depend upon the concentration of the specific binding protein, the concentration of ligand, and the affinity constant of the protein for the ligand. The sensitivity of these assays depend to a large extent on the affinity constant and the specific activity of the labeled ligand (minimal detectable mass). Specificity is primarily depended upon the quality of antibodies. It is conceivable that the sensitivity and specificity of EIAs may be improved by monoclonal antibodies. Another method to improve the sensitivity of EIAs may be to increase the availability of the hapten to the solution-phase antibody. The ionophore-hapten conjugates prepared for the PGE_2 EIA described in this paper contained 2 PGE_2 per crown ether molecule (Fig. 2). The PGE_2 molecules were bound close to the core of the ionophore. Since the ionophore must reside in the membrane to serve as an ionic binding site (carrier), it is quite possible that the molecular design of the ionophore-hapten conjugate was not optimal for interaction with antibody, which was free in solution. It is possible that the interaction between antibody and hapten could be improved by placing "spacer" groups synthetically between the ionophore and the hapten. Another approach that may improve the interaction between the antibody and the hapten may be to place branched spacers between the ionophore and hapten. The multiple ends of each spacer branch could be used to increase the ratio of conjugation between hapten and ionophore. Ideally, greater availability of hapten per ionophore could increase the binding between antibody and ionophore-hapten ligand and perhaps increase the magnitude of the voltage response to a given antibody concentration.

It may also be feasible to develop an EIA probe. The EIA tests described in this paper were performed on samples to which antibody was added at a sufficient concentration to cause a voltage response across the antibody-sensitive electrode (see Fig. 1). This means that EIA currently cannot be used for "on-line" measurements. Analyses could be performed on-line if a means was developed to isolate the antibody from the sample to be tested. This could be achieved be constructing a small chamber between the antibody-sensitive membrane and the external solution (see Fig. 6). The interior of this chamber would be isolated from the exterior by a dialysis-type membrane, which would allow the passage of the low molecular weight compounds (haptens) to be measured, while retaining immunoglobulins. If the volume of the chamber is small and the diffusion distance between the dialysis membrane and the antibody-sensitive membrane is minimized, then the intra- and extrachamber solutions would equilibrate rapidly. Fig. 5 shows a schematic drawing of a possible EIA probe.

Based on our model study with prostaglandin, it is likely that other immunogenic compounds or haptens can be measured by EIAs. The technique is a potentially important advance in the study of endogenous, biologically active substances and industrial monitoring of the production of biochemicals if "on line" EIA becomes feasible. In biological studies of local regulatory agents, such as prostaglandins, electroimmunoassay probes may eventually provide vital information about tissue, or even cellular concentrations of these agents.

SUMMARY

A technique is described in which an antibody sensitive electrode for anti-prostaglandin E_2 antisera was used to measure solution-phase PGE_2 in nM quantities. The electrode was constructed by incorporating a cation selective ionophore-hapten (PGE_2) conjugate into a polyvinyl chloride membrane. Transmembrane potential in a fixed potassium gradient was measured. The addition of anti-PGE_2 antisera changed membrane potential in a concentration-dependent manner. The effect of anti-PGE_2 antibodies on membrane potential was decreased by adding "free" PGE_2 to the buffer containing antisera. With this technique a competitive protein binding assay was developed and standard curves for solution-phase PGE_2 were generated over a concentration range of 1 to 1000 nM. The assay was relatively specific for PGE_2; PGD_2 and PGF_{2a} had only minor effects on transmembrane potential over the effective concentration range for PGE_2.

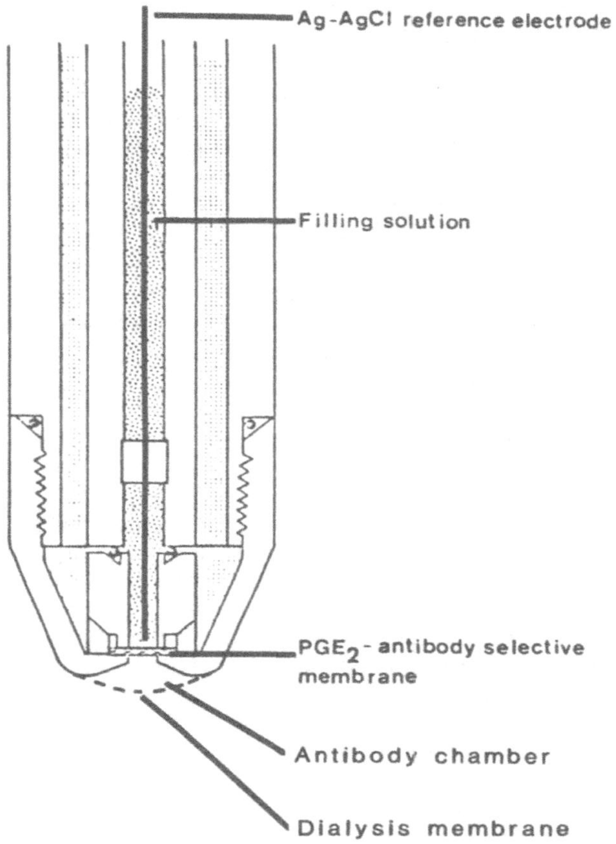

Fig. 6. Construction of an EIA probe. The figure shows a schematic of an electrode housing to which an "antibody" chamber has been added. The antibody chamber would be separated from the external bathing solution by a dialysis membrane through which small molecules (haptens) could pass, but immunoglobulins would be retained. The probe could be immersed in a solution containing unknown concentrations of hapten. The hapten would diffuse into the antibody chamber, compete for antibody binding sites with the hapten-ionophore conjugates incorporated into the antibody-sensitive membrane, and produce a quantifiable shift in transmembrane potential.

ACKNOWLEDGMENTS

The authors would like to thank Dr. Roy Williams for his input into the organic syntheses used in these studies. We are grateful to Dr. Frank Scully for performing NMR analysis. Mass spectral determinations were carried out at the Middle Atlantic Mass Spectroscopy Laboratory, a National Science Foundation shared instrument facility analysis. This research and preparation of this chapter were supported by Research Grant AM 32176 from the National Institutes of Health. Dr. Sanders was the recipient of and RCDA award (AM 01209) from NIH.

REFERENCES

Bodanszky, B. and Ondetti N.A., 1966, "Peptide Synthesis," John Willy and Sons, New York. p. 88.

Connell, G.R., Williams, R.L. and Sanders, K.M., 1983, Electroimmunoassay: A new competitive protein-binding assay using antibody-sensitive electrodes, Biophys. J., 44:123.

Connell, G.R. and Sanders, K.M., 1984, Design of ionophore-hapten conjugates for electroimmunoassay, Proc. West. Pharmacol. Soc., 27:337.

Feigenbaum, W.M. and Michel, R.H., 1971, Novel polyamides from macrocyclic ethers, J. Polym. Sci., 9:817.

Harper, H.A., Rodwell, V.W. and Mayes, P.A., 1979, "Review of Physiological Chemistry," Lange Medical Publications, Los Alamos, California. p. 199.

Lowrey, O.H., Rosebrough, N.J., Farr A.L. and Randall, R.J., 1951, Protein measurement with the folin phenol reagent, J. Biol. Chem., 193:265.

Sanders, K.M., 1976, Role of endogenous prostaglandin E in the regulation of cat intestinal function, Doctoral Dissertation, University of California, Los Angeles.

Solsky, R.L. and Rechnitz, G.A., 1979, Antibody-selective membrane electrodes, Science, 204:1308.

Solsky, R.L. and Rechnitz, G.A., 1981, Preparation and properties of an antibody-selective membrane electrode, Anal. Chim. acta, 123:135.

IMMUNO-POTENTIOMETRIC SENSORS WITH ANTIGEN COATED MEMBRANE

Shuichi Suzuki and Masuo Aizawa*

Department of Environmental Engineering
Saitama Institute of Technology
1690 Fusaiji, Okabe, Saitama 369-02, Japan

*Department of Chemical Engineering
Tokyo Institute of Technology
Ookayama, Meguro-ku, Tokyo 152, Japan

INTRODUCTION

In the last decade biosubstances such as enzymes and antibodies have been used in conjunction with electrochemical sensoring devices to form bio-selective sensors (biosensors), which include enzyme sensors, microbial sensors, and immunosensors (Aizawa, 1983; Suzuki, 1984). An enzyme sensor is the union of an enzyme, that is a biological catalyst which acts sensitively and specifically to almost all organic and inorganic componds in nature, with an electrochemical sensor. The result is a sensor which is useful for the assay of organic and inorganic compounds in such a manner as simple as a pH measurement with a glass electrode. Since an enzyme is water-soluble, it is immobilized (insolubilized) on solid matrix at or near the sensing device surface to serve as biocatalyst for specific molecular recognition. Due to marked progress in enzyme immobilization techniques, many enzyme sensors have been developed for ionic and low molecular weight compounds and evaluated particularly in clinical fields.

For biological and biomedical purposes, the possibilities for making sensitive and selective biosensors for a specific peptide and protein are an exciting prospect. A potentiometric immunosensor, which depends its selectivity on immunochemical affinity of an antigen to the corresponding antibody, has been developed for the determination of syphilis antibody in human sera(Aizawa, 1977a;1977b;1977c; 1979). Several other potentiometric immunosensors have also been reported (Janata, 1975; Yamamoto, 1978; Aizawa, 1980). These sensors consist of solid matrix (membrane and electrode)-bound antigen (or antibody); thus the antigen-antibody complex formation on the matrix is followed by measuring membrane (or electrode) potential as schematically represented in Fig.1.

If an antibody is attached to a membrane, the membrane reacts specifically with the corresponding antigen at the membrane surface as shown in Fig.2. Such an immunochemical reaction of the membrane-bound antigen with free antibody in solution may cause a change of charge density at the membrane-solution interface, generating a transmembrane potential change.

The membrane potential across a fixed charge membrane, $\Delta\phi$, is approximated by the following equation, when uni-uni valent electrolytes are separated.

$$\Delta\phi = (RT/F)\{\ln(C_1/C_2)-\ln(-\theta+\sqrt{\theta^2+4C_1^2})/(-\theta+\sqrt{\theta^2+4C_2^2}) \\ +(1-2t)\ln[(1-2t)\theta+\sqrt{\theta^2+4C_1^2}]/[(1-2t)\theta+\sqrt{\theta^2+4C_2^2}]\} \qquad (1)$$

where C_1 and C_2 ($C_1>C_2$) are electrolyte concentrations, θ is the charge density of the membrane phase, and t is the ion transport number in the membrane phase (Kobatake, 1965). The first and second terms result from the surface potential and the third term from diffusion potential. In case of extremely high electrolyte concentration, $C \gg |\theta|$, the membrane potential is approximated by diffusion potential. On the other hand, the surface potential may be predominant when electrolyte concentration is infinite.

The potential profiles across the antibody-bound membrane are presented in Fig.2 along with the conceptual illustration of the immunochemical reaction on the membrane surface.

Potentiometric immunosensors for syphillis, blood group typing, and human serum albumin (Aizawa, 1982) are involved in this chapter.

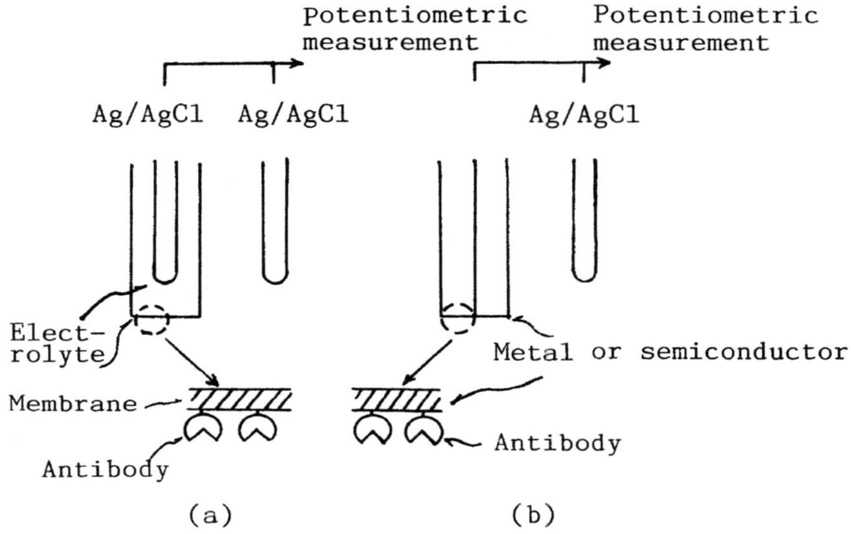

Fig. 1. Schematic representation of non-labeling type immunosensors.

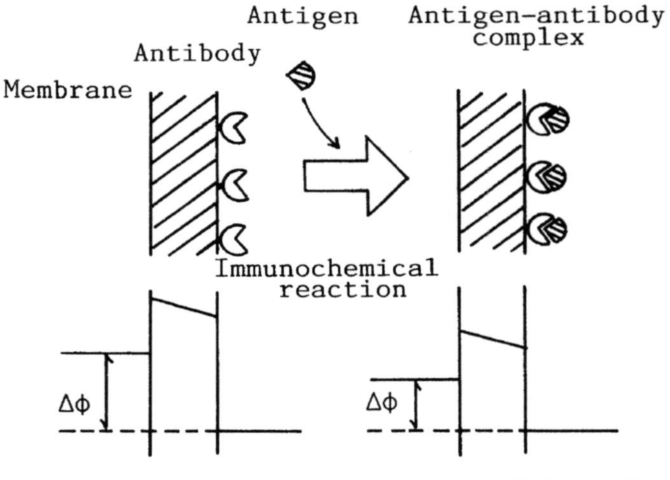

Fig. 2. Postulate potential profiles of an antibody-bound membrane before and after immunochemical reaction.

MATERIALS AND METHODS

Preparation of Antigen-Binding Membrane

(1) Cellulose triacetate (250 mg) was dissolved in 5 ml of mixed CH_2Cl_2/EtOH (9:1) solvent, One ml of the same solvent containing 25 mg of Ogata antigen for non-treponemal tests for syphilis was added to the cellulose triacetate solution and mixed with stirring. The Ogata antigen was obtained from Sumitomo Chemical Co. (Osaka) and contained 0.01% cardiolipin, 0.04% phosphatidylcholine, and 0.20 % cholesterol in ethanol. The ethanol was evaporated and replaced by a mixed solvent of CH_2CCl_2/EtOH (9:1) before use. The resulting solution was cast on a glass plate (20×20 cm^2). The cast membrane was dried at 25 °C for 20 min, and then allowed to stand at 25 °C under a reduced pressure for 3 h. The membrane was cut into small pieces (1.5× 1.5 cm^2) and immersed in physiological saline at 25 °C for more than 5 h to remove soluble substances.

(2) Freshly collected human blood using 0.1 % sodium citrate solution as anticoagulant was centrifuged at 3,000 rpm. The erythrocyte separated from serum was washed three times with 0.9 % NaCl solution, and was then hemolyzed by adding 20 vol. of water. The resultant solution was centrifuged at 15,000 g to separate the erythrocyte ghost membrane from soluble materials in the cell. The precipitated ghost membrane was washed twice with distilled water. The total lipid of the ghost membrane was extracted by immersing the ghost membrane in a $CHCl_3$: MeOH (1:2) solution for 2 days. After adding water,the $CHCl_3$ layer was separated and evaporated. The lipid was dried in a nitrogen gas stream and dissolved in a CH_2Cl_2:EtOH (9:1) solution. The lipid concentration used for membrane preparation was 2.3 $mg \cdot ml^{-1}$ for type A lipid and 3.2 $mg \cdot ml^{-1}$ for type B lipid, respectively.

Twenty milligrams of cellulose triacetate were dissolved in 0.5 ml of CH_2Cl_2:EtOH (9:1) solution containing each type of total erythrocyte lipid. The resultant solution was cast in a glass mold with a flat bottom (diameter = 19 mm). The solvent was evaporated at reduced pressure, and the 20-μm-thick membrane thus obtained was peeled off. The membrane was cut into small pieces (1.5×1.5 cm^2).

Preparation of Antibody-Binding Membrane

Cellulose triacetate (250 mg) was dissolved in 5 ml of dichloromethane, and 0.05 ml of 70 % glutaraldehyde followed by 0.15 ml of 1,8-diamino-4-aminomethyl octane. The solution was cast on a glass plate and allowed to stand at 25 °C to complete intermolecular cross-linking of 1,8-diamino-4-aminomethyl octane and glutaraldehyde. The membrane peeled off was placed in a 0.1 % solution of glutaraldehyde at 30 °C and pH 8.0 for 2 h, and was in contact with a phosphate buffer solution containing anti-HSA antibody around 4 °C for 16 h. The membrane was reduced with 0.1 M $NaBH_4$ for 3 min. After thorough washings, the membrane was used for a potentiometric immunosensor.

Measurement of Membrane Potential

Each membrane was mounted between a pair of glass compartments with the aid of silicone rubber gaskets. A concentration gradient was induced across the membrane by filling the compartments with NaCl solutions of different concentrations. The solutions in both compartments were stirred. The transmembrane potential was transmitted to an electrometer (Model HE-102A, Hokuto Denko, Tokyo) and a recorder through a pair of saturated calomel electrodes (SCE) or silver-silver cloride electrodes (Ag AgCl):

SCE_I or Ag $AgCl_I$	KCl or NaCl solution (C_1)	Membrane	KCl or NaCl solution (C_2)	SCE_{II} or Ag $AgCl_{II}$

where C_1 and C_2 ($C_1 > C_2$) are NaCl concentrations, and SCE_{II} or $AgAgCl_{II}$ is grounded.

A Potentiometric Immunosensor

Either antigen-binding or antibody-binding membrane was attached to a holder and a pair of reference electrodes were placed as illustrated in Fig.1 (a). The holder contained a fixed concentration of KCl solution (i.e., 4 mM). The sensor was in contact with a sample solution. Potential difference between the sensor and a reference electrode (SCE or AgAgCl) were measured through an electrometer.

EXPERIMENTAL RESULTS AND DISCUSSION

Potential Change Associated with the Immunochemical Reaction of Membrane-Bound Antigen

As indicated by eqn.(1), the membrane potential must be zero when $C_1 = C_2$, if the effective fixed charge is distributed homogeneously throughout the membrane phase. Membrane potential changes depend on the fixed charge density of the membrane phase and the ionic mobility when the electrolyte concentration ratio (C_1/C_2) is fixed. A plot of electrolyte concentrations versus membrane potential yields a sigmoid curve when C_1/C_2 is constant. In the high electrolyte concentration range, the membrane potential is determined by the diffusion potential. In contrast, the diffusion potential may become negligible with increasing Donnan potential is the low concentration range. The ultimate membrane potential due to the Donnan potential should be $-(RT/F)\ln(C_1/C_2)$.

An antigen-binding membrane was changed in a cell for membrane potential measurements. When both cell compartments I and II contained equivalent concentrations of electrolyte, no appreciable membrane potential generated. The electrolyte in compartment I was then replaced by a different concentration of electrolyte while the C_1/C_2 ratio maintained constant at a value of 2. Membrane potentials across an Ogata antigen-binding membrane is plotted against $\log C_1$ in Fig.3. As was expected, a sigmoid curve of $\Delta\phi$ versus $\log C_1$ was obtained. The plots of $(F/RT)\Delta\phi$ against $1/C_2$ gave a straight line as shown in Fig.4.

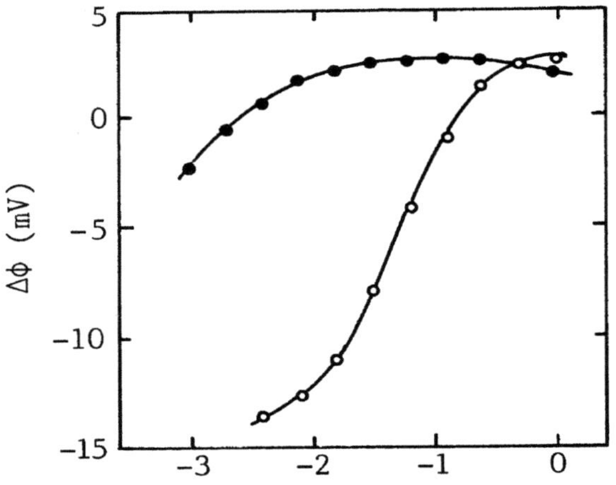

3

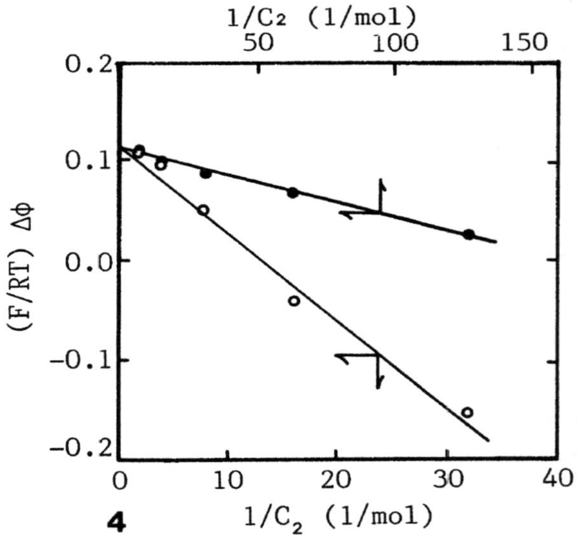

4

Fig.3. Dependence of membrane potential on NaCl concentration at $C_1/C_2=2.0$. Each membrane was reacted with an antibody-free serum(-o-) and antibody-containing serum(-•-) at 37 °C for 1 h.

Fig.4. Plots of $(F/RT)\Delta\phi$ against $1/C_2$ for NaCl solution at $C_1/C_2=2.0$. Each membrane was reacted with an antibody-free serum(-o-) and an antibody-containing serum(-•-).

The Ogata antigen-binding membrane was immersed in the antibody-containing solution at 37 °C for 1 h, and was extensively washed with physiological saline and water. Membrane potential was measured as indicated above. The relationship between $\Delta\phi$ and $1/C_2$ is involved in Fig.3. A marked shift in membrane potential was developed by the antigen-binding membrane when it was reacted with the antibody-containing serum. Since there would be both specific and non-specific adsorption of proteins onto the membrane. In order to minimize non-specific adsorption, the membrane was extensively washed with saline after the contact with serum. In the observation of trans-membrane potential, the effects of the non-specific adsorption could be neglected because the membrane reacted with the antibody-free serum exhibited no appreciable potential shift. Thus the membrane potential shift may be attributed to a potential change caused by the specific adsorption (immunochemical reaction) of antibody onto the membrane-bound antigen, probably at the membrane-solution interface.

From the slope and intercept of the plots of $(F/RT)\Delta\phi$ against $1/C_2$ in Fig.4, θ was determined to be 3.6×10^{-4} eq•l^{-1} for the antigen-binding membrane reacted with the antibody. According to the same procedure, θ was determined for the antigen-binding membrane reacted with the antibody-free serum. These results indicate that a decrease of negative charge resulted from the immunochemical reaction of the membrane-bound antigen with the free antibody in solution. The decrease of negative charge was calculated to be 3.6×10^{-2} eq•l^{-1}. The Ogata antigen consists of cardiolipin, phosphatidylcholine, and cholesterol. Since cardiolipin possesses the negative charge of phosphate groups, the antigen-binding membrane could assume the negative charge. The immunochemical reaction caused the antigen-binding membrane to decrease the negative charge by 3.6×10^{-2} eq•l^{-1}. A possible explanation is that the negatively charged groups of the membrane-bound antigen are hindered by the antibody.

The immunochemically-induced membrane potential change was also demonstrated with an erythrocyte antigen (blood group substance)-binding membrane. An erythrocyte antigen-binding membrane was designated as EL(A)M or EL(B)M, depending on whether it contained A or B blood group substance, respectively. The membrane potential was measured at 25 °C at a fixed NaCl concentration ratio of $C_1/C_2=2$ after incubating at 25 °C for 30 min with 0.9 % NaCl and antibody-containing serum. The NaCl concentration dependences of the membrane potential across EL(A)M and EL(B)M are presented in Figs.5 and 6.

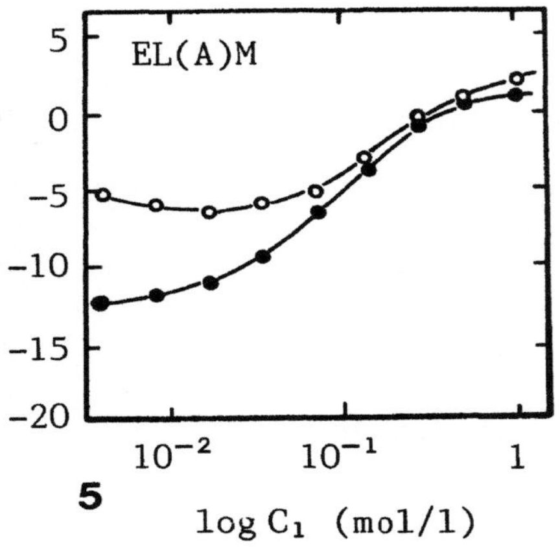

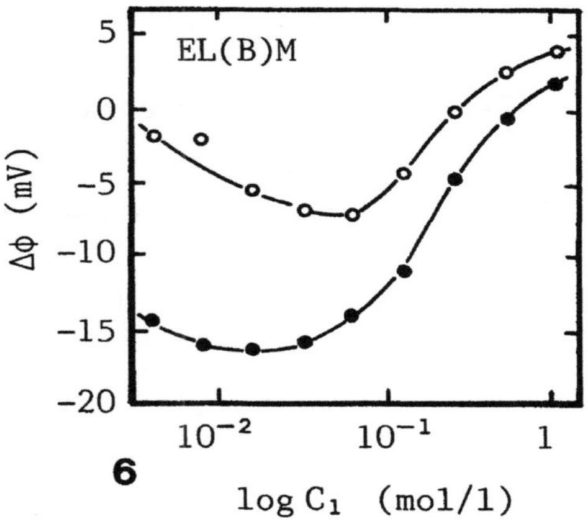

Fig. 5. NaCl concentration dependence of the membrane potential across EL(A)M. The membrane potential was measured at 25 °C at a fixed NaCl concentration ratio of $C_1/C_2=2$ after incubating at 25 °C for 30 min with 0.9 % NaCl(-o-) and anti-serum(-●-).

Fig. 6. NaCl concentration dependence of the membrane potential across EL(B)M. The conditions were the same as in Fig. 5.

The membrane potentials shifted markedly to negative values after the immunochemical reaction. These results clearly show that a distinct change in the membrane potentials resulted from the specific adsorption of antibody by membrane-bound blood group substance, and moreover that the immunochemical reaction induced an increase in the negative charge of the membrane phase.

The erythrocyte membrane lipid contains the blood group determinant of oligosaccharide as in represented in Table 1. As the total lipid extracted from each type of erythrocyte distinctly showed agglutination activity when in liposome form, it must have involved the blood group substance. Therfore, the membrane-bound total lipid was assumed to have the blood-group substance partly on the membrane surface. The membrane-bound blood-group substance may bind its corresponding agglutinin (antibody) on the membrane surface. Since the antibody is a polyelectrolyte, the charge density of the membrane surface could be markedly influenced by such a specific adsorption of antibody.

Both EL(A)M and EL(B)M contain the polargroups of lipids, such as phosphate, carboxyl, and amino groups. Since the pH values in this study were in the range of 5.5 - 6.0, the net charge of these membranes may have been negative. This interpretation is supported by the electrolyte concentration-dependence of the transmembrane potential. It suggested that either an anti-A or an anti-B antibody could specifically adsorb on the corresponding membrane-bound blood-group substance. The additional negative charge due to agglutination must result in a marked shift in transmembrane potential.

Fig.7 shows the electrolyte concentration dependence of membrane potential for the antibody-bound membrane. Electrolyte concentration gradient, C_1/C_2, was fixed at 2 and 8. The curves obtained were simulated by eq.(1). The membrane potential measurements indicate that the antibody-bound membrane is positively charged, which is interpreted by the positive charge of the amino groups of the membrane matrix and of the antibody under the conditions. The immunochemical reaction of membrane-bound antibody with free HSA in solution induced negative shift in membrane potential. Since HSA is negatively charged in solution, antigen-antibody complex formation may cause the antibody-bound membrane to reduce the membrane potential.

Table 1. Structure of Determinant of Blood Group Substance

Blood group	Structure
O	Gal-GlcNAc--- | Fuc
A	GalNAc-Gal-GlcNAc--- | Fuc
B	Gal-Gal-GlcNAc--- | Fuc

Gal:galactose, Fuc:fucose, GlcNAc:N-acetyl glucosamine, GalNAc:N-acetyl galactosamine.

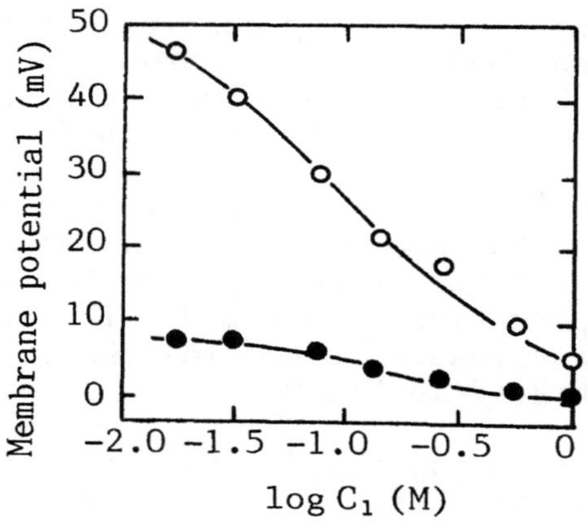

Fig.7. Electrolyte concentration dependence of transmembrane potential across an antibody-bound membrane. Electrolyte: KCl, Concentration gradient: C_1/C_2=8(●) and 2(○).

Response of Potentiometric Immunosensors

(1) Immunosensors for Syphilis. An immunosensor for syphilis was constructed using an Ogata antigen-binding membrane as represented in Fig.1. The membrane contained 37.5 μg of Ogata antigen per cm^2 of membrane. Figure 8 presents the relationship between the immunochemically induced potential change and the concentration of the antibody solution with which the immunosensor was contacted. The antibody solution was made up by the dilution of Positive Control Serum, containing the Wasserman antibody, which was used for non-treponemal serology tests for syphilis (supplied from DADE Division, American Hospital Supply Co., Miami, Fla.), with physiological saline. One arbitrary unit is equivalent to the antibody concentration of Positive Control Serum the reactivity of which was assayed at a dilution of 1:32 by Rapid Plasma Reagin (RPR) and Unheated Serum Reasin (USR) Tests. The immunosensor was extensively washed after each immunochemical reaction.

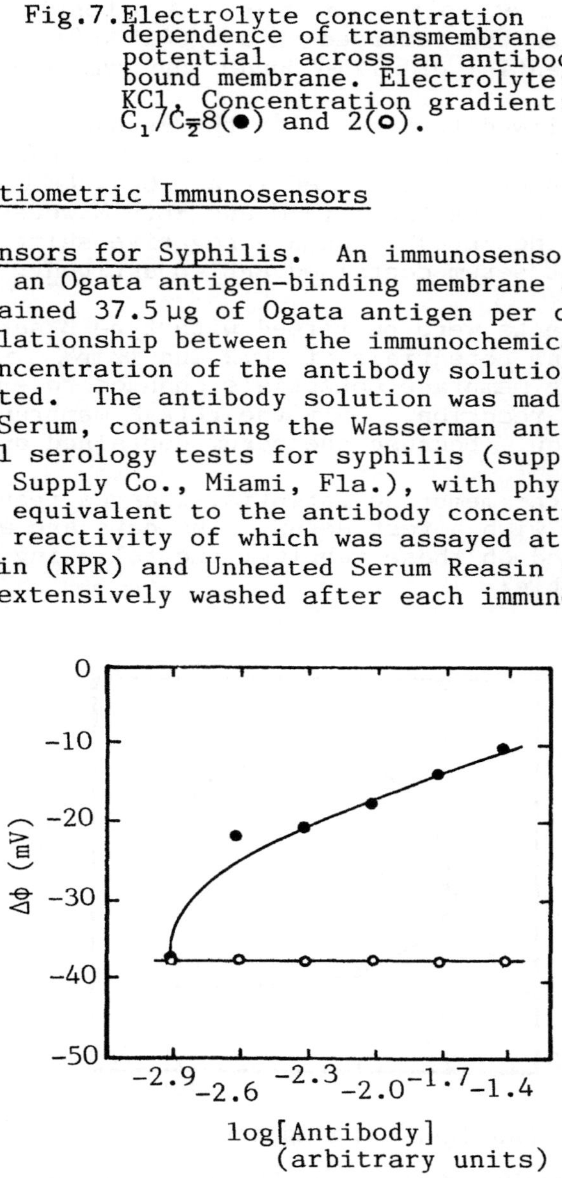

Fig.8. Relationship between the immunochemically induced potential change and the concentration of the antibody. One arbitrary unit is equivalent to the antibody concentration of Positive Control Serum. ○,in antibody-free serum; ●,in antibody-containing serum. C_1=1/64 M, C_2=1/512 M NaCl.

An appreciable potential change was obserbed even at an antibody concentration of 1/800 (log[antibody] = −2.9), which was comparable to an 800-fold dilution of Positive Control Serum. The immunochemically induced potential increased with an increase of the antibody concentration.

Membranes with different antigen contents were prepared in order to determine the dependency of the immunochemically induced potential. A membrane containing less than 25 μg antigen•mg^{-1} membrane gave no appreciable immunochemically induced potential change under the above experimental conditions. The immunochemically induced potential change correlated with the antigen content as demonstrated in Fig.9.

(2) Electrochemical typing of blood. Electrochemical typing of blood was performed with control sera for blood group tests. An EL(A)M was mounted in an immunosensor as illustrated in Fig.1. The immunosensor contained 1/512 M NaCl as an inner electrolyte solution. A potential of −7.2 mV was generated, when the sensor was immersed in a 1/64 M NaCl solution. An EL(B)M developed a membrane potential of −12 mV under the same conditions.

Type A serum was allowed to react with the EL(A)M and the EL(B)M of immunosensors at 25 °C for 30 min. After washing with a 0.9 % NaCl solution and water, the immunosensors were placed in a 1/64 M NaCl solution. The transmembrane potentials across the EL(A)M and EL(B)M were −7.2 and −33.6 mV, respectively. Only the EL(B)M showed a negative shift in membrane potential; thus, the test serum contained anti-B antibody.

The similar experiments were performed with type B serum. An EL(A)M and EL(B)M showed membrane potentials of −8.2 and 12 mV, respectively, before the reaction. The membrane potentials changed to −34 and 7.2 mV, respectively, due to the reaction. Only the EL(A)M membrane potential changed after the reaction , because the serum contained anti-A antibody.

Table 2 summarizes the membrane potential changes resulting from the contact of each membrane with a test serum. The data are average values for repeated tests. Based on these results, the following electrochemical typing of blood is possible:

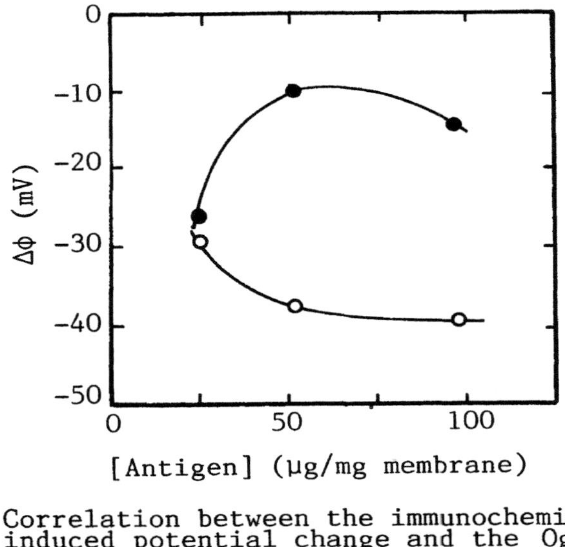

Fig. 9. Correlation between the immunochemically induced potential change and the Ogata-antigen content in the membrane. ○,in antibody-free serum; ●,in antibody-containing serum. C_1=1/64 M, C_2=1/512 M NaCl.

Table 2. Membrane Potential Changes Derived from Agglutination**

	EL(A)M*	EL(B)M*
Type A serum	0 ± 2 mV	−28 ± 2 mV
Type B serum	−22 ± 2 mV	0 ± 2 mV

*EL(A)M, EL(B)M: membranes containing type A and type B blood-group substances,respectively.
**The membrane was allowed to react with each serum for 30 min. The membrane potential was measured before and after reaction, while maintaining C_1/C_2 at 8.

a) Type A — EL(A)M, none; EL(B)M, negative shift
b) Type B — EL(A)M, negative shift; EL(B)M, none
c) Type AB — EL(A)M, negative shift, EL(B)M, negative shift
d) Type O — EL(A)M, none; EL(B)M, none

Any negative shift in membrane potential is attributed to the agglutination of membrane-bound blood group substance; therefore, blood can be typed by checking if the test serum causes the EL(A)M and EL(B)M to shift their membrane potentials.

(3) Immunosensors for HSA. An antibody-bound membrane was used to construct an immunosensor for HSA. The structure of the immunosensor is illustrated in Fig.1. An electrolyte solution (4 mM KCl) was contained in the immunosensor.

The immunosensor was immersed in a phosphate buffer containing 1 mg•ml^{-1} HSA at 37 °C for 1 h. After thorough washing, the immunosensor was placed in 32 mM KCl. The sensor output was reduced by 4.8 mV. The output change is caused by the immunochemical adsorption of HSA to the antibody-bound membrane of the sensor. The sensor output difference decreased with a decrease in HSA, reaching the minimum at a concentration of 10^{-6} g•ml^{-1}. Figure 10 shows the plots of the sensor output difference against HSA concentration, indicating that HSA is determined in the concentration range 10^{-5}–10^{-3} g•ml^{-1} with the immunosensor.

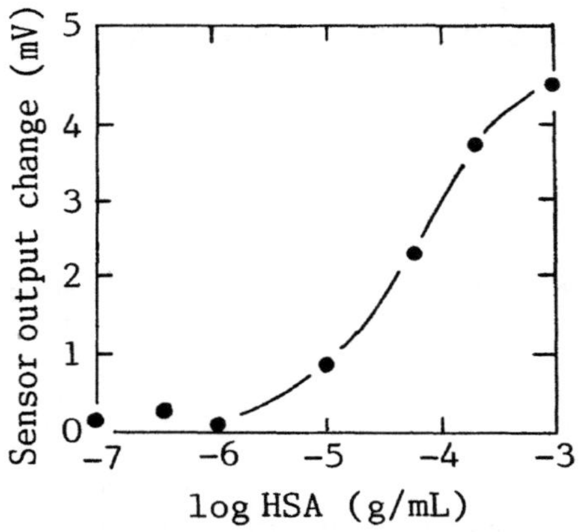

Fig.10. Calibration curve for HSA determination with a non-labeling type immunosensor.

DISCUSSION

It has been revealed that an antigen (or antibody)-binding membrane adsorbs the corresponding antibody (or antigen) at the membrane surface with a resulting change of the transmembrane potential. The transmembrane potential is believed to alter because of the adsorption of the antibody protein, perhaps in the form of a monolayer.

To elucidate the role of chemical substances, especially protein molecules, attached to the membrane surface on the generation of transmembrane potential, the surface of cellulose bromoacetate membranes was chemically modified by the covalent immobilization of several proteins. The transmembrane potential of these modified membranes was investigated as a guide to understanding their surface characteristics.

Cellulose triacetate (200 mg) was dissolved in 6 ml of CH_2Cl_2/C_2H_5OH (9:1). Cellulose bromoacetate was dispersed at the desired concentration in the solution. A cellulose bromoacetate membrane (ACM-Br) was prepared in the similar manner as a cellulose triacetate membrane.

An ACM-Br was chemically modified as follows. In the first step, amino groups were introduced using 1,6-diaminohexane to yield a membrane containing amino groups (ACM-NH_2). In the second step, the amino groups of the ACM-NH_2 were coupled to 1,2:3,4-diepoxybutane. The epoxy-attached membrane (ACM-EPOX) was obtained. The postulated reaction scheme of the chemical modification is schematically presented in Fig.11.

The ACM-EPOX was then incubated in a solution containing egg albumin or lysozyme to form a protein-binding membrane.

The concentration dependence of the transmembrane potential is shown for the ACM-Br_2, ACM-NH_2, and ACM-EPOX in Fig.12. The charge density of each membrane was calculated according eq.(1). Table 3 presents the results of θ and t_a for these membranes. The ACM-Br and ACM-EPOX exhibited a slightly negative charge. The introduction of amino groups to the ACM-Br produced a strongly positive charge. There was no appreciable variation of the transport number among these membranes.

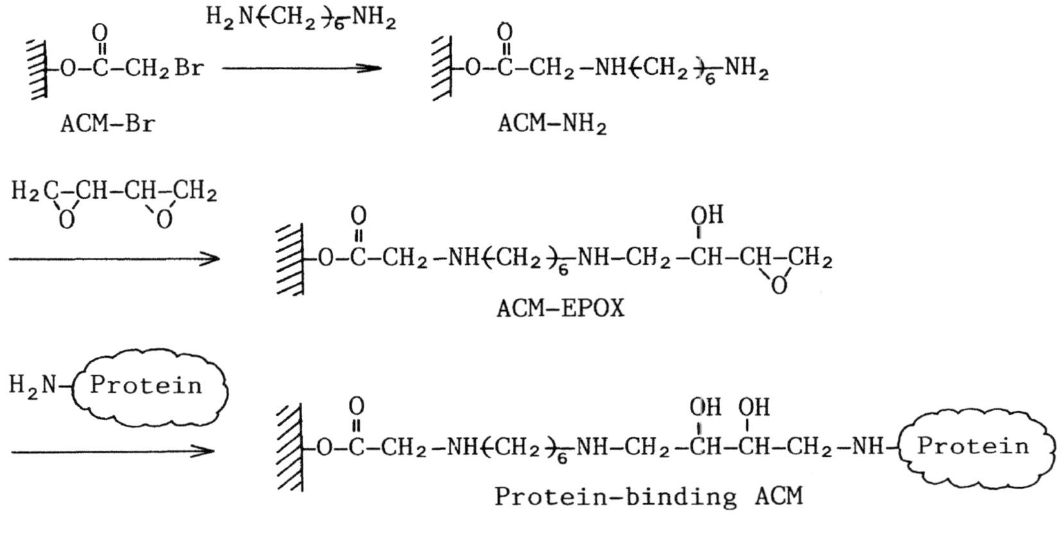

Fig.11. The postulated scheme for the chemical modification of a bromoacetyl cellulose membrane.

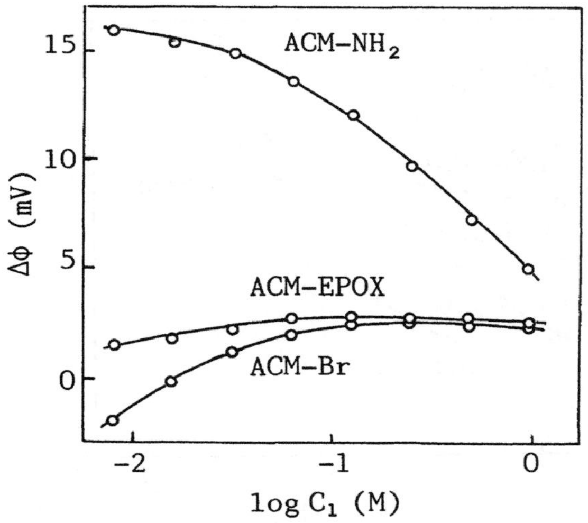

Fig.12. The transmembrane potential vs. $\log C_1$ curves for ACM-Br, $ACM\text{-}NH_2$, and ACM-EPOX. $C_1/C_2=2.0$, NaCl.

Table 3. Transport Numbers and Charge Densities of the Modified Membranes

Membrane	Transport number of Cl^-(t_a)	Charge density(θ) (eq./l)
ACM-Br	0.57	-3.54×10^{-3}
ACM-NH	0.57	1.86×10^{-1}
ACM-EPOX	0.58	-1.40×10^{-3}

Figure 13 involved the characteristics of the transmembrane potential across protein-binding membranes. The charge density of the ACM-Gly was estimated to be -3.2×10^{-2} eq•l^{-1}. A marked increase in negative charge was caused by the introduction of carboxyl groups. The lysozyme-binding membrane (ACM-Lys) had a positive charge and the egg albumin-binding membrane (ACM-Alb) had a negative charge. The calculated charge densities of these membranes are listed in Table 4. The charge densities of the ACM-Lys and the ACM-Alb were estimsted to be 4.1×10^{-9} eq•l^{-1} and -1.0×10^{-2} eq•l^{-1}, respectively. These values are reasonable, because lysozyme may be positive charged, and egg albumin negatively charged under the conditions of the transmembrane potential measurement. These results strongly suggest that the transmembrane potential should depend on the charged substances attached to the membrane surface.

It may be concluded that the potential generation of an antigen (or antibody)-binding membrane due to specific adsorption of the corresponding antibody (or antigen) results from a change in the charge density of the membrane surface.

SUMMARY

Antigen (or antibody)-binding membranes have found feasible application in constructing potentiometric immunosensors which can respond in membrane potential specifically to the corresponding antibody (or antigen). Potentio-

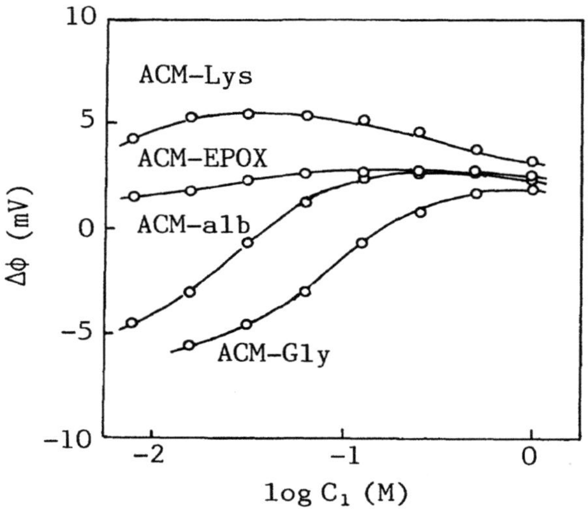

Fig.13. The transmembrane potential vs. $\log C_1$ curves for the protein-binding membranes. $C_1/C_2=2.0$,NaCl.

Table 4. Characterization of the Protein-Binding ACM

Membrane	pI[a]	$\Delta\phi$(mV)[b]	θ(eq./l)
Egg albumin-ACM	4.6	-13.5	-0.010
Lysozyme-ACM	11.0	19.4	0.041

[a]Isoelectric point of protein.
[b]$C_1=1/64$ M, $C_1/C_2=8.0$, NaCl.

metric immunosensors for syphilis antibody in serum, blood typing, and serum proteins have been developed and characterized. The mechanism of immunochemically induced membrane potenrial are discussed.

REFERENCES

Aizawa, M., Kato, S. and Suzuki, S., 1977, Immunoresponsive membrane, J. Membrane Sci. 2:125-132.
Aizawa, M., Suzuki, S., Nagamura, Y., Shinohara, R. and Ishiguro, I., 1977, An immunosensor for specific protein, Chem. Lett. 1977:779-782.
Aizawa, M., Kato, S., Suzuki, S., Nagamura, Y. and Ishiguro, I., 1977, Immunoresponsive membrane II., Kobunshi Ronbunshu, 34:813-818.
Aizawa, M., Suzuki, S., Nagamura, Y., Shinihara, R. and Ishiguro, I., 1979, An immunosensor for syphilis, J. Solid-Phase Biochem. 4:25-31.
Aizawa, M., Kato, S. and Suzuki, S., 1980, Electrochemical typing of blood using affinity membranes, J. Membrane Sci. 7:1-10.
Aizawa, M. and Suzuki, S., 1982, Chemical amplification in biosensors, Jap. J. Appl. Phys. 21(21-1):219-223.
Aizawa, M., 1983, Molecular recognition and chemical amplification of biosensors, in: "Chemical Sensors," T. Seiyama, et al., eds., Kodansha, Tokyo.
Janata, J., 1975, Immunoelectrode, J. Am. Chem. Soc. 97:2914-2915

Kobatake, Y., Takeguchi, N., Toyoshima, Y. and Fujita, H., 1965, Studies of membrane phenomina, J. Phys. Chem. 69:3981-3987.
Suzuki, S., ed., 1984, "Biosensors," Kodansha, Tokyo.
Yamamoto, Y., Nagasawa, Y., Sawai, M., Suda, T. and Tsubomura, H., 1978, Potentiometric investigations of antigen-antibody and enzyme-enzyme inhibitor reactions using chemically modified metal electrodes, J. Immunol. Methods 22:309-315.

THE OPSONIZED ELECTRODE

H. Allen O. Hill and Nicholas J. Walton

Inorganic Chemistry Laboratory, University of Oxford
South Parks Road, Oxford, OX1 3QR, England

Introduction

The superoxide anion is the one electron reduction product of molecular oxygen. For this reason it can be both produced from, and re-oxidised back to, dioxygen electrochemically in non-aqueous solvents. In aqueous solutions, however, electrochemical reduction of dioxygen always yields hydrogen peroxide. When superoxide _is_ produced, in aqueous media, for example, enzymatically, then it disproportionates rather rapidly to dioxygen and hydrogen peroxide. Human neutrophils are well known to produce superoxide from dioxygen when stimulated (see, for example, Babior et al., 1973, Johnston et al., 1975, Green et al., 1979). The principle behind the work described in this chapter is that if neutrophils could be induced to produce their cytotoxic superoxide at, or close to, the surface of an electrode then it may be possible to re-oxidise electrochemically that superoxide back to dioxygen _before_ disproportionation takes place. In this way an oxidation current response would detect the stimulation of the neutrophils. Such stimulation is always a result, however, of a membrane perturbation on the surface of the cell. The stimulus producing this can be soluble, as in the case of phorbol myristate acetate (De Chatelet, 1976) or, more commonly, it can be a surface coated with an immunoglobulin such as IgG. The latter generally takes the form of microscopic particles, for example, latex beads or invading bacteria, coated with IgG. We describe here, however, the use of a voltammetric electrode, coated with IgG, to stimulate human neutrophils. The surface of the electrode is opsonized, or "made appetising", towards the cells by non-specific adsorption of IgG onto its surface. The opsonized electrode then acts as stimulus to the neutrophils but is also able to re-oxidise electrochemically any superoxide produced by them back to dioxygen and so detect their respiratory burst. It is the unique combination of stimulus and detector in the guise of the opsonized electrode which allows the technique to succeed. If the superoxide were produced anywhere but _at_ the electrode surface then rapid disproportionation would ensure that very little indeed reached the electrode, resulting in very small currents.

We also describe the use of microvoltammetric electrodes, of 10μm diameter, to observe the respiratory burst of a _single_ neutrophil. Such microscopically small electrodes are directly comparable in size to that of the average human neutrophil with a diameter of approximately 8μm. It is for this reason that we suggest that a microvoltammetric electrode might be expected to re-oxidise the local flux of superoxide associated with only a single stimulated human neutrophil. There are very few techniques suitably sensitive to detect such an event. The electrochemical experiment at fixed potential is one of them. Methods such as measurement of dioxygen consumption or cytochrome c reduction can give only the ensemble response of a large number of cells. As will be shown, the opsonized electrode provides a novel method for the characterization of that immunological reaction taking place between an IgG-coated surface and a single human neutrophil which leads to the production of superoxide radicals.

Materials and Methods

Human neutrophils were prepared using a method similar to that of Boyum (1974). They were finally resuspended into either RPMI 1640 medium (macro-electrode experiments) or Hanks' balanced salt solution (micro-electrode experiments).

The electrochemical experiments using a macroscopic working electrode were performed in a glass cell which was water-jacketed and thermostatted at 37°C. The glass walls of the cell were silanized to prevent neutrophils aggregating on the glass surface. A side-arm with frit contained the platinum gauze counter electrode, and a saturated calomel reference electrode (Radiometer K401) in a second side-arm was connected to the working compartment by a Luggin capillary. The working electrode was a 6mm diameter pyrolytic graphite rotating disc. A rotation speed of 10Hz was used to induce gentle hydrodynamic mixing of the solution. The motor drive, motor speed controller and potentiostat were from Oxford Electrodes. The current-time traces were recorded on a Bryans 26000 series X-Y recorder operating in the time-base mode. A 1Hz low-pass filter plus 60uF capacitance across the inputs of the chart recorder were used to achieve a low experimental noise level at the expense of a small loss in unrequired frequency response.

The electrochemical experiments using a microscopically small electrode utilised a 2-electrode configuration throughout. The cell consisted simply of a small plastic sample tube which was thermostatted at 37°C. The secondary electrode was a loop of silver wire, lightly anodised to give a coating of AgCl, which rested at the bottom of the tube around its edges. The working electrode was a 10um diameter gold microvoltammetric electrode. The cell plus electrodes, together with fixed potential source (battery driven) and current amplifier (Keithley model 427) were all placed inside a Faraday cage to reduce noise pick-up. A gain of 10^{10} was used in current amplification. An amplifier rise-time of 300ms, together with a 1Hz low-pass filter, were used to remove high frequency noise. The current-time traces were finally recorded on a Bryans 60000 series X-Y recorder operating in the time-base mode.

The working electrode, large or small, was polished before each experiment using a slurry of alumina (0.1 μm particle size)/water and rinsed with deionized water. It was subsequently opsonized by dipping it for approximately two minutes into a solution of 30mg/cm^3 human IgG (Miles Biochemicals). The electrode was then washed thoroughly with deionized water before connection to the electrochemical cell. The fixed potential applied was +50 mV vs. SCE (or its equivalent in the case of the microelectrode experiment). This very modest oxidizing potential was considered easily sufficient to oxidize superoxide back to dioxygen (the thermodynamic potential is $E^o(O_2/O_2^-)$ = -574 mV vs. SCE) but nowhere near positive enough to oxidise hydrogen peroxide. In this way any oxidation current detected is unlikely to be due to hydrogen peroxide, one of the disproportionation products of superoxide. A rather more negative potential, still well positive of the thermodynamic potential, is undesirable from two aspects: firstly, at potentials more negative of -100mV vs. SCE the direct electrochemical reduction of dioxygen begins to take place. This would lead to a background current, of considerable magnitude, on top of which would be superimposed the small neutrophil-derived superoxide-oxidation current. Looking for small changes on top of large background responses, which may well be changing due to depletion effects, is difficult and undesirable. Secondly, the electrochemical reduction of dioxygen, resulting from use of a lower applied potential, would be undesirable simply because it would remove dioxygen from solution in the vicinity of the electrode. It is dioxygen which is the substrate for the NADPH oxidase of neutrophils, the enzyme which produces the superoxide. Clearly, in the extreme case, there could be no superoxide-oxidation current if the detecting electrode were actually consuming all the dioxygen arriving at it. For these reasons we suggest that the use of a potential of +50 mV vs. SCE is, in many ways, ideal.

The electrochemical experiments were actually performed as follows: after the working electrode was placed in the cell, containing buffer medium only, and connected to the electronics, the small background response was allowed to stabilise for 1-2 minutes. An aliquot of a suspension of neutrophils was then rapidly added from a Pasteur pipette. This alone gives sufficient mixing of the system in the case of the stationary microvoltammetric electrode experiment. The rotation of the electrode in the case of the macroscopic electrode experiment provides additional mixing and transport of cells to the electrode surface.

The Opsonized Macro-electrode

A typical result of adding an aliquot of a suspension of human neutrophils to the buffer medium in the electrochemical cell is illustrated in Figure 1(A). After a short lag-time of some 30-40s., the opsonized macro-electrode detected an oxidation current which increased fairly rapidly with time, reached a "plateau" level, and which then decreased again. This plateau current was found to be dependent, though not linearly, on the number of neutrophils added, as shown in Table I.

The oxidation current observed at any given time showed only a feeble dependence on the rotation speed of the electrode, i.e. the rate of transport of solution species towards the electrode surface. Variation over the range 2-50 Hz generally resulted in only perhaps a 10% variation in the observed oxidation current. It was confirmed that the oxidation current burst was not due to oxidizable impurities in the buffer medium by simply adding aliquots of the medium only to the cell: no current response was obtained.

Figures 1(B) and 1(C) show the current responses which resulted at a clean (B) and (bovine serum albumin)-treated (C) graphite electrode when a standard aliquot of human neutrophils were added to the electrochemical cell. There was, in both cases, no current response. This is in contrast to the large oxidation current burst observed when the graphite electrode was pre-treated with IgG (Figure 1(A)). This oxidation current burst was also observed at a gold rotating disc electrode which was pre-treated with IgG.

Table I: Plateau Oxidation Current Observed as a Function of the Concentration of Neutrophils

10^6 cells/cm^3	Current/nA
0	0
1.4	10
7.1	68
14.3	88

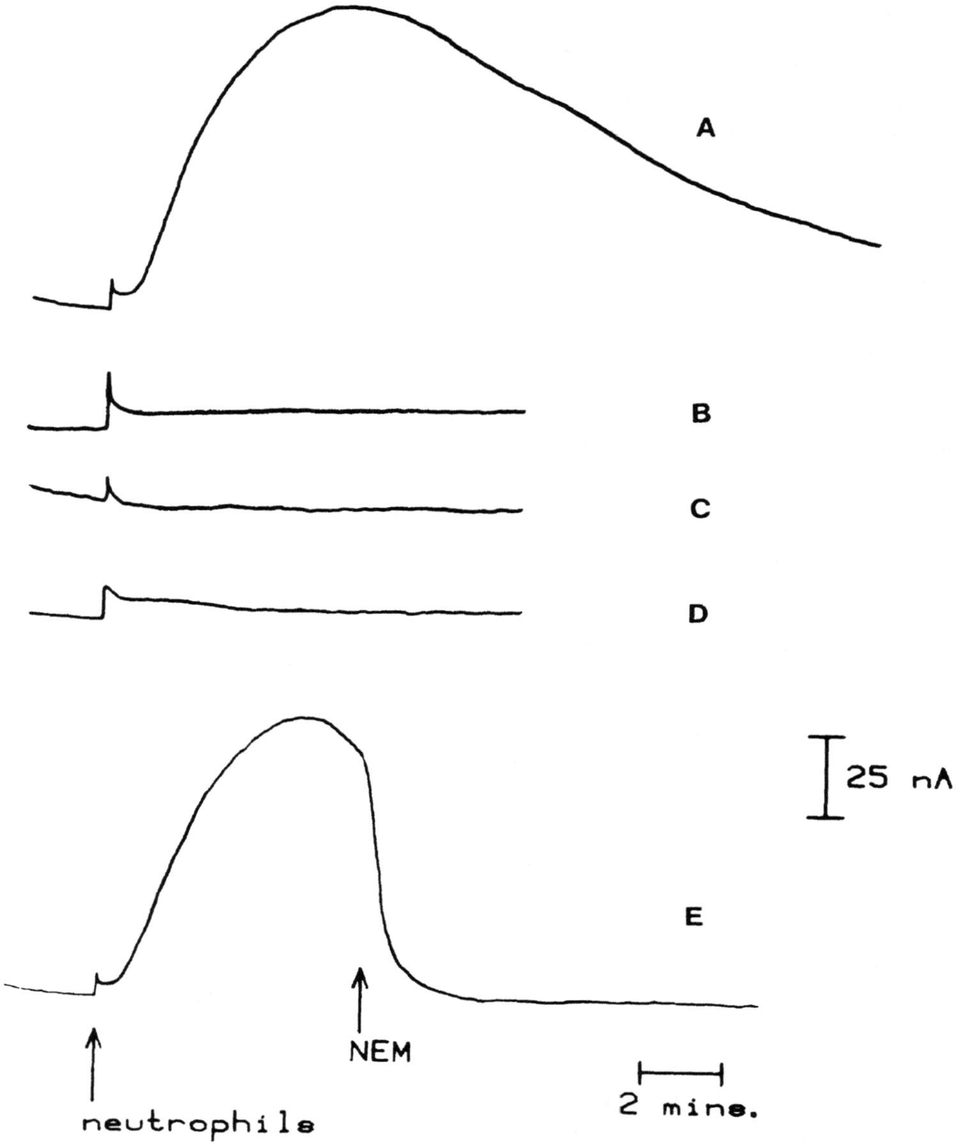

Figure 1: Current vs. time traces at a pyrolytic graphite electrode in RPMI 1640 medium with 5×10^7 neutrophils added as indicated.
(A) IgG-treated electrode
(B) clean electrode
(C) BSA-treated electrode
(D) IgG-treated electrode, medium containing 140 μM NEM
(E) IgG-treated electrode, with addition of NEM to 140 μM as shown.
(Reproduced with permission from Green et al., copyright 1984, FEBS

Addition of neutrophils to medium containing N-ethylmaleimide (NEM) produced no current response at an IgG-treated electrode, as shown in Figure 1(D). Furthermore, addition of NEM to the electrochemical cell during the course of an oxidation current burst (Figure 1(E)) led to a very rapid decline in observed oxidation current towards zero. As will be discussed later, the neutrophils are effectively "killed" by the reagent NEM.

The effect on the oxidation current response at an IgG-treated electrode of added superoxide dismutase (SOD) or catalase, present in the medium when neutrophils were added, is shown in Figure 2. Trace A shows the control experiment-medium only. Trace B, medium plus superoxide dismutase, shows a very large inhibition of the oxidation current burst. Its magnitude is decreased by > 90%. Trace C, medium plus catalase, shows no significant inhibition with respect to trace A. Interestingly, addition of superoxide dismutase to the electrochemical cell during the course of an oxidation current burst (in the same way as NEM was added in Figure 1(E)) generally gave an inhibition of less than 10%. This contrasts with > 90% when the same bulk concentration of the enzyme was present before the neutrophils were added.

The current-time profiles of Figure 1(A-C) suggest that added neutrophils recognise the immunoglobulin protein IgG that is adsorbed onto the electrode surface. In the absence of surface modification a clean electrode surface does not stimulate an oxidation current burst. Moreover, dipping a graphite electrode into a concentrated solution of BSA is expected to result in unspecific adsorption of this protein also. However, such a BSA-treated electrode fails to stimulate a current burst. The implication is that the observed oxidation current effect is not simply some result of general protein adsorption at the electrode surface but that it is specific to IgG itself. The electrode surface is effectively labelled immunologically by IgG. The neutrophils therefore recognise it and are stimulated by it. The mode of attachment of IgG to the electrode does not, however, appear to be specific since similar results can equally be obtained on graphite and gold surfaces.

The weak rotation speed dependence of the observed oxidation current, and its non-linear dependence on the concentration of neutrophils present, both suggest adsorption: the species being oxidized at the electrode is not freely diffusing in bulk solution. We suggest that recognition by neutrophils of IgG on the electrode leads to them binding to its surface; their stimulation by it causes them to produce an oxidizable species - superoxide. This species is probably produced in a volume of restricted diffusion bounded by the electrode on one side and the neutrophil on the other - a sort of vacuole. Under normal conditions a neutrophil will completely engulf an opsonized bacterium or latex particle but in the case of the opsonized electrode this is clearly impossible. It is to be expected, however, that neutrophils will bind to the opsonized electrode in an attempt to phagocytose it. There may be slow exchange of some neutrophils between surface and bulk suspension. The foregoing suggestions are supported by the observation that addition of SOD to an already-running system has rather a small effect. The enzyme is unlikely to have easy access to the "vacuolar" space in which, it is suggested, the superoxide is produced. It cannot therefore inhibit the observed oxidation current substantially.

The dramatic effect of superoxide dismutase, present in solution before addition of the neutrophils, on the oxidation current burst, identifies the oxidizable species as being the superoxide anion, O_2^-. Superoxide is the only known substrate for the enzyme. Furthermore, the lack of effect of catalase confirms that superoxide, and not hydrogen peroxide, is the electrode-oxidizable species. Superoxide does spontaneously dismute in aqueous solution to give dioxygen and hydrogen peroxide. However, the results shown in Figure 2 clearly suggest that, at the potential used, electrochemical oxidation of superoxide and not hydrogen peroxide, takes place. Furthermore, as discussed earlier, the choice of applied fixed potential is likely to be highly selective for the electro-oxidation of superoxide only.

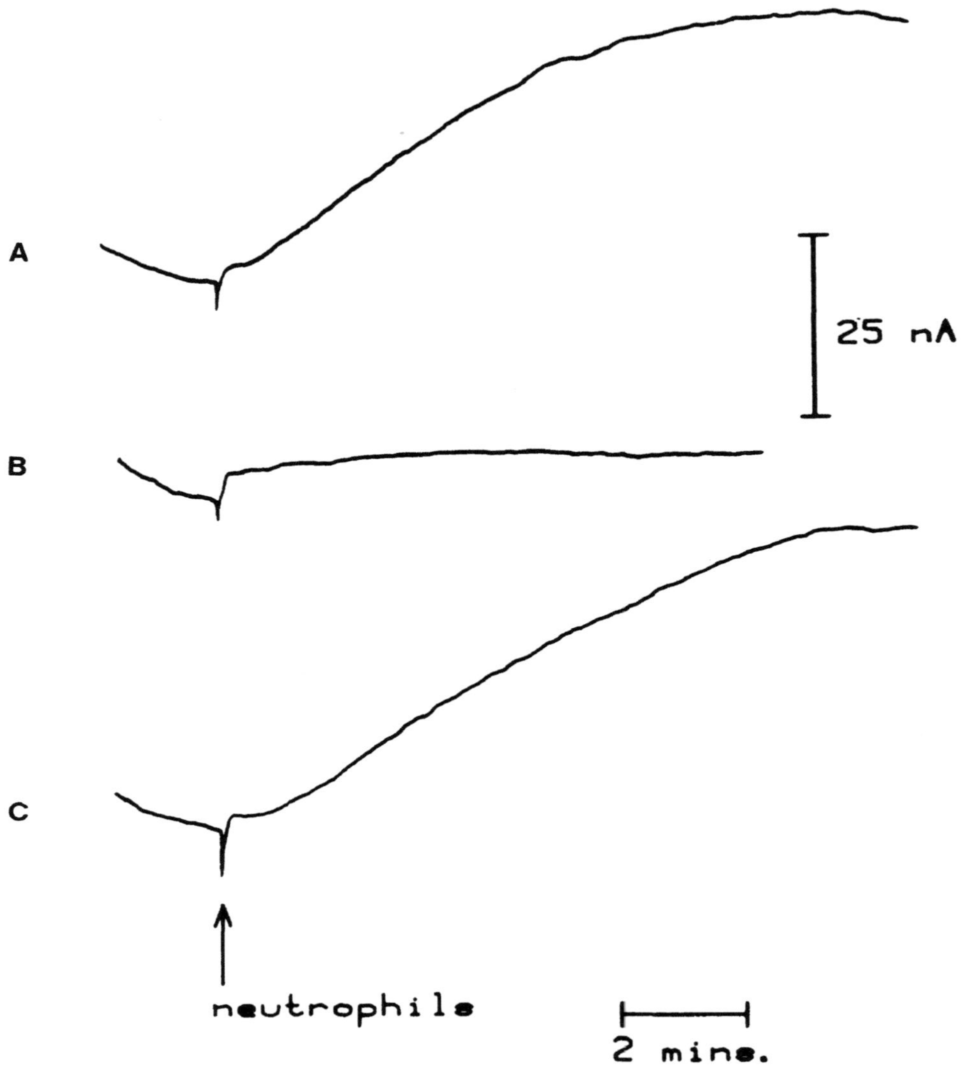

Figure 2: (A) Current vs. time trace at an IgG-treated pyrolytic graphite electrode in RPMI 1640 medium with 1.5 x 10^7 neutrophils added as indicated;
(B) +70 $\mu g/cm^3$ human CuZn SOD;
(C) +70 $\mu g/cm^3$ catalase.
(Reproduced with permission from Green et al., copyright 1984, FEB

Neutrophil metabolism, including the production of superoxide, is known to be irreversibly inhibited by the sulfhydryl reagent NEM (Yamashita, 1983). It is presumed that the reagent blocks metabolic pathways, including the NADPH oxidase ultimately responsible for superoxide production. Indeed it has been reported (Bellavite et al., 1983) that isolated NADPH oxidase is inhibited by the sulfhydryl reagent p-chloromercuribenzoate. It is not, therefore, surprising that additon of NEM to the electrochemical cell, either before or after neutrophil addition, leads rapidly to inhibition of the superoxide oxidation current burst, as is clearly shown by Figures 1(D) and 1(E). Quite simply, the neutrophils must be "alive" in order to produce superoxide and hence an oxidation current. These observations serve to emphasise that the observed oxidation current is not a result of any artifact such as, for example, subtle time-dependent changes in capacitance due to binding of cell membranes to the electrode. These would be expected with both dead cells and live cells. It is clear that the neutrophils must be enzymatically-active in order to give a superoxide-oxidation current at the opsonized electrode.

It is generally true that superoxide produced in aqueous solution at neutral pH is not a long-lived species. The fact that it can be detected so easily using the experiment described in this article is due entirely to the following: the IgG-treated electrode is both stimulus and detector combined. It is this dual role that the opsonized electrode plays that is fundamental to the successful operation of the system. We believe that it is only because the neutrophils are stimulated to produce their superoxide right next to the electrode surface that the electrochemical oxidation process has any substrate on which to act. The efficiency with which the electrode "collects" superoxide produced by the neutrophils, as compared with their bulk rate of dioxygen consumption, is discussed later in the article after the experiments conducted with microvoltammetric electrodes have been described.

The Opsonized Micro-electrode

It is reasonable to suggest that the current which is observed at an opsonized macro-electrode when human neutrophils are added is, in fact, an ensemble response of that cell population. Experiments which use an opsonized microscopically-small voltammetric electrode are, we suggest, likely to give a superoxide-oxidation current which might reasonably be associated with the respiratory burst of a single neutrophil. The neutrophils and the electrode are directly comparable in size so it is unlikely that more than one cell will bind to, and be activated by, the opsonized electroactive surface area of the electrode. This point is also discussed later in the context of the likely packing density of cells bound to an opsonized surface.

A control experiment, consisting of addition of an aliquot of buffer medium only to the electrochemical cell containing the opsonized gold microvoltammetric electrode, is illustrated in Figure 3(a). It shows only a small spike of current, which was short-lived. Figures 3(b) - 3(d) are representative examples of the result of adding a standard aliquot of human neutrophils to the cell containing the opsonized microelectrode. They each show a rapid rise in oxidation current immediately after the addition of the neutrophils. In all cases there was no observable time lag in the response. The results obtained were always somewhat variable in their current-time profile and were never precisely reproducible. We believe that this is significant in terms of an expected cell-to-cell variability - see discussion later. The majority of results, such as those reproduced in Figures 3(b) and 3(c), consisted of a rapid rise leading to a maximum observed current in 20-25 seconds followed by a much slower decay with time. There were a number of results, however, which had current-time responses of the type illustrated in Figure 3(d). In these cases an initial spike was followed by a rather slower rise in current, usually taking up to about three minutes to reach maximum, again followed by a slow decay of current with time. It is suggested that the initial spike can largely be assigned to a short-lived capacitative change of the type observed on adding buffer only, Figure 3(a). This feature is masked in the early parts of Figures 3(b) and 3(c). The magnitude of the maximum neutrophil-derived oxidation current was found to be, typically, within the range 1.05 - 1.85 pA for a

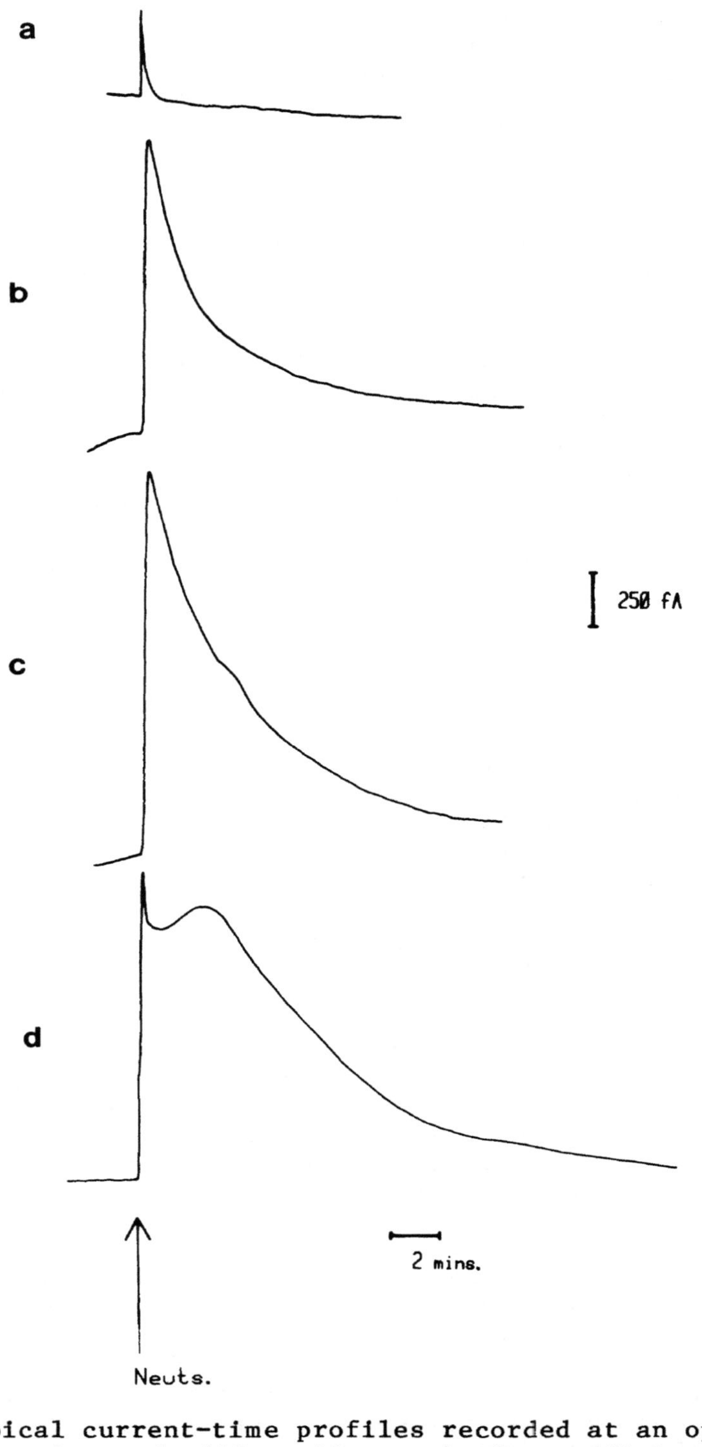

Figure 3: Typical current-time profiles recorded at an opsonized gold microelectrode (10 µm diameter) after addition, as indicated, of a 75 µl aliquot of:
(a) buffer;
(b-d) neutrophils (7.5 x 10^6 cells)
to 150 µl Hanks' medium.
Potential applied equivalent to 50 mV vs. SCE. Temperature 37°C.
(Reproduced with permission from Hill et al., copyright 1985, FEBS)

set of experiments using the same neutrophil preparation. The current decay following the maximum was always observed to be biphasic: a rapid decay at first was followed by a slower second phase, there being generally a clear transition point between the two. This can be seen by careful inspection of Figures 3(b) - 3(d). A logarithmic plot, as discussed later, brings this out rather clearly. In some cases not only is there a change of current decay rate but there is also observed to be a transient increase in current at the same time: in Figures 3(c) and 4(a) there is a "bump" in the current-time transient. When this was observed, and it should be made clear that it did not always occur, then it was consistently at a time of between two and three minutes after addition of the neutrophils to the electrochemical cell.

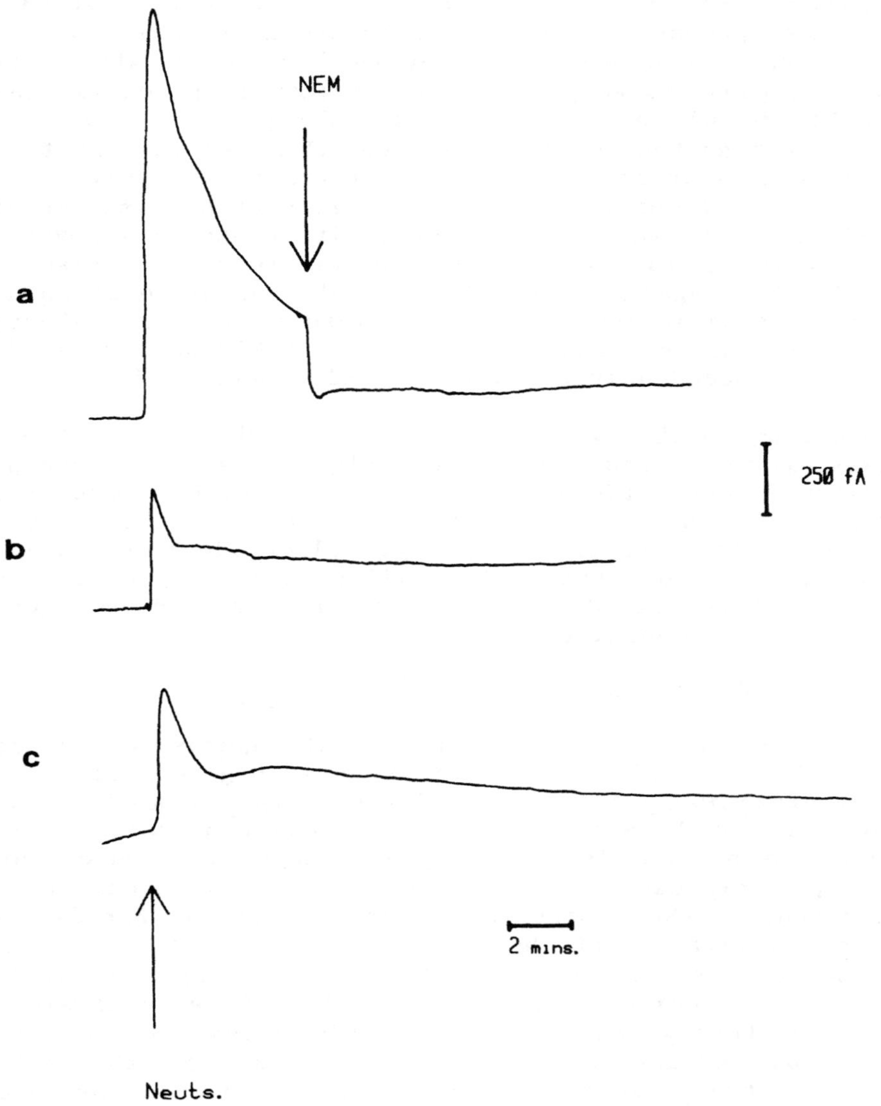

Figure 4: Current-time profiles recorded at an opsonized gold microelectrode on addition of neutrophils as indicated. Conditions as in Figure 3 plus:
(a) N-ethylmaleimide added to 100 μM as indicated;
(b) N-ethylmaleimide (100 μM) present in medium before addition of cells;
(c) human Cu-Zn superoxide dismutase (100 $\mu g.cm^{-3}$) present in medium before addition of cells.
(Reproduced with permission from Hill et al., copyright 1985, FEBS)

As in the case of the opsonized macroelectrode, some control experiments were carried out in order to define the type of electrode process being observed. In Figure 4(a) addition of neutrophils led to a rapid rise in oxidation current to a maximum of 1.5 pA. A rapid decay phase was followed by a "bump" at t = 2 minutes which was itself followed by a slower decay phase. Addition at this point of the reagent NEM caused the oxidation current to fall quickly to zero and to remain at that level subsequently. In the same way, addition of active neutrophils to medium in the electrochemical cell which already contained NEM (Figure 4(b)) led to little more than a short-lived spike due to a capacitative current. Inhibition of neutrophil metabolic pathways, and especially the superoxide-producing enzyme NADPH oxidase, by this reagent clearly results, once more, in rapid and complete loss of electrode response. The effect of added superoxide dismutase, present in the medium before neutrophils were added, is illustrated for the microelectrode experiment in Figure 4(c). The oxidation current response is very much smaller when compared with normal results such as Figure 3(b) or the first part of Figure 4(a), but signigicantly larger than the result of Figure 4(b). We suggest that there is simply a very greatly inhibited version of the normal results superimposed on the backgroup response. There is even a "bump" around three minutes after addition of neutrophils. The entire response appears to last for nearly the same length of time as those normal results carried out without added SOD. Once again the specificity of superoxide dismutase identifies the species being oxidised by the microvoltammetric electrode as the superoxide anion. The control experiments with NEM confirm again that such superoxide is produced only by viable, active neutrophils.

Before proceeding further with a discussion of the form of the current-time decay phases, it is instructive to investigate the relationship of the sizes of the observed superoxide-oxidation currents, at both macro- and micro-electrodes, to the rates of dioxygen consumption of the neutrophils which can be measured in a conventional oxygen-electrode experiment. The principle of the latter experiment is that all superoxide produced by the neutrophils will spontaneously dismute to give dioxygen and hydrogen peroxide according to the equation -

$$H^+ + O_2^- \longrightarrow \tfrac{1}{2}O_2 + \tfrac{1}{2}H_2O_2$$

Since one mole of dioxygen is consumed per mole of superoxide produced by the NADPH oxidase, but one half mole of dioxygen is regenerated by dismutation, then effectively one mole of dixoygen is <u>consumed</u> net per turnover of two moles by the enzyme. Since the enzyme performs, per turnover, a one-electron reduction of dioxygen then a dioxygen consumption rate of R moles O_2/s/cell can accordingly be equated with a current (electron flux) output of 2RF amperes/cell, where F is the Faraday constant. The rate of oxygen consumption of the neutrophils used for the experiments of Figure 1 was measured as 1.52×10^{-15} moles O_2/min./cell. We may now convert this to give a current of 4.8 pA/cell, which is effectively the average electron flux through the activated NADPH oxidase systems of each neutrophil in this bulk-phase experiment. In the same way, the neutrophils used for the microelectrode experiments of Figure 3 had an observed dioxygen consumption rate of 1.3×10^{-15} moles O_2/min./cell which converts to a current output of 4.2 pA/cell. The two figures are similar but, more importantly, are of the same order of magnitude as the currents experimentally measured at the microelectrode - typically 1.5 pA at maximum output. In the light of these figures it is therefore quite reasonable to suggest that the current due to the re-oxidation of superoxide generated near to the microvoltammetric electrode is due to one single neutrophil only. If more than one cell were contributing then the resulting current may well have exceeded the electron ouput figure per cell derived from bulk oxygen consumption measurements. There are a number of factors which may be contributing to the apparent 35% efficiency of collection of superoxide by the microvoltammetric electrode. One is the rate of spontaneous dismutation of superoxide in aqueous solution at pH 7. It is high and likely to lead to a superoxide-oxidation current less than that predicted, even at the opsonized electrode. The second factor is more subtle and has to do with the percentage cell surface area of each neutrophil which is activated for superoxide production in the electrochemical experiments and in the bulk oxygen-electrode experiments. In the electrochemical experiments the maximum proportion of the cell surface which could become activated is fifty per cent, and this only if the neutrophil were completely flattened out onto the electrode. In the bulk Clark electrode experiment used to measure dioxygen consumption, however, the addition of an excess of opsonized particulate stimulus may lead to the activation of a very much greater proportion of the cell membrane of each neutrophil. For this reason the electron ouput per cell calculated from these measurements may be rather an over-generous estimate of the oxidation current expected in an electrochemical experiment. The apparently modest 35% superoxide collection efficiency may therefore reflect, to a considerable extent, the relative cell surface areas of each neutrophil activated in the two different types of experiment.

It is interesting now to make a comparison of the maximum currents observed at opsonized macro- and micro-electrodes. The neutrophils used for the microelectrode experiments described gave maximum currents of around 1.5 pA. To compare this with the plateau currents of 90 nA observed at the opsonized macroelectrode it is necessary to make a correction for the different dioxygen consumption rates of the two sets of neutrophils: a microelectrode experiment might have been expected to give a maximum current of 1.75 pA using those neutrophils which gave a plateau current of 90 nA at the macroelectrode. Assuming that 1.75 pA is the electron flux collected from one cell only, then 90 nA would be produced by 5.1×10^4 neutrophils. Since the opsonized macroelectrode had a geometric surface area of 2.83×10^{-5} m^2, this implies an effective surface area allocation of 5.5×10^{-10} m^2/cell. Since each neutrophil has a diameter in the region of $0.8\text{-}1 \times 10^{-5}$ m, the area allocated to each one can be described by a square of side approximately two-to-three times the cell diameter. In other words, the apparent packing density of cells on the opsonized macroelectrode is not high. This may correspond to a limited number of surface sites on the electrode suitable for recognition and binding by neutrophils. If this packing is similar on the opsonized surface of the glass insulator surrounding the opsonized gold microelectrode then it would be unlikely for more than one neutrophil to be bound to the opsonized surface where it is electroactive.

The variability of results obtained with the opsonized microelectrode, illustrated in Figure 3, is a subject whose discussion can benefit from that of the previous paragraph. The apparent irreproducibility, in terms, say, of maximum current magnitude, would actually be expected: a neutrophil would not always be expected to be bound completely on the electro-active opsonized surface. The less the extent of such "overlap", the less the proportion one would expect to collect of the superoxide flux associated with that cell. Smaller currents are therefore anticipated, and observed, in a number of experiments. Another feature of variability is that of the time taken to reach maximum current. It is possible that cell activation may not be as rapid in all cases leading, in some cases, to a rather longer time to the peak of cell-derived current, as is the case in Figure 3(d). Finally, a heterogeneous population of neutrophils would be expected anyway to give responses at the opsonized microelectrode which showed some spread of behaviour, both in terms of extent of superoxide production and in terms of time response after activation.

The form of the current-time decays observed following peak current generation is interesting. In all the cases that we have observed it has been clear, just by visual inspection, that the decay phases are not simple exponential processes. For example, shown in Figure 3(c) is a changeover from fast to slower decay taking place three minutes after the addition of neutrophils. This is shown up very clearly in Figure 5 where the logarithm of the current during the decay phase is plotted against time. There is a marked change in the time-dependence of the current decay around three minutes from neutrophil addition. In Figure 3(c) not only is there a similar change in decay rate but at the same time there is also a transient increase in current. We suggest that such observations are important since it may be possible to relate them to the operation of those metabolic pathways within a cell which are known to be involved in the production of superoxide.

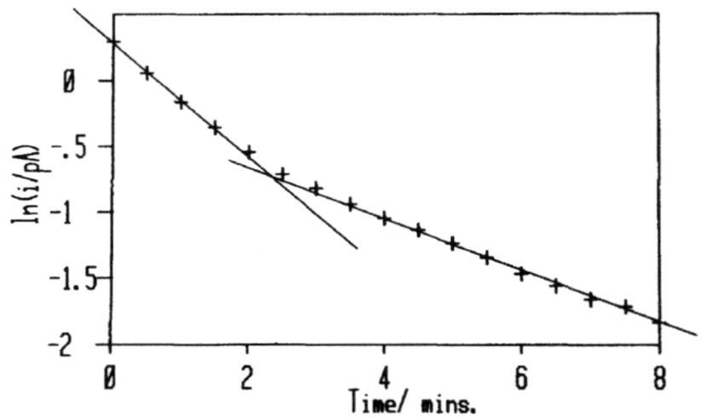

Figure 5: Plot of the natural logarithm of the superoxide-oxidation current observed in Figure 3(b) against time, commencing just after peak cell current observation.
(Reproduced with permission from Hill et al., copyright 1985, FEBS)

It is known that a pool of NADPH exists in resting neutrophils and that such a reserve of reducing equivalents serves as substrate for the oxidase enzyme which produces superoxide (Patriarca et al., 1971). We suggest that the first rapid phase of the observed superoxide-oxidation current decay may correspond to first order depletion of this pool. However, if the rate of superoxide production, as detected by the opsonized microelectrode, is determined by the concentration of intracellular NADPH then clearly a change to a slower decay rate, as observed, must imply the maintenance of an NADPH concentration higher than that anticipated from the initial first-order process. If the first-order process continues throughout the decay phase then the second, slower part can only be accounted for by the activation of a mechanism for production of NADPH, presumably recycled from the rapidly increasing concentration of $NADP^+$. We suggest that this process is the metabolism of glucose via the hexose monophosphate shunt. Indeed, a process of this kind, with a transiently increasing rate, can account for the frequent observation of a "bump" in the current-time decay. One alternative proposal, viz., that the rate constant for depletion of NADPH from the pool changes sharply at a certain point, leading to the biphasic behaviour, is rather improbable. Another possibility that would produce a biphasic decay is that the production of NADPH by the hexose monophosphate shunt is present throughout the decay period and its rate slowly decreases with time. It would be this process which is dominant at long times after the initial rapid depletion of the NADPH pool has taken place. A combined kinetic scheme such as this can give a biphasic decay but can never give the sort of transiently increasing current which has often been observed around the changeover time from fast to slow decay. The foregoing discussion is illustrated by Figure 6 which shows the summation of various mathematical

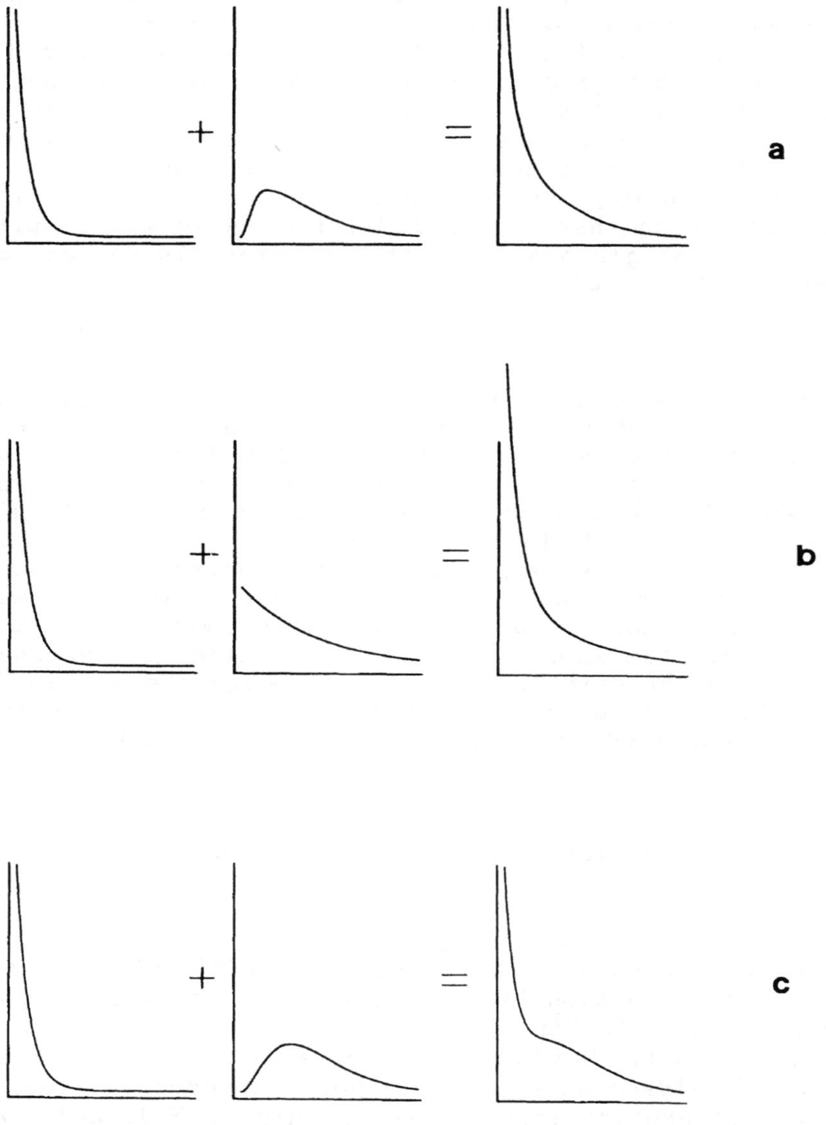

Figure 6: Summation of mathematical functions to simulate the decay phases of experimentally observed results. In all cases the left-hand function is based on exp(-3t). The second function added is:
(a) a normalised t^2 exp(-3t/2) function using the rapid growth phase of t^2 exp(-3t);
(b) exp(-3t/2);
(c) a normalised t^2 exp(-3t/2) function.
The scales are arbitrary.
(Reproduced with permission from Hill et al., copyright 1985, FEBS)

functions which might crudely model the rate processes taking place. Figure 6(a) shows that summing a simple exponential decay, corresponding to pool depletion, with a transiently-increasing function which has a slower exponential decay, does indeed give qualitatively the correct form of net current-time behaviour, viz., a biphasic decay. However, although the summation of two simple exponentials, as shown in Figure 6(b), also gives a biphasic decay it cannot, as noted, ever produce the increase in current observed at the changeover time. Accordingly, an ever-present and slowly decaying rate of production of NADPH cannot be a realistic possibility. As Figure 6(c) shows, the transient increase in net current will be observed if the time to maximum of the second function is slightly longer than in Figure 6(a). Clearly this time to the maximum value is critical in determining whether the net transient increase is observed or not. If the peak occurs early during the pool decay period then it will be masked. The time to the maximum value of this second function will correspond, metabolically, to the efficiency with which NADPH production by the hexose monophosphate shunt is activated by increasing $NADP^+$ levels. This may well be slightly variable from cell-to-cell leading to a "bump" in the current-time response in some cases but not in others. The reason why the second function of Figures 6(a) and 6(c), the NADPH-producing function, has been chosen to have a slowly decreasing magnitude is so that it can model the end of the respiratory burst; in the presence of glucose, activated neutrophils do not continue to produce superoxide indefinitely.

Summary

We have described in this chapter a rather different application of electrochemistry to the sensing of immunological reactions. We have used an electrode surface, opsonized with IgG, both to stimulate human neutrophils to produce superoxide and to oxidise electrochemically such superoxide back to dioxygen. Electrodes both macroscopic (Green et al., 1984) and microscopic (Hill et al., 1985) in their dimensions have been used, the latter yielding information, we suggest, on the respiratory burst of a single neutrophil. The intriguing success of the technique is, we believe, due to the singular function of the opsonized voltammetric electrode as both stimulus and detector: the labile species, superoxide, is necessarily produced right next to the detector itself. Although the system described is in many ways unique, we suggest that the principles involved may well find application in the study of other cell/surface recognition reactions.

Acknowledgements

We thank Dr. David Tew for his able contribution to this work.

References

Babior, B. M., Kipnes, R. S., and Curnutte, J. T., 1973, Biological defense mechanisms. The production by leukocytes of superoxide, a potential bactericidal agent, J. Clin. Invest., 52:741.

Bellavite, P., Serra, M. C., Davoli, A., Bannister, J. V., and Rossi, F. 1983, The NADPH oxidase of guinea pig polymorphonuclear leucocytes. Properties of the deoxycholate extracted enzyme, Mol. Cell. Biochem., 52:17.

Boyum, A., 1974, Separation of blood leucocytes, granulocytes and lymphocytes, Tissue Antigens, 4:269.

De Chatelet, L. R., Shirley, P. S. and Johnston, R. B. jr., 1976, Effect of phorbol myristate acetate on the oxidative metabolism of human polymorphonuclear leukocytes, Blood, 47:545.

Green, M. R., Hill, H. A. O., Okolow-Zubkowska, M. J., and Segal, A., 1979, The production of hydroxyl and superoxide radicals by stimulated human neutrophils - measurements by EPR spectroscopy, FEBS Lett., 100:23.

Green, M. J., Hill, H. A. O., Tew, D. G., and Walton, N. J., 1984, An opsonised electrode. The direct electrochemical detection of superoxide generated by human neutrophils, FEBS Lett., 170:69.
Hill, H. A. O., Tew, D. G. and Walton, N. J., 1985, An opsonised microelectrode. Observation of the respiratory burst of a single human neutrophil, FEBS Lett., 191:257.
Johnson, R. B. jr., Keele, B. B. jr., Misra, H. P., Lehmeyer, J. E., Webb, L. S., Boehner, R. L. and Rajagoplan, K. V., 1975, The role of superoxide anion generation in phagocytic bactericidal activity. Studies with normal and chronic granulomatous disease leukocytes, J. Clin. Invest., 55:1357.
Partriarca, P., Cramer, R., Moncalvo, S., Rossi, F. and Romeo, D., 1971, Enzymatic basis of metabolic stimulation of leucocytes during phagocytosis. The role of activated NADPH oxidase, Arch. Biochem. Biophys., 145:255.
Yamashita, T., 1983, Effect of maleimide derivatives, sulfhydryl reagents, on stimulation of neutrophil superoxide anion generation with concanavalin A, FEBS Lett., 164:267.

HOMOGENEOUS VOLTAMMETRIC IMMUNOASSAY OF HAPTENS

Sam A. McClintock and William C. Purdy

McGill University, Department of Chemistry
801 Sherbrooke St. W.
Montreal, Quebec, Canada H3A 2K6

INTRODUCTION

The analytical application of the immune defence system's very selective antigen-antibody reaction coupled with a suitable detection system has become a powerful technique for the analysis of a wide range of biologically important compounds. When a foreign species referred to as the antigen is recognised by an organism it starts immediately to produce an antibody that has a large binding constant for the various chemically distinctive elements of the antigen thus neutralizing its adverse effects. Although the normal function of antibodies is to destroy or neutralize foreign organisms it is this naturally occurring binding phenomena that is exploited in the immunoassay methodology and gives the technique its selectivity. However, the production of antibodies for small molecules such as therapeutic drugs is very inefficient. This problem can be overcome by attaching the small molecules known as haptens to a larger protein molecule; the resulting combination causes the production of antibodies some of which will have binding capacities as high as 10^{11} for the hapten. Other molecules having the same chemical groups in the determinant region will also bind with the antibody and it has only to assume the appropriate configuration temporarily in order for the binding process to be analytically useful. It should be kept in mind that although antibody and antigen are complex reagents they react in a simple equilibrium fashion. As in any competitive binding analysis three major components are involved: a hapten or antigen (the analyte), a "tagged" hapten which has been chemically modified in some way so that it can be measured in the presence of the analyte or "untagged" hapten and a binder which is usually the antibody. The "tagged" and "untagged" hapten can each bind to the antibody with an affinity K,

$$\text{Hap} + \text{Ab} = \text{Hap:Ab}$$

$$K1 = [\text{Hap:Ab}]/[\text{Hap}][\text{Ab}]$$

and

$$\text{Hap*} + \text{Ab} = \text{Hap*:Ab}$$

$$K2 = [\text{Hap*:Ab}]/[\text{Hap*}][\text{Ab}]$$

where Hap is the hapten, Hap* is the "tagged" hapten and Ab is the antibody. In any system at equilibrium containing the "tagged" hapten and the antibody some will be bound and some will be free (see FIG 1). If we add to this system a hapten whose concentration we wish to measure then as the system returns to equilibrium the amount of "tagged" free hapten will increase and the amount of bound "tagged" hapten will decrease. By measuring the change in concentration in either the bound or free "tagged" hapten the amount of "untagged" hapten or analyte can be calculated. If a separation of the free hapten from the bound hapten is required before a measurement can be carried out then it is called a heterogeneous assay, if no separation is required then it is referred to as a homogeneous assay.

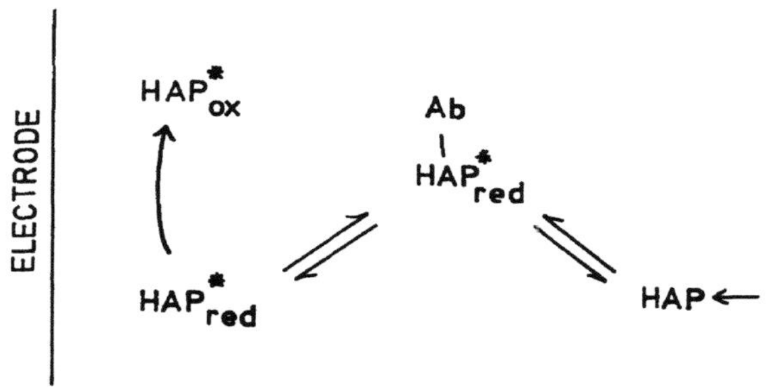

Figure 1. Schematic representation of voltammetric immunoassay of haptens; Ab, is the antibody and Hap* is the labelled hapten.

The most successful of the labelling systems used to date are radioactive labels first reported by Yalow and Berson (1959); however non-isotopic labels such as enzymes (Engvall and Perlmann, 1971; Rubenstein et al., 1972) and fluorescent labels (Watson et al., 1976) are gaining broader acceptance because of laboratory safety and reagent shelf life but perhaps more so because of the ready availability and adaptability of conventional analytical instrumentation. The excellent sensitivity and selectivity that can be achieved by modern electrochemical techniques such as liquid chromatography with electrochemical detection, flow injection analysis with electrochemical detection, differential pulse and square-wave voltammetry (Kissinger and Heineman, 1984) has prompted the investigation of these techniques (Heineman and Halsall, 1985) as alternate detection strategies to exploit the antibody-antigen reaction. In this report we will look at the approaches to homogeneous voltammetric immunoassay of haptens investigated and speculate on some of the possible approaches for future work.

Breyer and Radcliff (1951) first used electrochemical techniques to investigate the antigen-antibody reaction using azo-protein as an antigen. Its reaction with a specific antiserum was examined polarographically. The polarographic wave for azo-protein between 240 and 440 mV versus a S.C.E. was sharply reduced with the addition of specific antiserum. This is the first use of a quantitative electrochemical technique to study the antibody-antigen reaction.

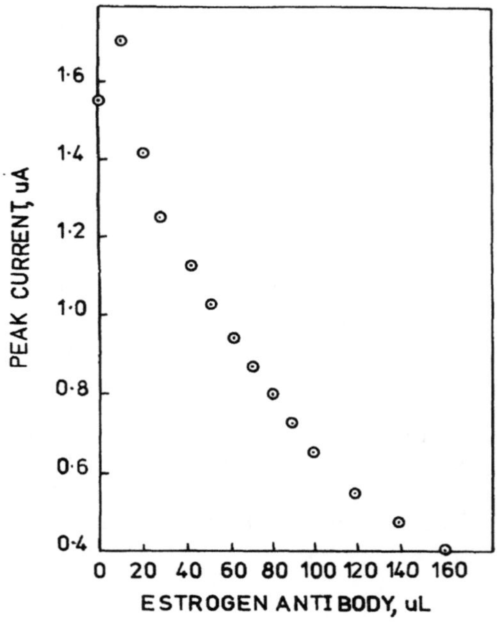

Figure 2. Plot of the differential pulse polarogram peak currents for 4.6 mL of phosphate buffer solution containing dinitroestriol (7.7 umol/L) vs. microlitres of estrogen-specific antibody added to the dinitroestriol solution. (From Wehmeyer et al., 1982, with permission).

Heineman and co-workers (Heineman et al., 1979; Wehmeyer et al., 1982) have investigated the use of differential pulse polarography as a technique for investigating the binding of a hapten marked with an electroactive "tag". They chose to use a dropping mercury electrode because it provides a new surface every few seconds on which to carry out the electrochemistry thus minimizing the problem of electrode fouling and deactivation by either solutions containing proteins or the products of the electrochemical reaction. They also chose to label an estrogen molecule, estriol, with a tag that can be reduced electrochemically. This is used to help differentiate it from the "untagged" estriol which shows little reductive behaviour. In this way a unique electrochemical region or "window" is created in which the measurements made will be highly selective for the "tagged" molecule. In this particular case 2,4-dinitroestriol was used as the labelled hapten. The plot in Figure 2 demonstrates the binding capacity of the 2,4-dinitroestriol to various amounts of estrogen specific antibody and Figure 3 demonstrates that the bound labelled hapten is reversibly displaced and is proportional to the amount of unlabelled hapten added.

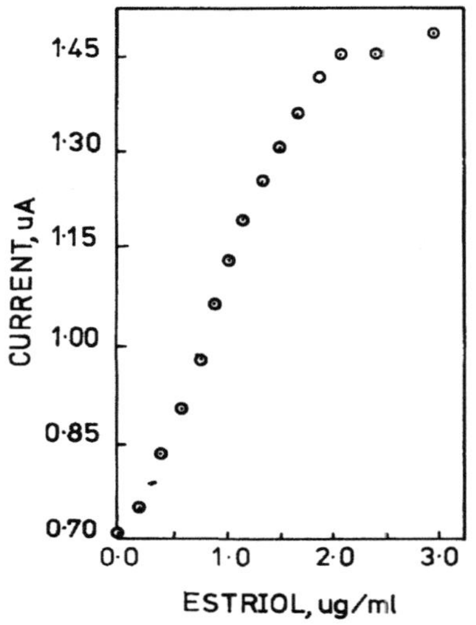

Figure 3. Plot of differential pulse polarogram peak currents of 5.5 mL of phosphate buffer solution containing dinitroestriol (5.6 umol/L) plus 150 uL of estrogen-specific antibody vs. estriol added to the solution. (From Wehmeyer et al., 1982, with permission).

Weber and Purdy (1979) investigated the use of flow injection analysis with amperometric detection in voltammetric immunoassay at a glassy carbon electrode. This technique takes advantage of the fact that no separation is required and so large numbers of samples can be processed in an automated fashion. They used the morphine immunoassay to demonstrate the feasibility of the process. Not only is morphine electroactive (Proksa and Molnar, 1978) but a non-electroactive compound, codeine, binds to the morphine antibody (Wainer et al., 1973). This provides a system for study without the necessity of performing organic synthesis to prepare "tagged" molecules. This system also has the advantage that morphine can be electrochemically labelled with a compound like ferrocene. Figure 4 shows the system response to morphine in the absence and the presence of morphine antibody while Figure 5 demonstrates the same process for the ferrocene "tagged" morphine. They also demonstrated that codeine introduced into one of the above systems displaces morphine increasing the morphine signal by a proportionate amount. There is theoretically no reason why this technique which is amenable to automation by Flow Injection Analysis cannot be extended to other biologically important compounds.

Generally the sensitivity of any immunoassay depends on the binding constant of the antibody-antigen complex, i.e. the larger the binding constant the more sensitive the assay. If a calibration curve is plotted of the ratio of bound hapten to free hapten versus total hapten concentration, this curve has a large slope initially. Since the sensitivity

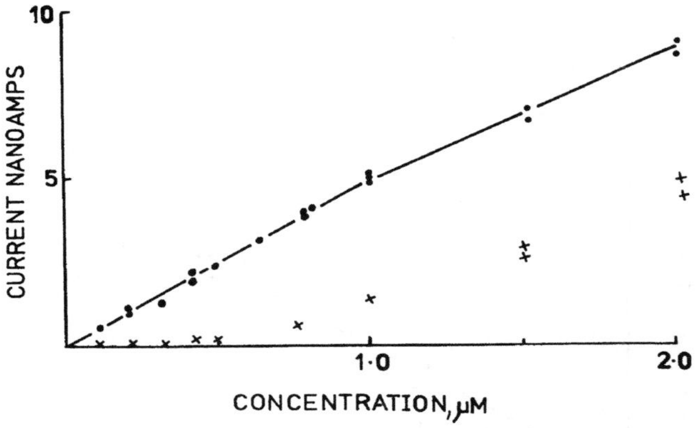

Figure 4. Peak current (nA) versus total concentration of morphine (uM) in the absence (o) and presence (x) of antibody to morphine. (From Weber and Purdy, 1979, with permission).

is defined as the slope of the analytical curve, it is this initial portion of the curve where the slope is measured to determine sensitivity, i.e. where the concentration of analyte goes to zero. At this point the only hapten is the labelled hapten and therefore the sensitivity of the assay can be increased by lowering the concentration of the labelled ligand. Using this criterion only, the maximum sensitivity can be achieved by minimizing the concentration of the labelled hapten. It is clear that this cannot be the only criterion used as it does not take into consideration the error in sample preparation or the error in the electrochemical measurement process. Ekins et al. (1968) recognizing lack of uniformity defined sensitivity as the quantity of analyte ligand which will change the measured quantity ratio of bound to unbound hapten by an amount equal to the standard deviation of the measured quantity when the analyte concentration is zero. Although Ekins and coworkers considered the problem for radioimmunoassay his approach is valid for electrochemical measurements. This quantity contains information from the sensitivity and the signal- to-noise ratio (measured quantity/standard deviation of the measured quantities), is related to the detection limit and as such is useful in the study of practical problems. However the precision involved in carrying out the assay does affect the detection limit. The measurement precision should be such that the label can be measured to a satisfactory degree of precision at a level near the lower limit of the range of values expected for the unknown concentration. It is for this reason that electrochemical labels, which can be detected at very low levels with excellent reproducibility, are under active investigation. The major problem to be overcome in this process is that of being able to discriminate between "tagged" and "untagged" hapten and to discriminate between this signal and the signals generated by naturally occurring electroactive species in biological samples.

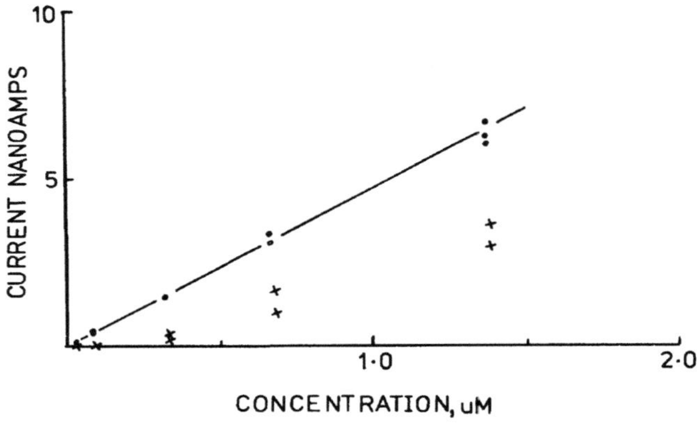

Figure 5. Peak current (nA) versus total concentration of ferrocenyl morphine (uM) in the absence (o) and presence (x) of antibody to morphine.

Homogeneous voltammetric assays function at least in part due to the difference in diffusion coefficients between the free labelled hapten and the labelled hapten bound to a much larger antibody (Steinhardt and Reynolds, 1969). The diffusion coefficient of globular proteins in aqueous media is 10^{-7} cm^2 sec^{-1} while small organic molecules have a diffusion coefficient of 10^{-5} cm^2 sec^{-1}. Thus even though the bound hapten produces an electrochemical signal due to its much larger diffusion coefficient, its signal will be reduced compared to the unbound hapten with a much smaller diffusion coefficient. The problem can also be minimized by selecting a tag that has a small potential window or at least a potential window relatively far away from the other components in the system; then signal discrimination between labelled hapten and background species will be greatly enhanced.

The feasibility of homogeneous voltammetric immunoassay of haptens has been demonstrated but the power of more recent developments in electrochemical detection technology has not been applied to the technique. In this section we will look at some of the newer electrochemical techniques and speculate on how they can be used to enhance the utility homogeneous voltammetric immunoassay of haptens.

The initial commercial success of electrochemical detectors for high pressure liquid chromatography has revived an interest in electroanalytical techniques generally as viable alternatives to existing procedures (Roston et al., 1982). Detector systems of this type are ideally suited to the homogeneous assay since no separation of bound and unbound species is required. Electrochemical detectors function in the following manner: a potential of sufficient magnitude is applied to a working electrode, over which the analyte passes, forcing an oxidation or reduction to take place. The resulting flow of electrons to or from the working electrode can be measured as current the magnitude of which is proportional to the amount of analyte passing through the detector. Recently attention has been focused on the inclusion of more than one working electrode in the electrochemical detector (Blank, 1976). These have become known as dual-working-electrode or multiple-working-electrode detectors. They can be classified in three distinct groups based on the position of the working electrodes in the cell; there are variations in the use of each type.

The most common type and original design (Blank, 1976) is the series-dual-electrode electrochemical detector in which one electrode is placed downstream from the other. In this configuration the first electrode acts as a "scavenger electrode". It can be maintained at a potential where background interfering species are selectively removed from the solution while permitting the labelled hapten to pass through to the second electrode thus improving both the detector's selectivity and sensitivity. This series approach can also be used to generate new species at the upstream electrode. If a particular label is easy to produce but not electroactive in a region for optimum detection then the labelled hapten could be converted electrochemically at the first electrode into a species which is more easily detected.

Electrochemical detectors have also been constructed by placing two electrodes parallel and adjacent to each other (Roston and Kissinger, 1982). If these electrodes are operated at appropriate potentials with respect to the oxidation/reduction potential of the labelled hapten then by electronically manipulating the output of the two electrodes in real time in the manner of Brunt et al. (1981), the signal-to-noise ratio of the analytical signal can be improved.

Two electrodes can also be placed parallel and opposed to each other in the walls of a thin-layer cell (McClintock and Purdy, 1983). For this type of detector to function successfully the species being detected must be part of an electrochemically reversible or quasireversible redox couple and the electrodes must be close enough so that species can diffuse between the electrodes. One electrode is held at a potential where oxidation takes place and the second is held at a potential where the products of the oxidation at the first electrode are reduced to the starting material, making it available for re-oxidation. Then for each conversion that takes place the current produced at each electrode will be increased. As with all the above techniques if the electroactive label and the operating potentials are carefully chosen the detection process in homogeneous voltammetric assay of haptens can be greatly enhanced.

Recently there has been renewed interest in electrochemical detectors where the potential is applied as a square-wave pulse (O'Dea et al., 1981; Eccles and Purdy, 1985). As with other pulse techniques current is sampled just before a voltage pulse is applied and at some time near the end of the pulse. Since charging current decays much more quickly than faradaic current the adverse effects of charging current are minimized while at the same time background current processes can be subtracted. As before the judicious choice of the electroactive label coupled with an appropriate pulse pattern will provide enhanced selectivity for that particular label.

Although carbon has been the most commonly employed electrode material in electrochemical detectors the next major advance in this area is likely to be achieved through chemical modification of electrode surfaces (Murray, 1984; Faulkner, 1984; Turk et al., 1986). Preliminary studies in our laboratories have shown that carbon surfaces can be modified by conducting polymer films to facilitate the electrochemistry of certain drugs (Turk et al., 1986). If an electrode surface can be tailored to be specific for certain electrochemical labels then the use of that label would improve both the sensitivity and selectivity of the electrode. All the electrode processes described could be further enhanced if a thin coating of a specific pore size could be placed at the electrode solution interface allowing only small molecules to penetrate; this would limit some of the problems associated with the fouling of electrode surfaces by large protein molecules.

One of the reasons for the slow development of homogeneous voltammetric immunoassay of haptens is the diverse skills that are required, from the production of appropriate antibodies to synthesizing electroactive labelled haptens, to the development of an adequate electrochemical detection system that will provide adequate selectivity and sensitivity. It is hoped that this report will stimulate some new investigations into a technique that has never fulfilled its true potential.

REFERENCES

Blank, C.L., 1976, Dual electrochemical detector for liquid chromatography. J. Chromatogr., 117:35-46.

Breyer, B. and Radcliff, F.J., 1951, Polarographic investigation of the antigen-antibody reaction. Nature, 167:79.

Brunt, K., Bruins, C.H.P., and Doornbos, D.A., 1981, Comparison between a differential amperometric detector in the reductive mode and a u.v. detector in high-performance liquid chromatography with vitamin K3 as test compound. Anal. Chim. Acta, 125:85-91.

Eccles, G.N. and Purdy, W.C., 1985, Development of a pulse cyclic voltammetric instrument. Anal. Lett. 18:657-672.

Ekins, R.P., Newman, G.B., and O'Riordan, J.L.H., 1968, Theoretical aspects of "saturation" and radioimmunoassay, in "Radio Isotopes in Medicine: In Vitro Studies", E.A. Goswitz and B.E.P. Murphy, eds., U.S. Atomic Energy Commission, Oak Ridge, Tenn., pp. 59-100.

Engvall, E. and Perlmann, P., 1971, Enzyme-linked immunosorbent assay (ELISA). Quantitative assay of immunoglobulin G. Immunochemistry, 8:871-874.

Faulkner, L.R., Feb. 27, 1984, Chemical microstructures on electrodes. Chem. Eng. News, 62(9):28-38, 43-45.

Heineman, W.R., Anderson, C.W., and Halsall, H.B., 1979, Immunoassay by differential pulse polarography. Science, 204:865-866.

Heineman, W.R. and Halsall, H.B., 1985, Strategies for electrochemical immunoassay. Anal. Chem., 57:1321A-1331A.

Kissinger, P.T. and Heineman, W.R., eds., 1984, "Laboratory Techniques in Electroanalytical Chemistry", Marcel Dekker, Inc., New York.

McClintock, S.A. and Purdy, W.C., 1983, Dual working-electrode electrochemical detector for liquid chromatography. Anal. Chim. Acta, 148:127-133.

Murray, R.W., 1984, Chemically modified mlectrodes, in "Electroanalytical Chemistry", A.J. Bard, ed., Marcel Dekker, Inc., New York, pp. 191-368.

O'Dea, J., Osteryoung, J., and Osteryoung, R.A., 1981, Theory of square wave voltammetry for kinetic systems. Anal. Chem., 53:695-701.

Proska, B. and Molnar, L., 1978, Voltammetric determination of morphine on stationary platinum and graphite electrodes. Anal. Chim. Acta, 97:149-154.

Roston, D.A. and Kissinger, P.T., 1982, Series dual-electrode detector for liquid chromatography/electrochemistry. Anal. Chem., 54:429-434.

Roston, D.A., Shoup, R.E., and Kissinger, P.T., 1982, Liquid chromtography/electrochemistry: thin-layer multiple electrode direction. Anal. Chem., 54:1417A-1434A.

Rubenstein, K.E., Schneider, R.S., and Ullman, E.F., 1972, Homogeneous enzyme immunoassay. New immunochemical technique. Biochem. Biophys. Res. Commun., 47:846-851.

Steinhardt, J. and Reynolds, J.A., 1969, "Multiple Equilibria in Proteins", Academic Press, New York.

Turk, D.J., McClintock, S.A., and Purdy, W.C., 1985, The electrochemical detection of certain tricyclic drugs at polymer electrodes: a preliminary report. Anal. Lett., 18:2605-2618.

Wainer, B.H., Fitch, F.W., Freid, J., and Rothberg, R.M., 1973, Measurement of the specificity of antibodies to morphine-6-succinyl-BSA [Bovine Serum Albumin] by competitive inhibition of [14C] morphine binding. J. Immunol., 110:667-673.

Watson, R.A.A., Landon, J., Shaw, E.J., and Smith, D.S., 1976, Polarisation of fluoro-immunoassay of gentamicin [in serum]. Clin. Chim. Acta., 73:51-55.

Weber, S.G. and Purdy, W.C., 1979, Homogeneous voltammetric immunoassay: a preliminary study. Anal. Lett., 12:1-9.

Wehmeyer, K.R., Halsall, H.B., and Heineman, W.R., 1982, Electrochemical investigation of hapten-antibody interactions by differential pulse polarography. Clin. Chem., 28:1968-1972.

Yalow, R.S. and Berson, S.A., 1959, Assay of plasma insulin in human subjects by immunological methods. Nature, 184:1648-1649.

IMMUNOASSAY BY DIFFERENTIAL PULSE POLAROGRAPHY AND ANODIC STRIPPING VOLTAMMETRY

Matthew J. Doyle and Kenneth R. Wehmeyer

The Procter and Gamble Company
Miami Valley Laboratories, P. O. Box 39175
Cincinnati, Ohio 45247

William R. Heineman and H. Brian Halsall

Department of Chemistry
University of Cincinnati
Cincinnati, Ohio 45221

INTRODUCTION

Immunoassay is a powerful method exhibiting high selectivity and sensitivity for biologically important molecules (Chait and Ebersole, 1981). The utility of immunoassay in diagnostic medicine is well documented, and a number of routine clinical methods have been established for the determination of a wide variety of common antigens (Landon and Moffar, 1976). However, immunoassay has only recently gained prominence within the analytical community (Monroe, 1984).

Typically, a heterogeneous competitive format is employed, with radioisotopes continuing to be the label of choice for many immunoassay protocols. However, in view of the economic, health and legislative concerns associated with the use of isotopic materials, alternative labeling schemes are increasing in popularity (Nakamura and Kito, 1980; O'Donnell and Suffin, 1979; Wisdom, 1976).

The low detection limit, wide dynamic range, rapid analysis time and relatively low cost of instrumentation have stimulated a renewed interest in electroanalytical methods. Both potentiometric and voltammetric approaches to electrochemically based assays have been investigated. The former employs ion-selective or chemically modified electrodes as potentiometric sensors and is discussed in greater detail elsewhere. The finite current technique of voltammetry comprises the second class of electrochemically based immunoassays, which exhibit greater sensitivity and a more rapid response time than present potentiometric approaches.

Voltammetric approaches have traditionally involved coupling an electroactive moiety directly to the antigen. Weber and Purdy (Weber and Purdy, 1979) developed a continuous flow, homogeneous voltammetric immunoassay for the determination of morphine in the presence of codeine at a glassy carbon electrode. Several investigators have assessed the interactions between azo-labeled protein (Breyer and Radcliff, 1953; Breyer and Radcliff, 1951), nitrophenyl-labeled lysine (Zikan, 1966; Zikan and Kotynek, 1968), and azo-labeled phenylarsonic acid (Schneider and Sehon, 1961) with their specific antisera via polarography. Additionally, Alam and Christian have reported the development of a voltammetric immunoassay for human serum albumin using lead (Alam and Christian, 1982), cobalt (Alam and Christian, 1984) and nickel (Alam and Christian, 1985) labels. However, clinical application of this approach is limited due to severe interferences, competing equilibria, and the non-specific attachment of the assay label. Anderson and coworkers (Heineman et al., 1979; Svoboda, 1983) have investigated the use of Hg labels for the determination of estriol and testosterone. However, these methods also suffer from non-specific protein interactions with the metal label. The products of enzyme labels have also been detected amperometrically as discussed in detail in other chapters.

In this chapter two additional approaches to the development of electrochemical immunoassay are reviewed. The first involves the direct labeling of a steroid with a nitro group and the detection of the nitro label by differential pulse polarography (DPP) in a homogeneous immunoassay (Wehmeyer et al., 1982). The second approach involves labeling a protein with a specific metal chelate and detecting the release of the metal label by differential pulse anodic stripping voltammetry (DPASV) in a heterogeneous format (Doyle et al., 1982; Doyle et al., 1983; Heineman and Halsall, 1985).

EXPERIMENTAL

Materials

Potassium phosphate (0.1 M, pH 7.4), sodium citrate (0.1 M, pH 1.5 or pH 7.5) and NaCl (0.15 M) solutions were prepared from reagent grade salts (MCB) using distilled/deionized H_2O of at least 10^6 Ω resistivity. The citrate and phosphate buffers were pre-electrolyzed over a mercury pool electrode maintained at approximately -1100 mV vs. SCE for at least 24 h to remove trace metal impurities prior to use.

Acetonitrile (MCB) was redistilled and stored over molecular sieves. Diethylenetriaminepentaacetic acid (DTPA, Aldrich), triethylamine (Aldrich), and isobutylchloroformate (Sigma) were of the highest available purity. Indium was obtained as anhydrous, ultrapure $InCl_3$ (Alfa Products).

Estriol, estradiol, and bovine immunoglobulin G (IgG) were of the highest available purity (Sigma). Estrogen-specific monoclonal antiserum was a gift from New England Nuclear. Human serum albumin (HSA, essentially globulin free) and rabbit IgG specific for HSA (αHSA, Sigma) were obtained in lyophilized form. Protein A from Staphylococcus aureous (SPA, Enzyme Center, Boston, MA) was employed as an immunoadsorbent. It was obtained as a 10% (w/v) suspension with a 1.47 mg/mL binding capacity for immune complexes. The immunoadsorbent was prepared as follows: 2.0 mL of SPA solution were mixed with 35 μL of rabbit IgG (82 mg/mL) specific for HSA. The SPA has a high molecular weight (~10^6) and is insoluble. The SPA-αHSA complex retains specific affinity for HSA and HSA-DTPA-Indium (HDI) and the resulting SPA-αHSA-HDI/HSA complexes are easily separated from solution by centrifugation (Kessler, 1976; MacSween and Eastwood, 1978).

Apparatus

Polarographic analyses were performed using a Model 174A polarographic analyzer (Princeton Applied Research Corp) with a Model 2000 x-y recorder (Houston Instruments), a Model 303 static mercury drop electrode, a saturated Ag/AgCl reference electrode, and a platinum auxiliary electrode. DPP conditions for both the homogeneous and heterogeneous assays were: range of scan, 0.0 to -1500 mV vs saturated Ag/AgCl; pulse amplitude, 25 mV; scan rate, 10 mV/s; drop time, 1.0 s. Solutions were purged approximately 10 min with nitrogen passed through vanadous chloride deoxygenating towers. They were then blanketed with nitrogen while recording polarograms.

DPASV analyses were performed using a 5 min deposition time and a potential of -800 mV for preconcentrating indium. The potential range between -800 mV and 0.0 mV was then scanned positively at 2 mV/s with a 25 mV pulse amplitude and a 0.5 s clock time. Typically, 5.0 mL sample volumes were examined; however, smaller volumes (1.0-2.0 mL) can be routinely tested. All peak potentials are reported relative to a saturated Ag/AgCl reference electrode.

Label Synthesis

Estriol was nitro-labeled in the 2 and 4 positions by the procedure of Konyves and Olsson (Konyves and Olsson, 1964). The product was purified by silica gel column chromatography as previously described (Wehmeyer et al., 1982). Labeled estriol was dissolved in an ethanol/water 50% (v/v) solution and used as a standard solution.

A modification (Doyle et al., 1982) of the four-step mixed anhydride procedure of Krejcarek and Tucker (Krejcarek and Tucker, 1977) was utilitized to couple DTPA to HSA. A 17-fold molar excess of In^{3+} was added to the cooled HSA-DTPA (HD) solution and the mixture stirred for 35 min. The pH was adjusted to 4.50 with 1.0 M NaOH and the solvent was sequentially replaced by ultrafiltration using a 10,000 MWCO membrane, against 200 mL of 0.15 M NaCl and then 400 mL of 0.1 M citrate buffer (pH 5.50) and dialyzed once against 1 L of 0.1 M citrate buffer (pH 5.50) to remove any unbound In^{3+} or noncovalently attached DPTA. The HDI complex was shown to be monodisperse by gel chromatography and electrophoresis, immunoreactive <u>via</u> Ouchterlony radial immunodiffusion, and to contain 3-5 mol In^{3+}/mol HDI as determined by plasma emission spectroscopy (Doyle et al., 1982).

Procedure for Heterogeneous Assay

The assay protocol is outlined in Figure 1. Briefly, 110 μL of HDI (1.038 mg/mL stock) and 36 μL of SPA-αHSA were added separately, in order, to a 5.0 mL HSA sample, mixed in 12 x 75 mm polystyrene test tubes, and incubated at 37 °C for 1 h. The tubes were centrifuged at 2000 x g for 10 min, and the supernate was discarded. This supernate contained excess protein, and only SPA-αHSA bound antigen (i.e., HSA or HDI) remained within the tubes. The pellets were rinsed with 0.1 M citrate (pH 7.50) and resuspended in 0.1 M citrate (pH 1.50) which released In^{3+} bound in the HDI complexes. The tubes were centrifuged a second time at 2,000 x g for 15 min to pellet the SPA-αHSA-Antigen complexes. The free In^{3+} concentration in the supenatant liquid was then determined by DPASV.

PRINCIPLE

Homogeneous Assay

This assay involves the determination of haptens by detection of an electroactive label by DPP. As in a conventional competitive assay, the unlabeled hapten and the hapten labeled with an electroactive group compete for a limited number of antibody binding sites (Scheme I). The free labeled hapten present at equilibrium is determined by detection of the reduction of the electroactive label at a

$$\begin{matrix} E \\ E\text{-}NO_2 \end{matrix} \; + \; Ab \; \rightleftharpoons \; \begin{matrix} Ab{:}E \\ Ab{:}E\text{-}NO_2 \end{matrix}$$

Scheme I

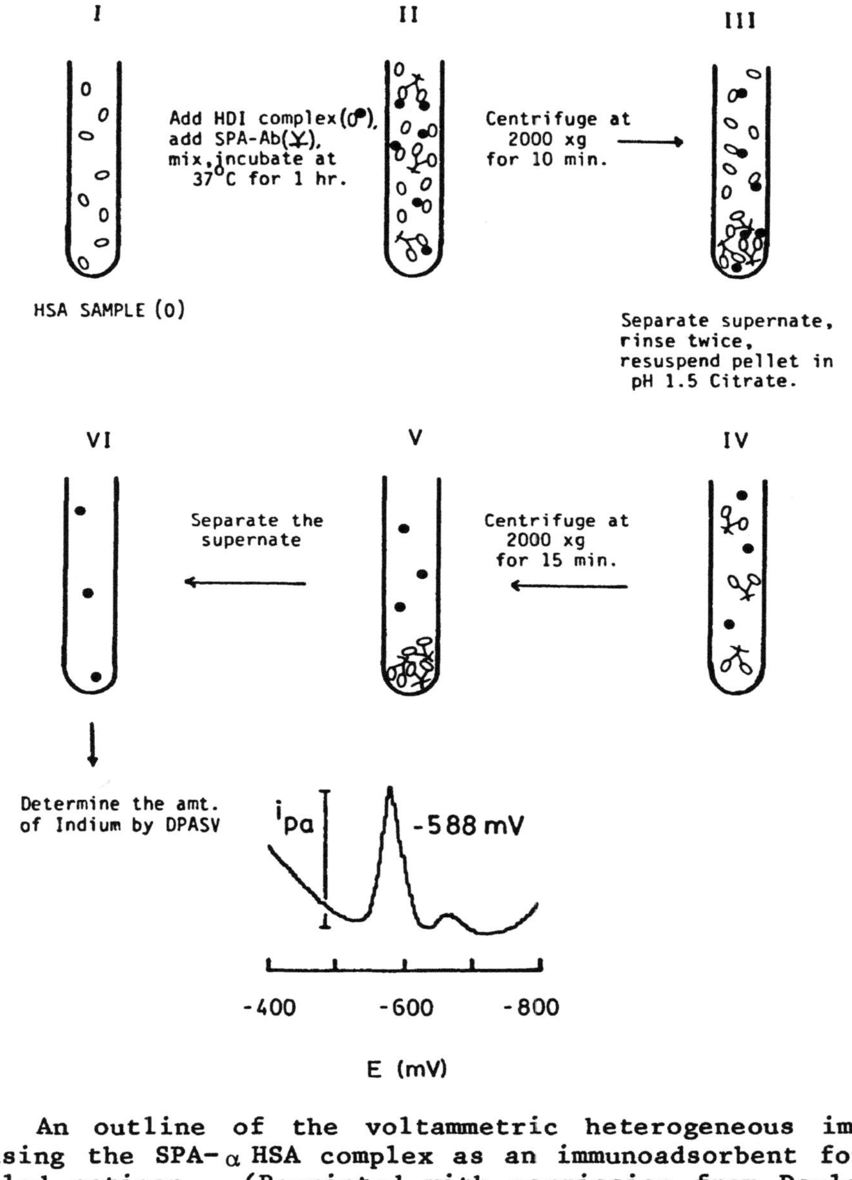

Figure 1. An outline of the voltammetric heterogeneous immunoassay protocol using the SPA-α HSA complex as an immunoadsorbent for labeled and unlabeled antigen. (Reprinted with permission from Doyle et al., 1982. Copyright 1982, American Chemical Society.)

dropping mercury electrode (DME). The cathodic current from the reduction of the label is proportional to the concentration of unlabeled hapten present. The separation of bound electroactively labeled hapten from free electroactively labeled hapten is not necessary, since the reduction of the antibody-bound labeled hapten is attenuated. Estriol was chosen as a model compound to demonstrate the assay. Dinitroestriol (DNE) was the labeled compound.

Heterogeneous Assay

The heterogeneous electrochemical immunoassay for protein antigens is based upon the determination of a releasable metal ion (In^{3+}) label. Human serum albumin (HSA) was chosen as a model antigen for these studies; however, the universal nature of the labeling protocol makes this approach broadly applicable for the determination of a number of macromolecules. DTPA, a pentadentate ligand, was covalently attached to HSA serving as a "site-specific" chelate for the In^{3+} label. Both labeled and unlabeled HSA are allowed to compete for a limited number of antibody binding sites which have been insolubilized on an immunoadsorbent [I)-Ab] as depicted in Scheme II below:

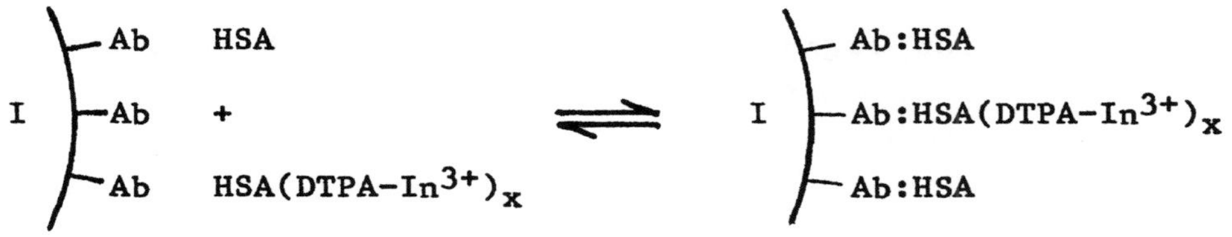

Scheme II

Following equilibration, the antibody "bound" complexes are separated from excess "free" antigen and the In^{3+} label is then released from the complex by acidification (Scheme III). Levels of released In^{3+} are then determined in an interference-free manner by DPASV.

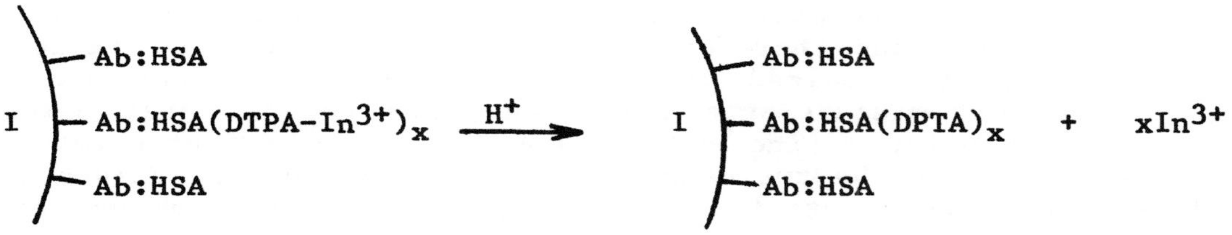

Scheme III

RESULTS AND DISCUSSION

Electrochemical Detection

The electroanalytical techniques of differential pulse polarography and differential pulse anodic stripping voltammetry are highly sensitive methods for the trace determination of electroactive organic compounds (10^{-7} to 10^{-8} M) and metal ions (10^{-10} to 10^{-11} M), respectively (Kissinger and Heineman, 1984). Both methods are considered "controlled-potential" techniques, which involve manipulating a solution-working electrode interfacial potential difference relative to a reference electrode. At sufficiently negative or positive potentials, electrons may be transfered between an electrode and an analyte species in solution. The resulting current comprises the analytical signal. In both methods peak current (i_p) vs. concentration response is linear over a wide range (10^3 to 10^5).

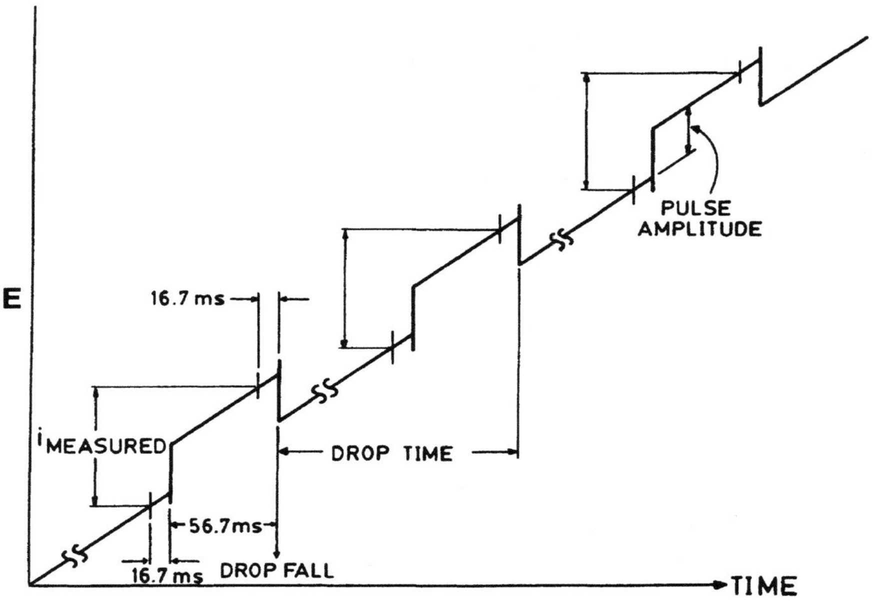

Figure 2. Excitation signal employed for differential pulse polarography.

DPP differs from the more classical polarographic techniques in both the applied waveform and current sampling mode. The excitation signal for DPP consists of small amplitude pulses (typically 5-100 mV) superimposed upon a linearly varying DC voltage ramp. Each pulse is applied during the last 50 ms of drop life when the electrode area changes and consequently the non-faradaic current contributions, are small. The DPP waveform results in a discrimination against the charging current and therefore an increased signal/noise ratio. The drop time is synchronized to terminate each drop at the end of the pulse. Current measurements are taken prior to, and just before the end of each pulse. The output signal of the potentiostat is the difference between the two measured currents as shown in Figure 2. The differential signal output results in peak-shaped current vs potential plots.

Anodic stripping voltammetry is ideally suited to the determination of metal ions. The electrode is maintained at a potential sufficient for the reduction of the metal ion, for a specified period of time and under stirred solution conditions. During this "preconcentration" phase of the experiment metal ions are reduced at the electrode surface, typically a mercury drop or film, resulting in the formation of an elemental amalgam. Following deposition, the electrode potential is scanned in the positive direction resulting in the oxidation (stripping) of the preconcentrated metal. The subsequent anodic current is a measure of the amount of deposited metal. Like DPP, differential pulse anodic stripping voltammetry imparts discrimination against the capacitive component of the stripping signal by judicious current sampling (Figure 2). The applied waveforms and current sampling regimen are similar for both DPP and DPASV. The current response and hence, sensitivity, during DPASV is a function of deposition time, stirring rate, and electrode surface area. Analyte preconcentration in the mercury electrode results in limits of detection that can be several orders of magnitude lower than those achieved by DPP.

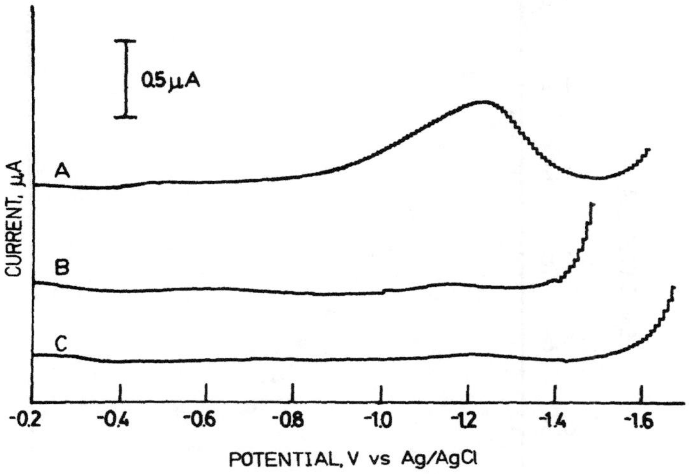

Figure 3. Differential pulse polarogram of 0.1 M phosphate buffer solution containing (A) 46 μM estriol, (B) 100 μL of estrogen-specific antibody per 5.0 mL of buffer, and (C) no additive. (From Wehmeyer et al., 1982, with permission.)

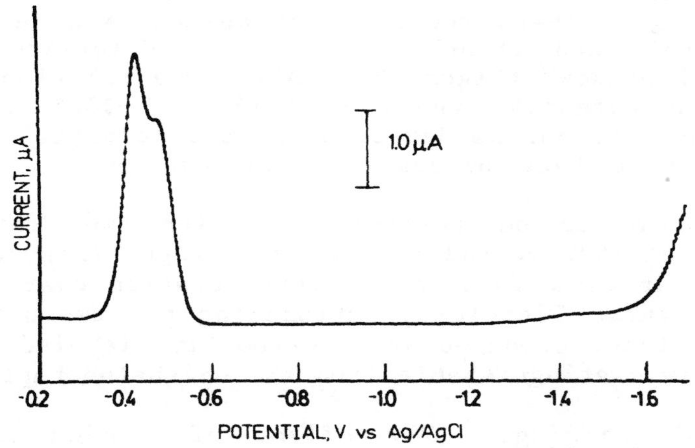

Figure 4. Differential pulse polarogram of 0.1 M phosphate buffer solution containing dinitroestriol, 7.3 μM, recorded with a 2 mV/s potential scan rate. (From Wehmeyer et al., 1982, with permission.)

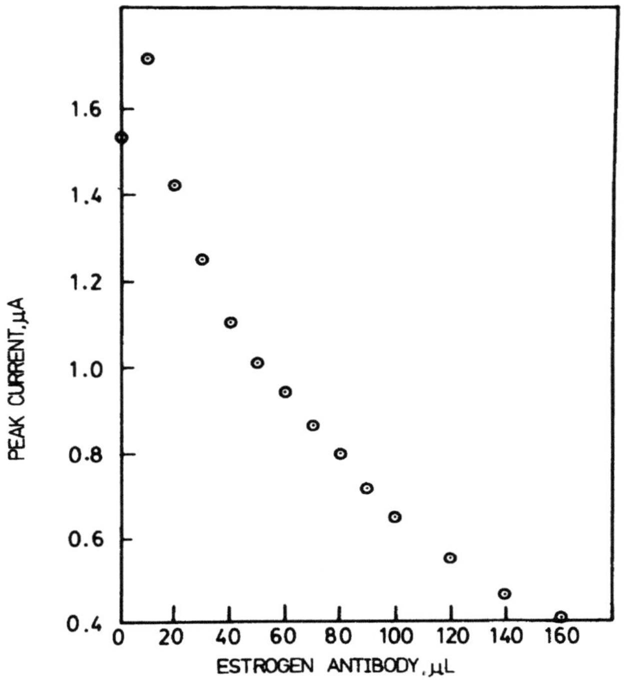

Figure 5. Plot of the polarographic peak current for 4.6 mL of phosphate buffer solution containing dinitroestriol (7.7 μ M) vs microliters of estrogen-specific antibody added to the dinitroestriol solution. (From Wehmeyer et al., 1982, with permission.)

Homogeneous Immunoassay

Electrochemical Evaluation. The homogenous assay is based on the DPP detection of an electroactive nitro label. In order to detect the nitro label in an interference-free manner the other assay components must be electroinactive in the potential region where this group is reduced. Polarograms were recorded for solutions of estrogen-specific antibody, estriol, and phosphate buffer to determine the available "electrochemical window" (Figure 3). All three solutions were electroinactive in the potential range -200 mV to -1000 mV. Thus, the nitro-labeled analyte can be detected in this potential "window" region without interference from the assay components.

DNE was shown to be electroactive with two distinct reduction waves appearing at -422 mV and -481 mV vs Ag/AgCl (Figure 4). The peak current was linear as a function of concentration over the range of 60 ng/ml to 3.7 μg/mL. Therefore, introducing the two nitro groups into the A ring of estriol produced an electroactive labeled hapten that was electrochemically distinguishable from the unlabeled hapten.

Antibody-DNE Binding. The effect of sequential additions of estrogen antibody on the reduction of DNE is shown in Figure 5. The binding of DNE by antibody is characterized by a concomitant decrease in peak current. Two explanations for the decrease in peak current are plausible. First, antibody binding of DNE may sterically sequester the electroactive nitro groups thereby preventing electron transfer with the electrode surface. Alternatively, it is possible that the DNE-Ab complex retains its electroactivity but due to its bulkiness, diffuses more slowly than free DNE. Diffusion coefficients of globular proteins of M_r=150,000-160,000 in aqueous media are about 10^{-7} cm^2/s (Sober,

1970), whereas those of small organic compounds such as estriol are on the order of 10^{-5} cm^2/s (Miller et al, 1964). In DPP the peak current is directly proportional to the square root of the diffusion coefficient of the electroactive species and therefore the decrease could be attributable to a change in diffusion coefficient upon binding. In either case, the effect of binding is clearly measurable without separation of the bound-labeled from the free-labeled species. The increase in peak current observed for the addition of the first aliquot of antiserum is probably due to an adsorption phenomenon at the electrode surface.

Immunoassay of Estriol. An antigen-antibody reaction must be reversible in order to serve as a functional competitive-type analytical method. The displaced labeled hapten should be proportional to the concentration of unlabeled hapten in the standard or sample analyzed. The reversible displacement of DNE from antibody-binding sites is demonstrated by the sequential increase in peak current with successive aliquots of unlabeled estriol (Figure 6). The quantitative detection of this reversible displacement is the basis of an immunoassay based on DPP. Estrone and estradiol were also able to displace antibody-bound DNE, as would be expected from the reported cross reactivity of the antiserum with these steroids. Progesterone, however, did not displace DNE from the antibody-binding sites, indicating that the reaction is specific for estrogens.

Nonspecific Interactions. Bovine IgG was used to probe for nonspecific interactions between DNE and the globular IgG proteins. The addition of successive aliquots of bovine IgG resulted in a decrease in the normalized peak current i_{pn} (Figure 7). The effect of bovine IgG appeared to approach a limit and its further addition resulted in no significant decrease in i_{pn}. Further, the addition of estriol did not result in an increase in the i_{pn}. In contrast, the incremental addition of estrogen specific antiserum to the same solution resulted in corresponding decreases in the i_{pn}. Subsequent addition of estriol to this solution resulted in an increase in i_{pn}.

The bovine IgG-induced decrease in i_{pn} may be the result of protein "fouling" of the electrode (Reilley and Stumm, 1962) or more likely, a nonspecific hydrophobic binding of the DNE by bovine IgG. The diffusion coefficient of the resulting DNE-bovine IgG complex would be greatly reduced. However, regardless of the cause of this decrease, the fact that the addition of unlabeled estriol does not increase the peak current indicates that there is no specific binding similar to that observed with the estrogen-specific antisera.

Urine and Plasma Analysis. Levels of estriol in female serum and urine are well below the detection limit of the present assay (Thompson and Have, 1977). However, urine and plasma samples were analyzed by spiking these fluids with high levels of estriol. It was found that binding and displacement of DNE from Ab sites by estriol was readily observable in urine. In plasma, however, the binding and release of DNE was not apparent possibly due to a high amount of nonspecific binding by serum albumins. The effect of nonspecific binding by serum proteins could be avoided by first removing interfering protein (e.g. ultracentrifugation, precipitation).

Heterogeneous Immunoassay

Electrochemical Evaluation The electrochemical character of the metal-chelate-protein tracer was evaluated by DPP. The 0.1 M citrate (pH 5.50) supporting electrolyte was shown to be electroinactive in the

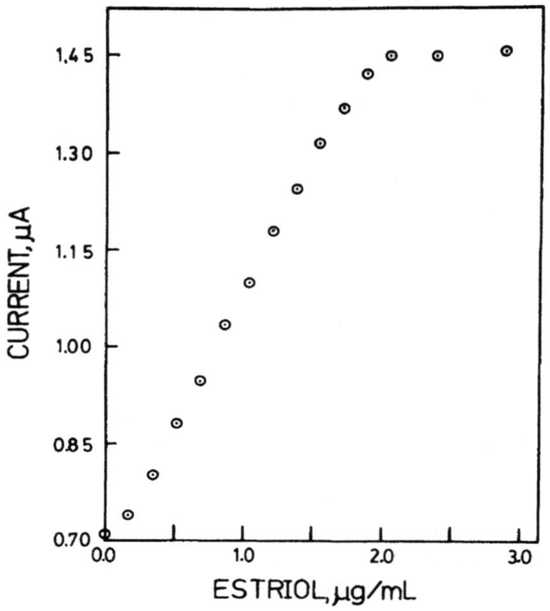

Figure 6. Plot of polarographic peak current of 5.5 mL phosphate buffer containing dinitroestriol (5.6 μM) plus 150 μL estrogen-specific antibody vs [estriol] added to the solution. (From Wehmeyer et al., 1982, with permission.)

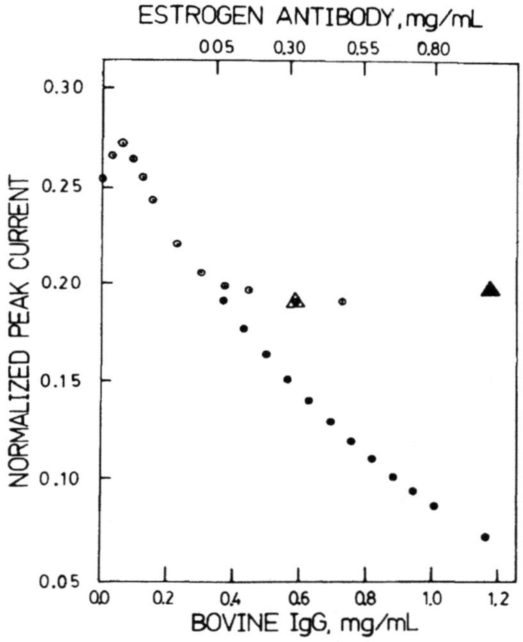

Figure 7. Plot of normalized peak current [(peak current)/(concentration DNE)] from differential pulse polarograms of 5.0 mL of phosphate buffer solution containing DNE, bovine IgG, estriol, and (or) estrogen-specific antiserum. ⊙ DNE 7.0 μM plus bovine IgG as shown on bottom scale; Δ, DNE 7.0 μM plus bovine IgG (0.7 g/L) and estriol (74 μM): ●, DNE 8.8 μM plus bovine IgG (0.3 g/L) and estrogen-specific antisera as shown on top scale; ▲, DNE 8.8 μM plus bovine IgG (0.3 g/L), estrogen-specific estriol (0.9 g/L), and estriol (6.0 μM). (From Wehmeyer et al., 1982, with permission).

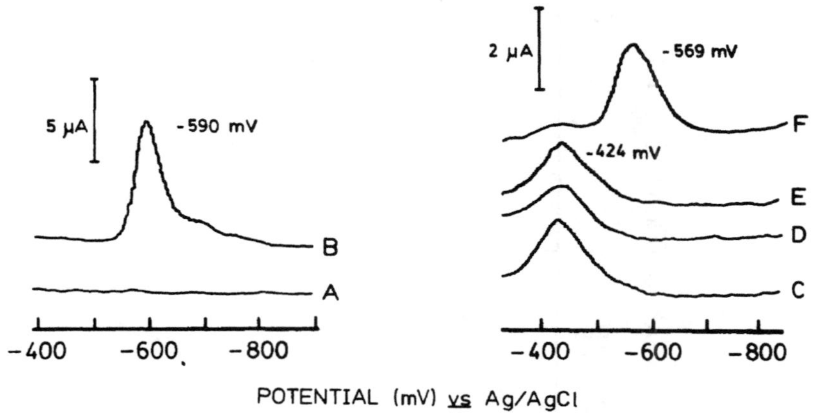

Figure 8. Differential pulse polarograms of: (A) 0.1 M citrate pH 5.50 supporting electrolyte, (B) 1.0 x 10^{-4} M $InCl_3$ in citrate, (C) 1.50 x 10^{-5} M HSA, (D) 1.40 x 10^{-5} M HSA-DTPA, (E) 1.08 x 10^{-5} M HDI, and (F) 1.08 x 10^{-5} M HDI following acidification with 1.0 M HCl to pH 3.80. (Reprinted with permission from Doyle et al., 1982. Copyright 1982, American Chemical society.)

potential region scanned (Figure 8A). Reduction of In^{3+} occurs as a three-electron process (Losev and Moledov, 1976) resulting in a pronounced and characteristic cathodic wave at -590 mV (Figure 8B). HSA, HD, and HDI exhibit reduction waves at approximately -424 mV due to the reduction of surface disulfides (Stankovich and Bard, 1978) of which albumin has seven (Figure 8C-E). Following acidification of the HDI complex to pH 3.8, a reduction wave appeared at -569 mV characteristic of free In^{3+} at this pH (Figure 8F). No wave at -569 mV appeared when either HSA or HD alone was acidified. Assay conditions were optimized with respect to pH to achieve maximal current response and maintain the highest degree of selectivity for the determination of indium (Doyle et al., 1982).

A standard curve for the determination of HDI by DPP is shown in Figure 9. The lower limit is approximately 1.0 μg/mL. The response was linear at lower HDI concentrations, however, current attenuation occurred at higher concentrations due to protein adsorption on the electrode surface. This problem can be surmounted by either diluting the analytical solution to a point where adsorption poses no problem (i.e., the linear section of the curve) or preferably by removing all interfering protein from solution prior to electrochemical investigation.

Minimization of Protein Electrosorption. With an immunoadsorbent such as SPA, which is essentially insoluble yet specific for a single antigen when complexed with the appropriate antiserum, separation of protein is easily achieved by centrifugation. These principles were used in a competitive heterogeneous immunoassay, which removed HDI and HSA from solution, and retained them so that the indium could be detected at the electrode surface in an interference-free manner. The kinetics of antigen dissociation in the presence of salt are slow (Hardie and van Regenmortal, 1977) compared to the equilibria governing indium release from DTPA upon acidification. Thus, the

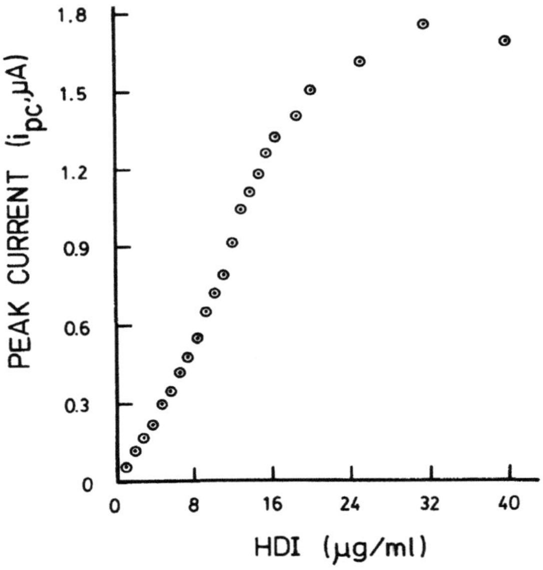

Figure 9. A plot of the cathodic peak current (for the -490 mV indium wave) vs. the total amount of HDI complex (μg/mL) added to 6.0 mL of 0.1 M citrate. The solution was acidified to pH 1.00 to release In^{3+} prior to recording the polarogram. (Reprinted with permission from Doyle et al., 1982, Copyright 1982, American Chemical Society.)

possibility of antigen leaching is unlikely and any residual protein remaining in the supernate would be of such low concentration that current attenuation would not be a problem. Additionally, the total protein concentration of each tube is identical at this point; only the ratios of bound HSA/HDI differ. Thus, any depression in current response due to adsorption would be reflected equally throughout the assay. Other approaches designed to remove interfering protein complexes (e.g., reversed-phase chromatography) could also be developed.

The controlled release of indium, simply by adjusting solution pH, coupled with the removal of protein, enables one to detect the tracer in an essentially interference-free environment. This approach eliminates any possibility of competing equilibria, which have been encountered by others (Alam and Christian, 1982; Alam and Christian, 1984; Alam and Christian, 1985; Svoboda, 1983) and would seriously compromise the utility of the assay.

<u>Anodic Stripping Voltammetry</u> The detection limit for the determination of HDI by DPASV is approximately 0.1 μg/mL. Following the electrochemical immunoassay protocol outlined in Figure 1, removal of the interfering protein was achieved as decribed above. A characteristic anodic wave corresponding to released indium appeared at -588 mV as seen in Figure 10. The wave at -633 mV, which also appears in blank analysis, is attributed to trace cadmium associated with the commercial immunologic reagents. The -588 mV wave is resolvable in the presence of the -633 mV wave. Since HSA will effectively displace HDI bound to the SPA-Ab adsorbent phase, decreasing amounts of released indium will be observed as the amount of HSA in the original analytical solution increases.

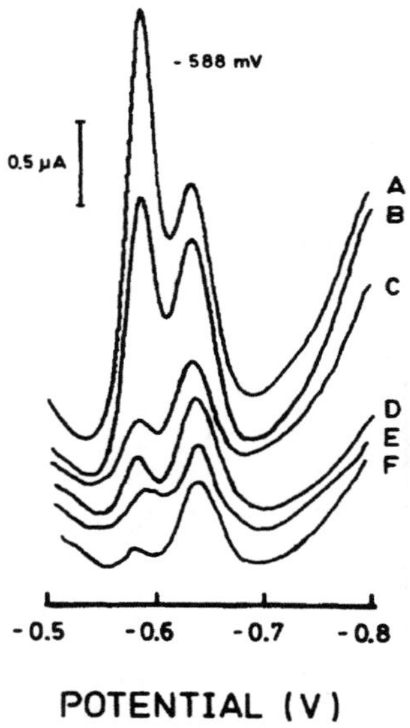

Figure 10. Differential pulse anodic stripping voltammograms of a typical series of HSA standard solutions of increasing concentration (A–F) in 0.1 M citrate following the electrochemical immunoassay procedure. (Reprinted with permission from Doyle et al., 1982. Copyright 1982, American Chemical Society.)

Consequently, a progressive decrease in anodic peak current (i_{pa}) resulted as HSA standards of increasing concentration were examined (Figure 10A–F). A plot of peak current (-588 mV wave) vs HSA concentration generated a smooth standard curve, which can be utilized for unknown sample analysis (Figure 11).

Comparison with Microbiuret Method Analytical methods employed for the quantitation of serum proteins vary widely depending on analyte concentration. The microbiuret assay was chosen as the most appropriate technique to compare with the electrochemical immunoassay method, primarily because their applicable concentration ranges overlap. Further, the microbiuret assay is frequently used for HSA quantitation when precision is required. The microbiuret assay measures the UV absorption of a complex formed between peptide nitrogens and Cu^{2+} in an alkaline $CuSO_4$ media (Itzhaki and Gill, 1964). The biuret reaction is relatively nonspecific and suffers from interferences including bile pigments and ammonia (Bradford, 1976) which may be present as a pathological consequence.

The electrochemical immunoassay technique for HSA is highly specific and free from endogenous interferences. A correlation plot obtained for the analysis of a common set of solutions by electrochemical immunoassay and the microbiuret assay yielded a correlation coefficient of 0.998 over a 20–100 μg/mL range.

SUMMARY

The concept of a homogeneous electrochemical immunoassay based on DPP detection of nitro-labeled antigens has been demonstrated. The assay offers a rapid means for the determination of analytes present at

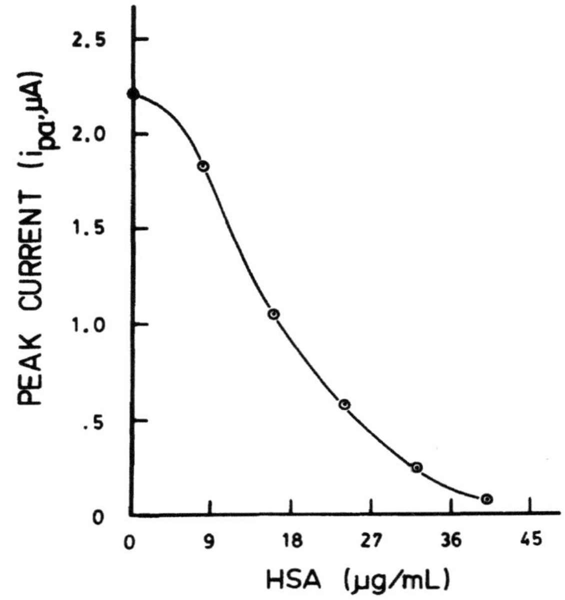

Figure 11. A plot of the anodic peak current (µA, for the -588 mV indium wave) vs. the concentration of HSA (µg/mL) in a series of standard solutions. This plot is a typical electrochemical immunoassay standard curve obtained for the model albumin system. (Reprinted with permission from Doyle et al., 1982. Copyright 1982, American Chemical Society.)

the µg/mL level. The direct analysis of urinary samples was shown to be feasible. However, the determination of haptens in other fluids, such as plasma, would require the removal of proteins capable of non-specific binding interactions.

The heterogeneous electrochemical immunoassay procedure described here is widely applicable to protein antigens due to the universal nature of the DTPA coupling reaction. Since we first reported this work (Doyle et al., 1982; Doyle et al., 1983) a similar heterogeneous immunoassay protocol has been made commercially available by LKB using a Eu-DTPA complex as a tracer (LKB News, 1984). The europium label is released following acidification and detected by fluorescence spectroscopic methods. The popularity of this class of immunoassays should continue to grow as chelate-coupled macromolecules become readily available. In addition, the utilization of a "sandwich" immunoassay format and longer deposition times should result in a series of assays with even better limits of detection.

ACKNOWLEDGMENT

Financial support provided by NSF Grants CHE-8217045 and CHE-8401525 and NIH Grant 5-R01-AI16753 is gratefully acknowledged.

REFERENCES

Alam, I. A. and Christian, G. D., 1982, Voltammetric determination of lead labeled albumin and of albumin antiserum by immunoassay, Anal. Lett. 15:1449-1456.

Alam, I. A. and Christian, G. D., 1984, Voltammetric immunoassay of human albumin and goat antiserum to human albumin by cobalt labeling, Fr. Z. Anal. Chem. 318:33-36.
Alam, I. A. and Christian, G. D., 1985, Voltammetric immunoassay of human albumin and goat antiserum to human albumin by nickel labeling, Fr. Z. Anal. Chem. 320:281-284.
Bradford, M., 1976, A rapid and sensitive method for the quantitation of microgram quantities of protein utilizing the principle of protein-dye binding, Anal. Biochem. 72:248-254.
Breyer, B. and Radcliff, F. J., 1951, Polarographic investigation of the antigen-antibody reaction, Nature 167:79.
Breyer, B. and Radcliff, F. J., 1953, Polarographic study of the antigen-antibody complex, Aust. J. Exp. Biol. 31:167-172.
Chait, E. M. and Ebersole, R. C., 1981, Clinical analysis: a perspective on chromatographic and immunoassay technology, Anal. Chem. 53:682A-692A.
Doyle, M. J., Halsall, H. B. and Heineman, W. R., 1982, Heterogeneous immunoassay for serum proteins by differential pulse anodic stripping voltammetry, Anal. Chem. 54:2318-2322.
Doyle, M. J., Halsall, H. B. and Heineman, W. R., 1983, A releasable label for immunoassay by anodic stripping voltammetry, Trends Anal. Chem. 2:XI-XII.
Hardie, G. and van Regenmortel, M. H. V., 1977, Isolation of specific antibody under conditions of low ionic strength, J. Immunol. Methods 15:305-314.
Heineman, W. R., Anderson, C. W. and Halsall, H. B., 1979, Immunoassay by differential pulse polarography, Science 204:865-866.
Heineman, W. R. and Halsall, H. B., 1985, Strategies for electrochemical immunoassay, Anal. Chem. 57:1321A-1331A.
Itzhaki, R. F. and Gill, D. M., 1964, The micro-biuret method for estimating proteins, Anal. Biochem. 9:401-410.
Kessler, S. W., 1976, Cell membrane antigen isolation with the staphylococcal protein A-antibody adsorbent, Immunol. 117:1482-1490.
Kissinger, P. T. and Heineman, W. R., Eds., 1984, "Laboratory Techniques in Electroanalytical Chemistry," Dekker, New York.
Konyves, I. and Olsson, A., 1964, Nitrated derivatives of estriol, Acta Chem. Scand. 18:483-487.
Krejcarek, G. E. and Tucker, K. L., 1977, Covalent attachment of chelating groups to macromolecules, Biochem. Biophys. Res. Commun. 77:581-585.
Landon, J. and Moffar, A. C., 1976, The radioimmunoassay of drugs, , Analyst 101:225-243.
LKB News, Spring, 1984. The breakthrough in immunotechnology,
Losev, V. V. and Moledov, A. I., 1976, Vol. VI, in: "Encyclopedia of Electrochemistry of the Elements," A. J. Bard, ed., Marcel Dekker, New York.
MacSween, J. M. and Eastwood, S. L., 1978, Recovery of immunologically active antigen from staphylococcal protein A-antibody adsorbent, Immunol. Methods 23:259-267.
Miller, T. A., Lamb, B., Prater, K., et al., 1964, Tracer diffusion coefficients of aromatic organic molecules, Anal. Chem. 36:418-420.
Monroe, D., 1984, Enzyme immunoassay, Anal. Chem. 56:921A-931A.
Nakamura, M. and Kito, W. R., 1980, Nonradioisotopic immunoassays for therapeutic drug monitoring, Lab. Med. 11:807-817.
O'Donnell, C. M. and Suffin, S. C., 1979, Fluorescence immunoassays, Anal. Chem. 51:33A-40A.
Reilley, C. N. and Stumm, W., 1962, Adsorption in polarography, in: "Progress in Polarography," P. Zuman and I. M. Kolthoff, eds., Interscience, New York.

Schneider, H. and Sehon, A. H., 1961, Determination of the lower limits for the rate constants of hapten-antibody reaction by polarography, Trans. N. Y. Acad. Sci. 24:15.

Sober, H. A., ed., 1970, "CRC Handbook of Biochemistry: Selected Data for Molecular Biology," 2nd ed., The Chemical Rubber Co., Cleveland.

Stankovich, M. T. and Bard, A. J., 1978, The electrochemistry of proteins and related substances part III. bovine serum albumin, J. Electroanal. Chem. 86:189-199.

Svoboda, G. J., 1983, Electrochemical studies of organomercurials and microvoltammetric electrodes: applications to voltammetric immunoassays, Ph.D. Thesis, Duke University.

Thompson, J. and Have, G., 1977, Evaluation of a radioimmunoassay for serum estriol using commercial reagents, Clin. Path. 68:474.

Weber, S. G. and Purdy, W. C., 1979, Homogeneous voltammetric immunoassay: a preliminary study, Anal. Lett. 12:1-9.

Wehmeyer, K. R., Halsall, H. B. and Heineman, W. R., 1982, Electrochemical investigation of hapten-antibody interactions by differential pulse polarography, Clin. Chem. 28:1968-1972.

Wisdom, G. B., 1976, Enzyme immunoassay, Clin. Ghem. 22:1243-1255.

Zikan, J., 1966, Polarographic determination of the extent of association of antibodies against the 2,4-dinitrophenyl group with hapten, Collec. Czech. Chem. Commun. 31:4260-4267.

Zikan, J. and Kotynek, O., 1968, Interactions of antibodies specific towards the 2,4-dinitrophenyl group and of their fragments with hapten, Biopolymers 6:681-690.

SEPARATION-FREE (HOMOGENEOUS) AMPEROMETRIC IMMUNOASSAY USING AN APOENZYME AS THE LABEL

That T. Ngo

Department of Developmental and Cell Biology
University of California at Irvine
Irvine, California 92717, USA

INTRODUCTION

From an operational standpoint enzyme mediated immunoassays can be divided into two major categories (Ngo and Lenhoff, 1981 and 1982; Blake and Gould, 1984; Oellerich, 1984; Ngo, 1985): (1) separation-free, also known as homogeneous system in which the physical separation of unbound ligand-indicator conjugates from that bound to the antibody is not required; (2) separation-required, also known as heterogeneous system, in which an antibody bound ligand indicator conjugates must first be separated from the unbound ones before measuring the activity of the indicator conjugate in either the free or the antibody-bound fraction.

In general, the separation-required enzyme mediated immunoassays provide a greater assay sensitivity, i.e. lower detection limits than separation-free systems (Ngo and Lenhoff, 1982; Blake and Gould, 1984; Ngo, 1985) . The latter systems, however, were simpler to perform and automate (Blake and Gould, 1984). The basis of all separation-free enzyme-mediated immunoassays is the ability of an anti-ligand antibody to modulate the activity of a covalently linked: (a) ligand-enzyme conjugate (Rubenstein et al., 1972), (b) antibody enzyme conjugate in a consecutive enzymic reaction system (Litman et al.,1980), (c) ligand-enzyme substrate (Burd et al.,1977; Ngo et al., 1981) , (d) ligand-prosthetic groups conjugate (Morris et al., 1981; Ngo, 1985b), (e) ligand-enzyme modulate conjugate (Ngo, 1980a) and (f) ligand-apoenzyme conjugate (Ngo and Lenhoff, 1983; Ngo et al., 1985). Currently most separation-free enzyme mediated immunoassays are measured by a spectrophoto-metric method. Only a few systems using electrochemical detectors have been developed (Guilbault, 1983; Wehmeyer et al., 1983; Heineman and Halsall, 1985). Owing to the advantages in using electrochemical sensors to measure immunochemical reactions, there have been increasing interests in applying various electrochemical sensors in enzyme mediated immunoassays. Ion selective electrodes have been employed to follow enzyme-mediated immunochemical assays using urease (Meyerhoff and Rechnitz, 1979), adenosine deaminase (Gebauer and Rechnitz, 1981), horseradish peroxidase (Boitieux et al., 1978), chloroperoxidase (Fonong and Rechnitz, 1984). Direct potentiometric immunoassays using membranes containing covalently linked ligand-ionophore conjugates have also been developed (Solsky and Rechnitz, 1979 and 1981; Connell et al., 1983; Keating and Rechnitz, 1984). Direct or enzyme-mediated voltammetric immunoassay techniques are also rapidly gaining popularity (Mattiasson and Nilsson, 1977; Aizawa et al., 1979; Wehmeyer et al., 1982; Doyle et al., 1984; Heineman and Halsall, 1985).

In this chapter I described a separation-free (homogeneous) amperometric immunoassay using apoenzyme as the label for measuring a model hapten, N-2,4-dinitrophenyl-6-aminocaproic acid (DNP-ACA) and hydrogen peroxide sensitive electrode as the sensor.

MATERIALS AND METHODS

Chemicals

2,4-Dinitrofluorobenzene, bovine serum albumin, flavin adenine dinucleotide (FAD), glycerol, horseradish peroxidase, Sephadex G-50C and N-2,4-dinitrophenyl-6-aminocaproic acid (DNP-ACA) (Sigma Chemical Co.); 3-dimethyl-aminobenzoic acid (DMAB) and 3-methyl-2-benzothiazolinone hydrazone (MBTH) (Aldrich Chemical Co.,); rabbit anti-DNP serum and highly purified glucose oxidase from Aspergillus niger (Miles Labs, Inc.).

Preparation of DNP-linked apoglucose oxidase (DNP-AG)

DNP-linked apoglucose oxidase was prepared from acid treatment of DNP-linked hologlucose oxidase. Such a treatment removed most of the enzyme bound FAD and provided an apoenzyme preparation in high yield. First the holoenzyme was covalently linked to 2,4-dinitrophenyl groups (DNP) by reacting the holoenzyme with 2,4-dinitrofluorobenzene in sodium bicarbonate buffer (Ngo and Lenhoff, 1983). The DNP-linked hologlucose was then treated with 10% sulfuric acid and bovine serum albumin coated charcoal to dissociate and thoroughly remove FAD from the holoenzyme according to the following procedure which is adapted from Ngo and Lenhoff (1980).

Glycerol was added to DNP-linked hologlucose oxidase to a final concentration of 35% by volume. The solution was stirred and cooled to 0 C and subsequently was acidified to pH 1 with 10 percent sulfuric acid solution by dropwise addition while maintaining the temperature at 0 C and continuously stirring for 1 hour. The solution was then applied onto a column of Sephadex G-50C (2.2 x 55 cm) which was equilibrated with 35 percent glycerol pH 1 (acidified with 10 percent sulfuric acid solution). Fractions were collected into test tubes containing 1 ml 0.1 M Tris, pH 10.

The acid glycerol solution was also used as the eluent. A flow rate of 4 ml/min was maintained with a peristaltic pump. Protein fractions were pooled and mixed with 5 ml of a suspension of charcoal particles which had been coated with bovine serum albumin (see below for preparation procedure) and stirred for one hour at 0°C. The suspension was then centrifuged at 20,000 rpm for ten minutes. The supernatant layer was saved and passed through a Millipore filter (0.45 µ). The pH of the filtrate was adjusted to 6.5 and sodium azide was added to the apo-glucose oxidase solution to give a concentration of 0.1 percent. This solution was stored at 4 and was stable for at least seven months.

The charcoal coated with bovine serum albumin was prepared according to Ngo and Lenhoff (1980) from a suspension of 1 g of charcoal powder in 5 ml of 10 percent BSA in 0.5 M sodium phosphate pH 7.0. The suspension was stirred at room temperature for 30 minutes and then centrifuged at 2,000 g for 15 minutes. The pellet was resuspended in 10 ml 0.1 M sodium phosphate pH 8.5 and stored at 4 until use. The DNP-AG showed no activity without the addition of FAD.

Assay for Glucose Oxidase

Colorimetric assay: This was carried out with a horseradish peroxidase coupled reaction using DMAB, MBTH and hydrogen peroxide as substrates (Ngo and Lenhoff, 1980a). The absorbance of the reaction product, the indamine dye (purple-blue chromopore), was measured at 590 nm in a Beckman Acta C III spectrophotometer.

Amperometric Assay: The rate at which glucose oxidase catalyzed the formation of hydrogen peroxide from glucose and oxygen was monitored amperometrically by using a YSI-Clark 2510 electode and a YSI model 25 oxidase meter and the amperometric reading was also displayed on a Fisher chart recorder. Glucose oxidase was assayed in a 5 mL solution with the electrode immersed in it. The assay solution consisted of 0.1 M sodium phosphate buffer pH 6.5 containing 0.15 M NaCl and 0.3 M glucose. It was prepared at least 1 day before use to ensure adequate time for the glucose to mutarotate. The reaction was initiated by adding 110 uL of a solution containing either the DNP-labeled enzyme or DNP-labeled apoenzyme with FAD.

RESULTS

The FAD of DNP-labeled hologlucose oxidase treated with sulfuric acid dissociated from the holoenzyme, and was separated from the denatured DNP-labeled apoglucose oxidase (DNP-AG) by gel permeation chromatography on a desalting gel (Ngo and Lenhoff, 1983 and Ngo et al., 1985). The enzyme activity of the acid denatured DNP-AG was restored by incubating it with FAD at around pH 7 (Ngo and Lenhoff, 1983 and Ngo et al., 1985). By increasing the concentration of FAD, we observed colorimetrically an increasing amount of glucose oxidase activity until a maximal level of reconstitutable enzyme activity was achieved (Fig. 1).

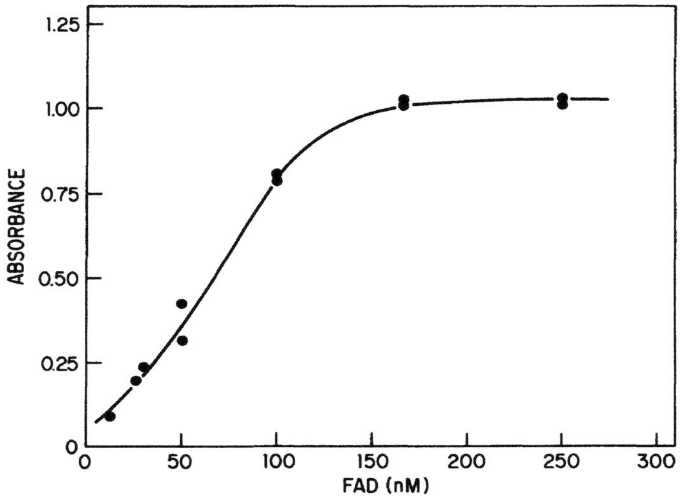

Fig. 1. Reconstitution of glucose oxidase activity from DNP-AG and FAD as monitored by colorimetric method.

Solution (10 µl) containing varying concentrations of FAD (25-500 nM) in 0.1 M sodium phosphate pH 6.5 were added to 10 ul DNP-AG and incubated at 25 for 30 minutes. Assay solution (4 ml) was added to each of the above solutions and incubated at 25° for another 5 minutes before the absorbance (at 590 nm) was read. The assay solution was made up in 0.1 M sodium phosphate pH 6.5. It contained 75 mM glucose, 3mM DMAB, 0.075 mM MBTH, 31.25 nM horseradish peroxidase and 37.5 mM NaCl.
(Reprinted from Ngo and Lenhoff,1983, with permission).

Similar results were obtained using amperometric assay (Fig. 2). From these experiments we calculated the concentration of FAD binding site. The level of reconstitutable glucose oxidase activity is directly dependent on the concentration FAD binding site (Fig. 3).

Addition of increasing amounts rabbit anti-DNP serum to solution containing a fixed concentration of acid denatured DNP-AG and preincubating the solution at 25°C for 30 minutes before adding an excess of FAD, resulted in decreasing levels of reconstitutable glucose oxidase activity being formed on the addition of excess FAD (Fig. 4). Hence, the modulation effect of the antibody was inhibitory. In contrast, when anti-DNP serum was added to DNP-labeled hologlucose oxidase which has not previously been subjected to treatment with sulfuric acid or to DNP-labeled hologlucose oxidase which was reconstituted from DNP-AG and FAD, there was no change in enzyme activity upon adding such anti-serum.

When a fixed concentration of DNP-AG was added to solutions containing different concentrations of the analyte, i.e. DNP-ACA and after a brief mixing of the solution, adding a fixed amount of anti-DNP serum to each tube, incubating the tubes for a fixed period and finally adding an excess of FAD, the results showed that the reconstituted hologlucose oxidase activity increased with increasing concentration of DNP-ACA (Fig. 5).

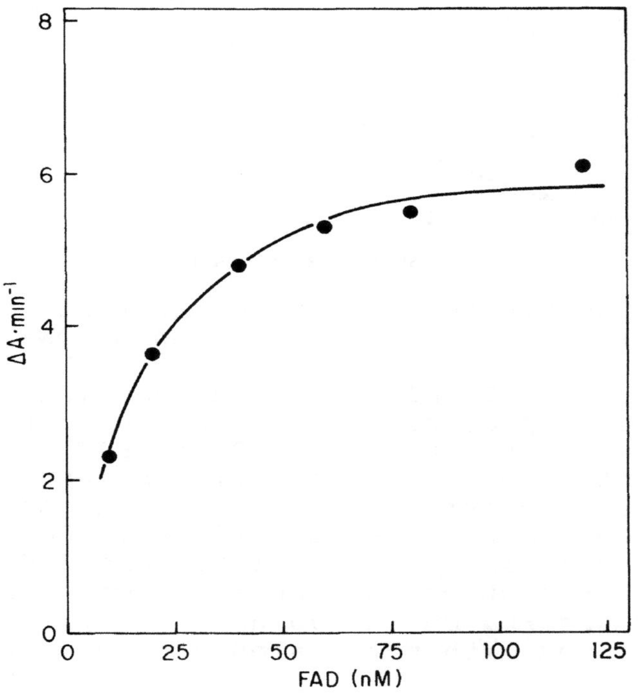

Fig. 2. Amperometric assay for reconstituting glucose oxidase activity from DNP-AG and FAD. Solutions (20 µL) containing varying concentration of FAD (50-1500 nM) in 0.1 M sodium phosphate pH 6.5 were added to 100 µL of DNP-AG and incubated at 25°C for 15 minutes. Immediately, 100 µL of each of the above solutions was added to 5 ml glucose solutions (0.3 M glucose, 0.15 M NaCl in 0.1 M sodium phosphate, pH 6.5) which has been preincubated in a reaction chamber at 25°C, and with the tip of the electrode pre-immersed in the solution. (Reprinted from Ngo et al., 1985, with permission).

The experiment indicated that DNP-ACA competed effectively with DNP-AG for the binding sites of antibodies to DNP; hence, fewer antibodies to DNP were available to complex with DNP-AG. As a result, the fraction of DNP-AG which was not

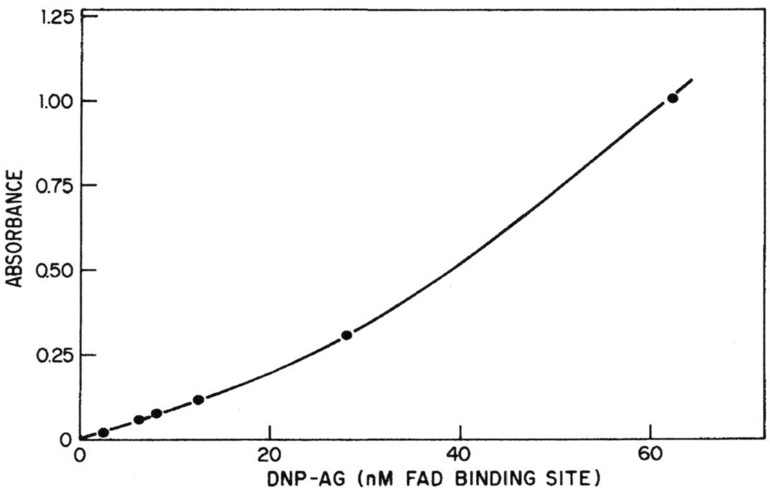

Fig. 3. Relationship between reconstitutable glucose oxidase activity and the concentration of DNP-AG.

Solution of 10 µl containing varying amounts of DNP-AG, expressed in terms of FAD binding sites, were added to 10 µl of 1000 nM FAD (saturating concentration, see Fig. 1), and incubated at 25°C for 15 minutes; then 4 ml of assay solution (its composition was described in legend to Fig. 1) were added and the solutions were incubated at 25°C for another 15 minutes before measuring their absorbance at 590 nm.
(Reprinted with permission from Ngo anf Lenhoff, 1983).

complexed with anti-DNP antibodies, was able to combine with FAD to form a catalytically active holoenzyme, which was subsequently measured amperometrically for its hydrogen peroxide produced (Fig. 5).

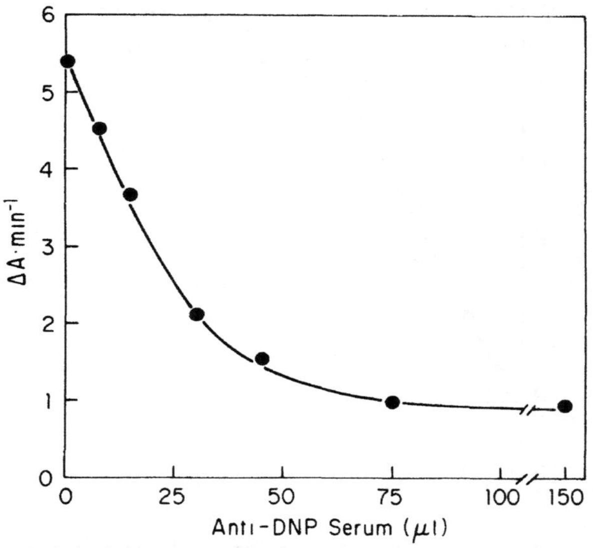

Fig. 4. The inhibition by anti-DNP serum of the reconstitution of hologlucose oxidase from DNP-AG and FAD as measured by amperometric glucose oxidase assay.

Varying amounts of anti-DNP serum were added to 60 μL DNP-AG. The solutions were adjusted to 310 μL with 0.1 M sodium phosphate pH 6.5, and incubated at 25°C for 30 minutes. To these solutions were added 100 μL of 1 uM FAD, and they were further incubated at 25°C for 15 minutes. Next, 300 μL of the solution was removed and assayed for glucose oxidase activity amperometrically (see legend to Fig. 2).
(Reprinted with permission from Ngo et al., 1985).

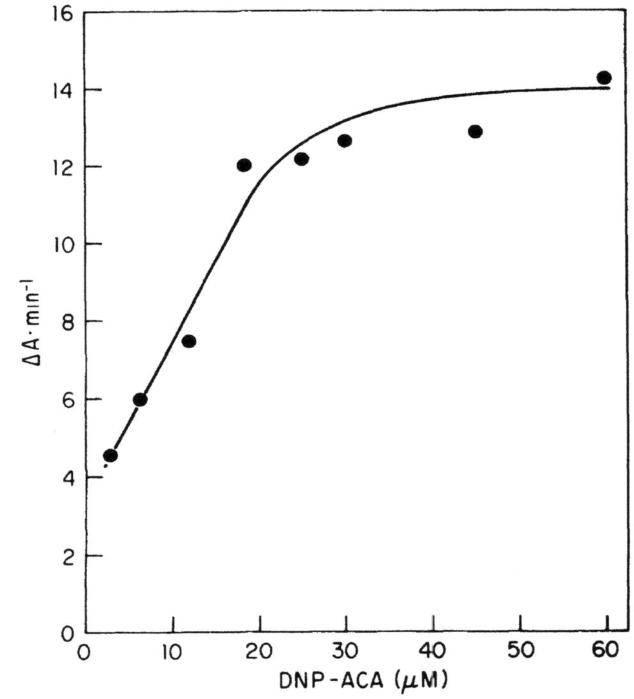

Fig. 5. Standard curve for measuring DNP-ACA by separation-free amperometric enzyme immunoassay.

Solutions of 50 µL containing varying amounts of DNP-ACA were added to 60 µL DNP-AG solutions. To these solutions were added 40 µL anti-DNP serum, and they were incubated at 25°C for 30 minutes. Then 60 µL of 1 uM FAD were added to each of these solutions, and they were further incubated at 25°C for 15 minutes. Next 200 µL was removed from each of the solutions and were assayed for glucose oxidase activity a perometrically (see legend to Fig. 2).
(Reprinted with permission from Ngo et al., 1985).

DISCUSSION

The development of a separation-free amperometric enzyme mediated immunoassay using apo-glucose oxidase as the label was dictated by three factors. First, the activity of holo-glucose oxidase reconstituted from apo-glucose oxidase or ligand-linked apo-glucose oxidase and FAD can be assayed amperometrically with ease by measuring the rate of hydrogen peroxide production (Ngo and Lenhoff, 1980 and Ngo, et al.,1985). Second, both free apo-glucose oxidase and ligand-linked apo-glucose oxidase can be prepared in high yield by a relatively simple method (Ngo and Lenhoff, 1980). Third, a separation-free colorimetric enzyme mediated immnoassay for haptens has been successfully developed using the same apo-glucose oxidase as the label (Ngo and Lenhoff, 1983).

The rationale and experimental sequences involved in the development of the assay are depicted in Fig. 6. In step 1, the holoenzyme was first covalently linked with the analyte, DNP. The second step involved acid denaturation of the DNP-linked holo-glucose oxidase which resulted in its prosthetic group (FAD) being dissociated from it; then,through a desalting gel chromatography of the acidified solution, the FAD was separated from the denatured, enzymatically inactive DNP-linked apo-glucose oxidase, DNP-AG.

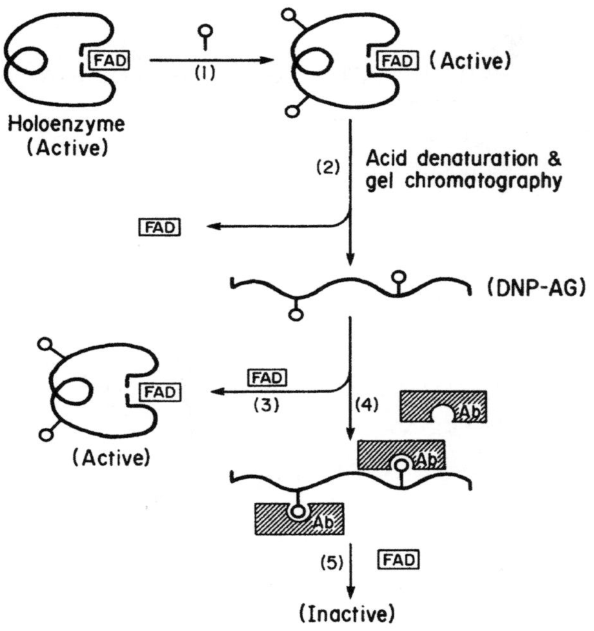

Fig. 6. Rationale and processes involved in the development of a separation-free apoenzyme-labeled amperometric immunoassay.
Holoenzyme: holo-glucose oxidase; o:2,4-dinitrophenyl group (DNP); DNP-AG: DNP-linked apo-glucose oxidase; and Ab: antibody to DNP group. (Reprinted from Ngo and Lenhoff, 1983, with permission)

The reconstitution of holo-glucose oxidase activity from DNP-AG and FAD, as monitored colorimetrically using a peroxidase coupled system and amperometrically using a hydrogen peroxide sensitive electrode was, respectively shown in Fig. 1 and 2.

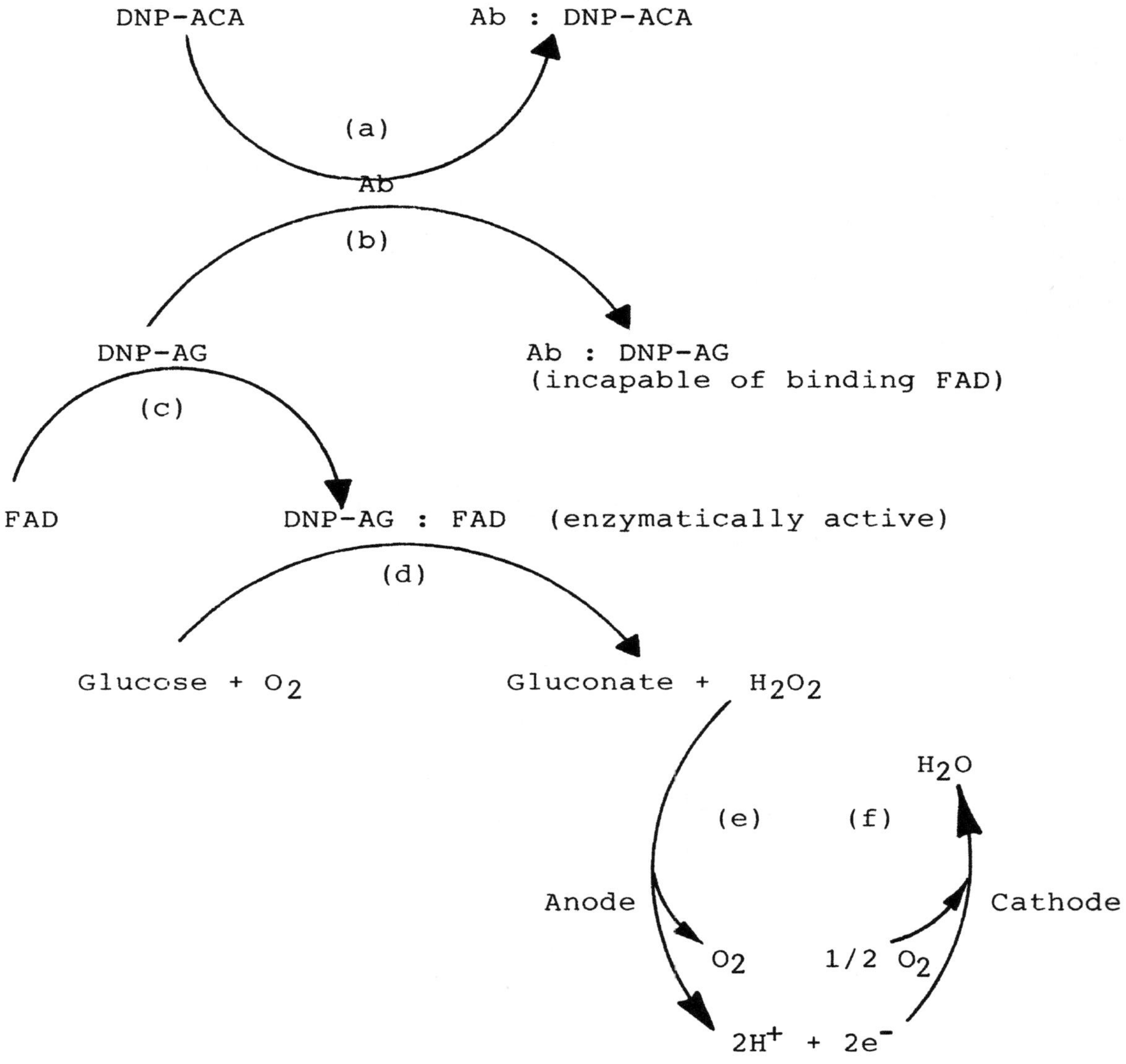

Fig. 7. Overall reactions of the apoenzyme-labeled separation free amperometric immunoassay.
All abbreviations were defined in the text.

The concentration of apo-glucose oxidase cannot be conveniently determined by measuring the amount protein in the solution because of the presence of relatively large amount of bovine serum albumin in the apo-glucose oxidase solution. The addition of albumin to apo-glucose oxidase solution was necessary for the stability of the apoenzyme. Even if a procedure capable of measuring the total protein concentration of apo-glucose oxidase was developed, the results would not be too meaningful because the important parameter was not the protein concentration but the concentration of the functional FAD binding site. This FAD binding site concentration was determined by using FAD titration method (Fig. 3).

The inhibition of the reconstitution of holo-glucose oxidase activity from DNP-AG and FAD by antibody to DNP was followed amperometrically for the rate of enzymatically production of hydrogen peroxide (Fig. 4). The binding of the antibody to the DNP group of DNP-AG presumably restricted the movement of the polypeptide chain of the apoenzyme and therefore prevented it from assuming a proper conformation for binding FAD and from becoming a catalytically active species in the presence of excess of added FAD. Further evidence supporting such a mechanism was provided by the finding that DNP-ACA could counteract the inhibitory action of the antibody to DNP (Fig. 5). It is also possible that the binding of antibodies to DNP-AG may cause some steric hindrance that prevents FAD from approaching the prosthetic group binding sites.

Figure 7 summarizes the overall steps involved in the separation-free amperometric enzyme immunoassay described here. In the presence of a large quantity of DNP-ACA (the analyte), most of the antibodies (Ab) to DNP would be tied up as Ab:DNP-ACA via reaction (a). As a result, less of the antibody (Ab) would be available to combine with a fixed amount of DNP-AG to form the enzymatically inactive complex Ab:DNP-AG [reaction (b)]. Therefore, in the presence of more analyte, there would be more of the free, uncomplexed DNP-AG capable of forming with FAD the enzymatically active DNP-conjugated holoenzyme, DNP-AG:FAD via reaction (c). Accordingly, there would be an increase in the enzymatic formation of gluconate and of hydrogen peroxide which was determined by measuring the amount of current generated upon oxidation of H O at 700 mV polarizing voltage at the platinum anode of the electrode [reaction (e)]. The electrochemical circuit is completed by the reduction of oxygen to water at the silver cathode of the electrode [reaction (f)].

Conversely, if the concentration of the analyte (DNP-ACA) is decreased, reaction (1) would decrease, and DNP-AG and the anti-DNP antibody (Ab) would combine [reaction (b)], leaving less DNP-AG to combine with FAD [reaction (c)] to form enzymatically active DNP-AG. FAD, and, therefore, less hydrogen peroxide would be produced.

REFERENCES

Aizawa, M., Morioka, A., Suzuki, S. and Nagamura, Y., 1979, Enzyme immunosensor, III. Amperometric determination human chorionic gonadotropin by membrane-bound antibody, Anal. Biochem. 94: 22-28.

Blake, C. and Gould, B.J., 1984, Use of enzymes in immunoassay techniques. A Review, Analyst, 109: 533-547.

Boitieux, J-L., Desmet, G. and Thomas, D., 1978, Immobilization of anti HbsAg antibodies on artificial proteic membranes, FEBS Letters, 93: 133-136.

Boitieux, J-L., Desmet, G. and Thomas, D., 1979, An "Antibody Electrode", Preliminary report on a new approach in enzyme immunoassay, Clin. Chem. 25: 318-321.

Burd, J.F., Wong, R.C., Feeney, J.E., Carrico, R.J. and Boguslaski, R.C., 1977, Homogeneous reactant labeled fluorescent immunoassay for therapeutic drugs exemplified by gentamicin determination in human serum, Clin. Chem.,23: 1402-1408.

Connell, G.R., Sanders, K.M. and Williams, R.L., 1983, Electroimmunoassay. A new competitive protein-binding assay using antibody-sensitive electrodes. Biophys. Journal, 44:123-126.

Doyle, M.J., Halsall, H.B. and Heineman W.R., 1984, Enzyme-linked immunoadsorbent assay with electrochemical detection for 1-acid glycoprotein, Anal. Chem. 56: 2355-2360.

Fonong, T. and Rechnitz, G.A., 1984, Homogeneous potentiometric enzyme immunoassay for human immunoglobulin G, Anal. Chem. 56: 2586-2590.

Gebauer, C.R. and Rechnitz, G.A., 1981, Immunoassay studies using adenosine deaminase enzyme with potentiometric rate measurement, Anal. Letters, 14(B2): 97-109.

Guilbault, G.G., 1983, Immobilized biological and immuno-Sensors, Anal. Proc., 20: 550-552.

Heineman, W.R. and Halsall, H.B., 1985, Strategies for electrochemical immunoassay, Anal. Chem. 57: 1321A-1331A.

Keating, M.Y. and Rechnitz, G.A., 1984, Potentiometric digoxin antibody measurements with antigen-ionophore based membrane electrodes, Anal. Chem. 56: 801-806.

Litman, D.J., Hanlon, T.M. and Ullman, E.F., 1980, Enzyme channelling immunoassay: A new homogeneous enzyme immunoassay technique, Anal. Biochem., 136: 223-229.

Mattiasson, B. and Nilsson, H., 1977, An enzyme immuno-electrode, FEBS Letters, 78: 251-254.

Meyerhoff, M.E. and Rechnitz, G.A., 1979, Electrode-based enzyme immunoassays using urease conjugates, Anal. Biochem., 95: 483-493.

Morris, D.L., Ellis, P.B.,Carrico, R.J.,Yeager, F.M., Schroeder, H.R., Albarella, Boguslaski, R.C., Hornby, W.E. and Rawson, D., 1981, Flavin adenine dinucleotide as a label in homogeneous colorimetric immunoassays, Anal. Chem., 53: 658-665.

Ngc, T.T.,1985, Enzyme Mediated Immunoassay: An Overview, in "Enzyme-Mediated Immunoassay", T.T. Ngo and H.M. Lenhoff, Ed., Plenum Press, New York, pp. 3-32.

Ngo, T.T., 1985a, Enzyme modulator as label in separation-free immunoassay: Enzyme modulator mediated immunoassay (EMMIA),in "Enzyme-Mediated Immunoassay", T.T. Ngo and H.M. Lenhoff, Ed. Plenum Press, New York, 57-72.

Ngo, T.T., 1985b, Prosthetic group labeled enzyme immunoassay in "Enzyme-Mediated Immunoassay", T.T. Ngo and H.M. Lenhoff, Ed. Plenum Press, New York, pp.73-84.

Ngo, T.T., Bovaird, J.H. and Lenhoff, H.M., 1985, Separation-free amperometric enzyme immunoassay, Applied Biochem. Biotechnol., 11: 63-70.

Ngo, T.T., Carrico, R.J., Boguslaski, R.C. and Burd, J.F., 1981, Homogeneous substrate labeled fluorescent immunoassay for IgG in human serum, J. Immunolog. Methods, 42: 93-104.

Ngo, T.T. and Lenhoff, H.M., 1980, Amperometric etermination of picomolar levels of flavin adenine dinucleotide by cyclic oxidation-reduction in apo-glucose oxidase system, Anal. Letters, 13(B13): 1157-1165.

Ngo, T.T. and Lenhoff, H.M., 1980a, A sensitive and versatile chromogenic substrate for peroxidase and peroxidase coupled reactions, Anal. Biochem., 105: 389-397.

Ngo, T.T. and Lenhoff, H.M., 1980b, Enzyme modulators as tools for the development of homogeneous enzyme immunoassays, FEBS Letters, 116: 285-288.

Ngo, T.T. and Lenhoff, H.M., 1981, Recent advances in homogeneous and separation-free enzyme immunoassays, Applied Biochem. Biotechnol., 6: 53-64.

Ngo, T.T. and Lenhoff, H.M., 1982, Enzymes as versatile labels and signal amplifiers for monitoring immunochemical reactions, Mol. Cell. Biochem. 44: 3-12.

Ngo, T.T. and Lenhoff, H.M., 1983, Antibody-induced conformational restriction as basis for new separation-free enzyme immunoassay, Biochem. Biophys. Res. Comm., 114: 1097-1103.

Ngo, T.T., Bovaird, J.H. and Lenhoff, 1985, Separation-free amperometric enzyme immunoassay, Applied. Biochem. Biotechnol., 11: 63-70.

Oellerich, M., 1984, Enzyme-immunoassay: A Review, J. Clin. Chem. Clin. Biochem., 22: 895-904.

Rubenstein, K.E., Schneider, R.S. and Ulman, E.F., 1972, "Homogeneous" enzyme immunoassay, New immunochemical technique, Biochem. Biophys. Res. Comm., 47, 846-851.

Solsky, R.L. and Rechnitz, G.A., 1979, Antibody-selective membrane electrodes, Science, 204: 1308-1309.

Solsky, R.L. and Rechnitz, G.A., 1981, Preparation and properties of an antibody-selective membrane electrode, Anal. Chim. Acta, 123: 135-141.

Wehmeyer, K.R., Halsall, H.B. and Heineman, W.R., 1982, Electrochemical investigation of hapten-antibody interactions by differential pulse polarography, Clin. Chem., 28: 1968-1972.

Wehmeyer, K.R., Doyle, M.J., Halsall, H.B. and Heineman, W.R., 1983, Immunoassay by electrochemical techniques in Immunochemical Techniques, Part E, "Methods in Enzymology", Vol. 92, J.J. Langone and H.V. Vunakis, Ed., Academic Press, New York, pp. 432-444.

DIGOXIN HOMOGENEOUS ENZYME IMMUNOASSAY WITH AMPEROMETRIC DETECTION

D. Scott Wright

Warner Lambert
Pharmaceutical Research Division
2800 Plymouth Road
Ann Arbor, Michigan 48105

H. Brian Halsall and William R. Heineman

Department of Chemistry
University of Cincinnati
Cincinnati, OH 45221

INTRODUCTION

Enzyme immunoassays (EIAs) are methods which use the chemical amplification feature of an enzyme and the high specificity of an antibody to provide a highly sensitive and selective means for quantitating an analyte. The applications of EIAs are numerous and ever-expanding (Jarvis, 1979). The sensitivity of EIAs is as high as that of radioimmunoassays (Maggio, 1980) which have detection limits ranging from 1-500 fmol/mL assay fluid (Oellerich, 1980). EIAs can be classified as either heterogeneous or homogeneous.

Heterogeneous EIAs are based on the separation of free enzyme-labeled analyte from enzyme-labeled analyte-antibody complex prior to measuring the enzyme activity in either resulting fraction. Homogeneous EIAs do not have this separation step. The activity of the enzyme label is sufficiently diminished by antibody binding that the free enzyme-label activity can be measured in the presence of the complex. Several excellent reviews (Monroe, 1984; Schuurs and Van Weeman, 1977; Wisdom, 1976) and books (Maggio, 1980; Engvall and Pesce, 1978) describe the pros and cons of both EIA types along with variations in methodology.

Homogeneous EIAs are well-suited for therapeutic drug monitoring because they are fast, convenient, and capable of measuring small amounts (ng/mL) of drugs in biomatrices. Success of the homogeneous approach depends upon the extent of antibody-mediated change in enzyme activity. Proposed mechanisms of the antibody-mediated change include steric exclusion of the enzyme's substrate, a conformational change in the enzyme, and prevention of conformational changes in the enzyme which are necessary for catalytic activity (Schuurs and Van Weeman, 1977). The extent of the antibody-mediated change together with the susceptibility of the assay to endogenous interferences, the association constant of the antibody-drug interaction, and the detectability of the enzyme, determine the sensitivity of homogeneous EIAs (Maggio, 1980).

This paper demonstrates the feasibility and advantages of using an amperometric detection scheme for homogeneous EIAs. The scheme includes high-performance liquid chromatographic (HPLC) column switching (Erni et al., 1981; Apffel et al., 1981) which involves a selective transfer of a chromatographic zone(s) from one column to another. Many homogeneous EIAs use enzyme systems which produce NADH as the detectable component for evaluating drug concentrations. The rate of NADH production is commonly measured by UV spectrophotometry at 340 nm; however, the rate can be measured amperometrically since NADH is electroactive. Hydrodynamic amperometric detection has been used in conjunction with Syva's EMIT Phenytoin assay to monitor the drug at the μg/mL level (Eggers et al., 1982). The high sensitivity of amperometric detection does, however, permit the monitoring of a substance present in serum at lower levels. An EMIT digoxin assay served as a model system. Digoxin, a cardioactive drug used in treating chronic heart disease, is clinically important because of its widespread use, low therapeutic level (ng/mL), and narrow therapeutic range, 0.9-2.0 ng/mL (Koch-Weser, 1972).

The digoxin assay involves the reaction of digoxin (DIG) and enzyme-labeled digoxin (DIG-G6PDH) with a limiting amount of antibody (Ab) to digoxin as shown below:

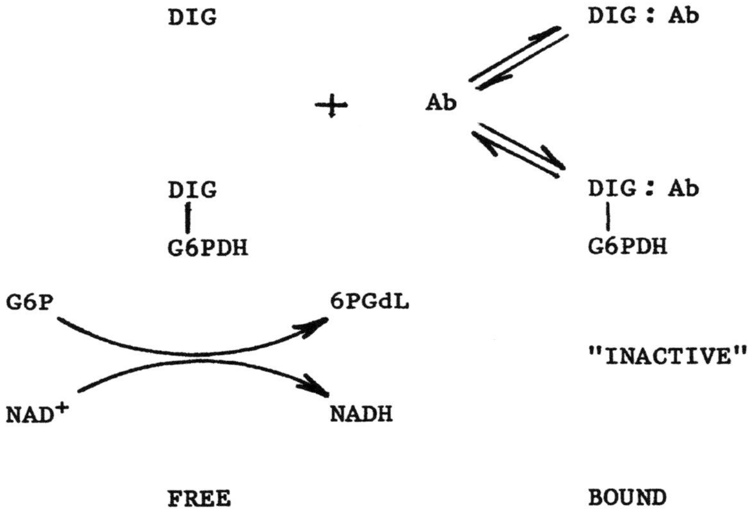

The total amount of enzyme-labeled digoxin distributes into two fractions: a free fraction (DIG-G6PDH) which is enzymatically active and an antibody-bound fraction (Ab:DIG-G6PDH) which is relatively "inactive". The free enzyme-labeled digoxin in conjunction with a substrate (G6P) is capable of converting NAD+ to NADH. The rate of NADH production can be directly related to the digoxin concentration. As the amount of digoxin in a sample increases, the amount of free enzyme-labeled digoxin necessarily increases, resulting in a greater rate of NADH production. Detection of NADH in the immunoassay reaction mixture was accomplished by injection of an aliquot into an HPLC column switching system in which a heart-cut was performed. The NADH zone was routed from a precolumn (microparticulate gel filtration) to an analytical column (reversed phase C-18) which was in-line with an amperometric detector; earlier and later-eluting zones which contain electrochemical interferences were routed from the precolumn to waste.

This simple and rapid on-line clean-up has enabled repetitive injections of sample-immunoassay reaction mixtures into the chromatograph while maintaining the working electrode as a stable sensor for NADH.

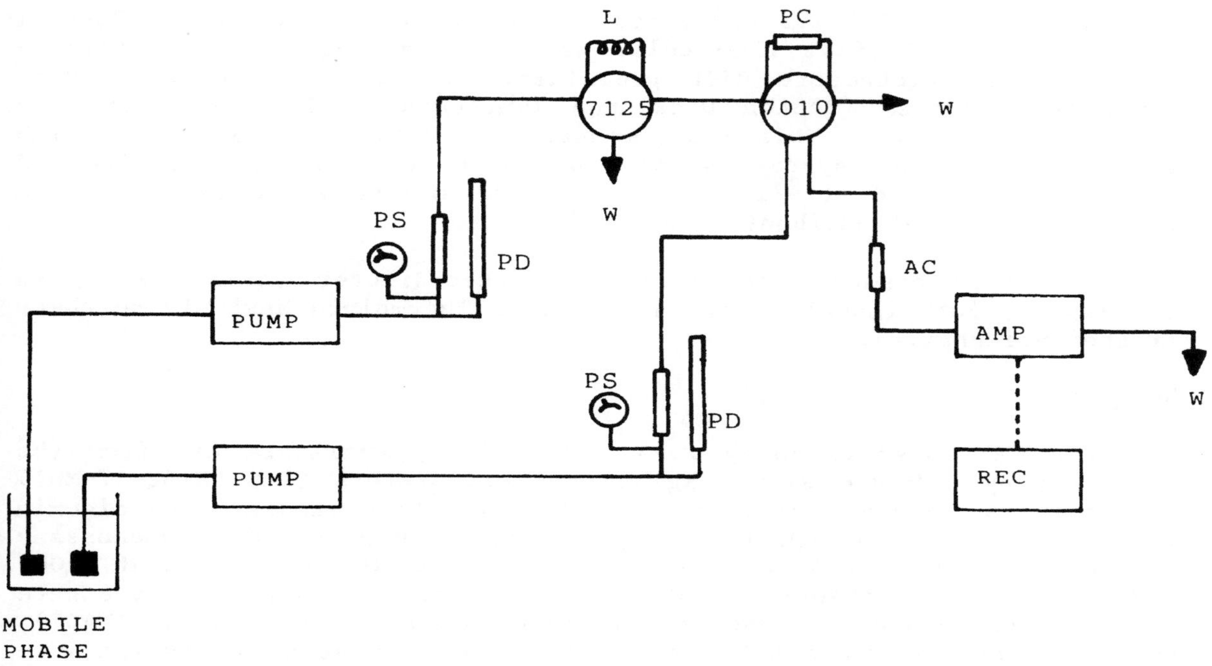

Figure 1. Block diagram of the HPLC column switching system with amperometric detection. L, sample loop; 7125 and 7010, sample injection valves; W, waste; PC, precolumn; AC, analytical column; AMP, amperometric detector; REC, stripchart recorder; PD, pulse damper; PS, presaturator column.

MATERIALS AND METHODS

Apparatus

A block diagram of the HPLC column switching system is shown in Figure 1. Two high-pressure solvent delivery pumps (Milton Roy Laboratory Data Control model 396-31 and model 396-57) were used to provide flow rate stability upon column switching. Both pumps delivered mobile phase from a common reservoir at 1.1 mL/min. Each flow-line was supplied with a standing-air column pulse damper, a high pressure gauge, and a presaturator column. Both presaturator columns contained pellicular C-18. Two Rheodyne sample injection valves were used for column switching; model 7125 (20 μL sample loop) syringe-loader and model 7010. The switching columns included a precolumn (10 μm Lichrosorb DIOL, 25 cm x 4.6 mm Knauer) which was positioned where the sample loop would normally be in the 7010 injection valve, and an analytical column (10 μm RSIL C-18 HL, 2 cm x 2.0 mm Upchurch). A stop watch was used to time the manual switching of the injection valves.

Instrumentation for amperometric detection included a BAS (Bioanalytical Systems, Inc.) LC-4B amperometric detector, a BAS TL-3 thin-layer amperometric flow cell (5 mil spacer), and an RC-2A reference compartment. The cell is a three electrode design: carbon paste (paraffin oil) working electrode, silver/silver chloride reference electrode (RE-1), and stainless steel auxiliary electrode. A BAS RYT strip-chart recorder was used.

A Beckman model 153 Analytical UV Detector and Analytical Optical Unit (8 μL analytical flow cell) were used during the characterizations of the Lichrosorb DIOL precolumn. An ARIA HT digoxin radioimmunoassay system (Becton Dickinson Immunodiagnostics) was used for method correlation. This radioimmunoassay (RIA) is an automated high sample throughput system which is based on the competition of radioactive-labeled (^{125}I) digoxin with unlabeled digoxin for a limiting number of antibody binding sites.

Ancillary hardware included a pipetter-diluter (model 1500), an Isotemp Dry Bath (model 145), and a Sybron Thermolyne Maxi-Mix vortexer (Fisher Scientific).

Reagents

The EMIT Digoxin Assay is a commercially available kit from the Syva Company. Immunoassay reagents and calibrators (lot 6H132-17-M01) were prepared as prescribed in the kit instruction manual. Six digoxin calibrators (0 to 7.5 ng/mL in human serum) were provided. Immunoassay reagents are as follows: Serum Pretreatment Reagent - 0.5 M NaOH; Reagent A - digoxin antibodies (sheep), enzyme substrate glucose-6-phosphate, coenzyme nicotinamide adenine dinucleotide (NAD^+), and preservatives in a 0.055 M Tris-HCl buffer; Working Reagent A - prepared by diluting 1.25 mL of Reagent A to 25 mL with pH 7.4 phosphate buffer; Reagent B - enzyme-labeled (glucose-6-phosphate dehydrogenase) digoxin and preservatives in 0.055M Tris-HCl buffer.

Lichrosorb-DIOL (Merck, Darmstadt, Batch VV758) is a glycerol propyl silyl phase bonded to an irregularly shaped silica gel particle (10 μm mean particle size, 100 A mean pore size). This packing was slurried in methanol, and packed in the Knauer precolumn under 7200 psi. The reversed-phase C-18 packing (RSIL C18 HL, Batch F11, Alltech Associates, Inc.) is an irregularly shaped particle with a carbon load of 16%; the mean particle size and mean pore size are 10 μm and 60 Å, respectively. This packing was slurried in trichlorethylene, and packed (7200 psi) in the Upchurch analytical column. Vydac SC pellicular C-18 (30-40 μm, Batch 503, Rainin Instrument Co.) was dry-packed in the presaturator columns.

NADH (N-8129, Batch 83F-73251) and immunoglobulin G (goat origin, lot 121F) were purchased from Sigma Chemical Co., uric acid (UX90 2271, lot 16) from MCB, and human serum albumin (lot 36) from Miles Laboratories, Inc. as a lyophilized preparation. The mobile phase (0.1 M phosphate buffer at pH 6.6) was prepared from Na_2HPO_4 and Na_2HPO_4 (Fisher Scientific Co.). Phosphate buffer (0.1 M pH 7.4) was used to prepare solutions of NADH, uric acid, immunoblobulin G, human serum albumin, and working reagent A; this buffer was also a diluent for the pipetter-diluter.

Procedures

Enzyme Immunoassay The homogeneous enzyme immunoassay is a sequential antibody saturation method (Zettner and Duly, 1974; Bastiani and Wilcox-Thole, 1982). The enzyme activity (rate of NADH production) was determined by a two-point kinetic measurement. First, 200 µL of the digoxin calibrators and serum sample unknowns were added to individual test tube. Then 50 µL of Serum Pretreatment Reagent were added to all tubes; each tube was vortexed for 10 s and then placed in a rack for 5 min. This pretreatment reaction dissociates digoxin from serum binding proteins and eliminates the activity of endogenous enzymes (Oellerich, 1980; Bastiani and Wilcox-Thole, 1982). Next, 1000 µL of working Reagent A were added to all tubes: each tube was vortexed for 10 s and placed in the dry bath (34.0°C) for 15 min. Afterward, 50 µL of Reagent B and 500 µL of pH 7.4 phosphate buffer were added to a tube by the pipetter-diluter; the tube was vortexed for 10 s and then returned to the dry bath. At two min and seven min after Reagent B addition, a 50 µL aliquot of the reaction mixture was diluted with 500 µL of pH 7.4 phosphate buffer using the pipetter-diluter. After vortexing, the diluted aliquot was injected in the HPLC column switching system for amperometric detection. Additional tubes were processed every 12 min in the same manner.

HPLC Column Switching HPLC column switching provided an on-line clean-up of the serum-based reaction mixtures; direct injection of the mixtures was made. A heart-cut approach was used to separate the NADH chromatographic zone from both early and late-eluting zones which contained electrochemical interferences. The diluted aliquot of the enzyme reaction mixture was loaded by syringe into the 20 µL sample loop of valve 7125. The sample loop contents were injected into the Lichrosorb-DIOL precolumn; the precolumn separated components within the enzyme reaction mixture by size exclusion. High molecular weight components (working electrode passivators) of the mixture eluted first and were routed to waste. Lower molecular weight components eluted later. When the NADH zone was at the precolumn exit (162 s after sample injection), the precolumn was switched (valve 7010) inline with the analytical column. After transfer of the NADH zone to the analytical column (20 s), the precolumn was switched out of line with the analytical column to route late-eluting interferences (uric acid, for example) to waste. The precolumn was flushed with mobile phase for two min before injecting additional reaction mixtures.

Comparison with Radioimmunoassay Forty-eight human serum samples from patients receiving digoxin therapy were first assayed by the ARIA HT system, and then within 12 days the same samples were assayed by the amperometric homogeneous enzyme immunoassay. In between the assays, the samples were stored refrigerated at 4°C. A correlation plot was constructed for the paired observations using linear regression and correlation analyses.

RESULTS AND DISCUSSION

NADH Detection

NADH is the enzyme-catalyzed reaction product which can be measured and related to the amount of digoxin present in a blood sample. A cyclic voltammogram of a 4 mM NADH solution, Figure 2, shows that NADH can be oxidized at a carbon paste (paraffin oil) electrode and suggests that the anodic current can serve as a measure of NADH concentration. The anodic peak potential is 650 mV. This oxidation occurs via a single-step two electron and one proton loss (Elving et al., 1982; Bresnahan el al., 1980). The nicotinamide portion of the molecule is the electroactive center.

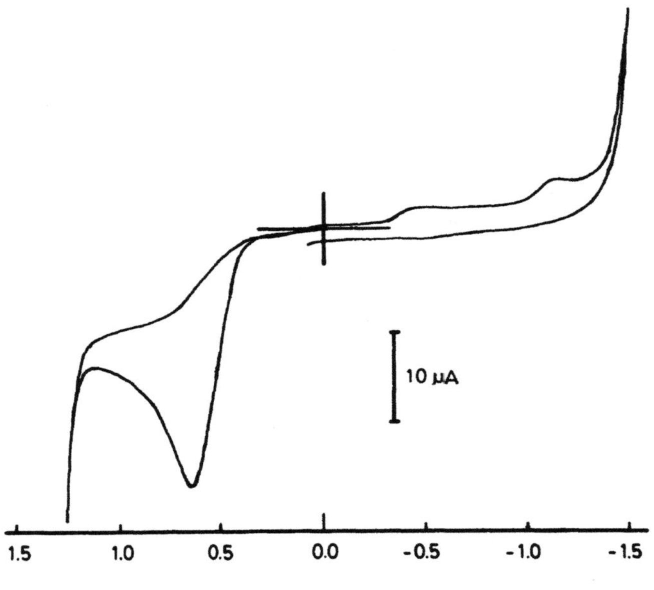

Figure 2. Cyclic voltammogram for 4 mM NADH in 0.1 M phosphate buffer pH 7.4 at a carbon paste electrode vs Ag/AgCl. Potential scan initiated at 0.0 V in the positive direction. Scan rate, 20 mV/s.

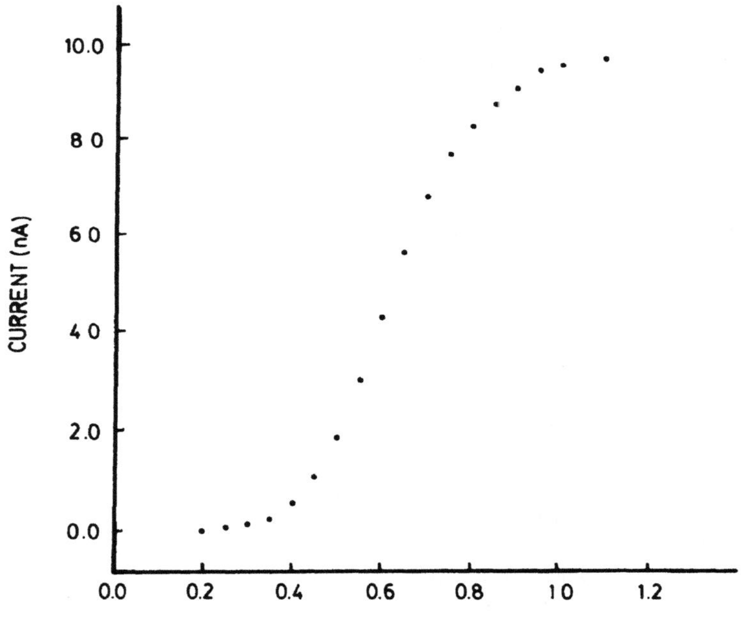

Figure 3. Hydrodynamic voltammogram for 20 µ M NADH in 0.1 M phosphate buffer pH 7.4.

An NADH hydrodynamic voltammogram, Figure 3, was produced to select the constant working potential required for amperometric detection. Each point represents the potential dependent current response after the 20 μ L injection of a 20 μ M NADH solution into the HPLC column switching system. Conventionally, one would choose to operate approximately 100 mV out on the limiting current plateau (i.e. 1200 mV). To improve selectivity a much lower constant operating potential was chosen (750 mV vs. Ag/AgCl).

An NADH calibration plot at this potential was linear over the range of 2.31×10^{-7} M to 2.95×10^{-4} M with a correlation coefficient of 0.999. The detection limit (2 S/N) for NADH was 6.20×10^{-8} M.

Removal of Electrochemical Interferences

HPLC column switching provided a convenient and precise on-line sample clean-up, which removed electrochemical interferences prior to amperometric detection of NADH. The serum biomatrix was the most severe interference source, since it contained substituents which readily passivated the working electrode. Working electrode passivation was observed after several blank human serum injections into a unidimensional HPLC system (C-18 or Lichrosorb-DIOL column). For example, after one serum injection (1/2 dilution factor) into the C-18 system a 39% reduction in NADH peak height resulted. After a total of seven serum injections, a 70% reduction resulted. Similarly, a 57% reduction resulted after 13 serum injections into a Lichrosorb-DIOL HPLC system. Similar results were obtained for injections of a human serum albumin (HSA) solution into both HPLC systems. Neither the C-18 packing nor the Lichrosorb-DIOL packing alone protected the working electrode from passivation.

Uric acid, a serum component having a maximal normal level of 6 mg/100 mL (Orten and Neuhaus, 1982), was a major electroactive interference. A hydrodynamic voltammogram for uric acid displayed a $E_{1/2}$ of 430 mV vs. Ag/AgCl. Consequently, NADH could not be resolved from uric acid by means of the selected working potential.

Immunoassay reagents were also sources of electrochemical interference. Immunoassay antibody has been shown to passivate a glassy carbon working electrode (Eggers et al., 1982).

Column switching times were necessarily important in producing an effective sample clean-up. Selection of the times was made after characterizing the Lichrosorb-DIOL precolumn and conducting a switching time survey. Lichrosorb DIOL has been reported (Schmidt et al., 1980; Roumeliotis and Unger, 1979; Buchholz et al., 1982) to be well-suited for size-exclusion of biopolymers. The diol portion of the packing minimizes adsorption of biomacromolecules to the packing material, while the small particle size allows a rapid and highly efficient separation. The size-exclusion mechanism enabled separation of NADH from the very large biopolymers of serum and immunoassay reagents.

The precolumn was unidimensionally characterized by evaluating the retention times of NADH and the known electrochemical interferences. Spectroscopic detection was used for this characterization, since the electrode passivators (IgG, HSA), uric acid, and NADH absorb radiation at 280 nm, IgG and HSA being electroinactive at 750 mV vs Ag/AgCl. The following retention times were observed: IgG, 109 s; HSA, 116 s; NADH, 173 s; and uric acid, 187 s, with higher molecular weight species (IgG = 150,000; HSA = 69,000) eluting before the lower molecular weight species (NADH = 709; uric acid = 168). Figure 4 shows a chromatogram

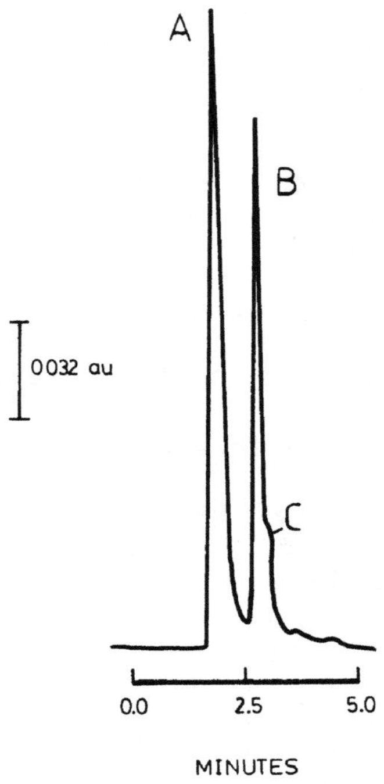

Figure 4. Chromatogram of a blank human serum NADH mixture. Lichrosorb-DIOL column; 1.1 mL/min; absorbance monitored at 280 nm. (A) HSA and IgG, (B) NADH, and (C) uric acid.

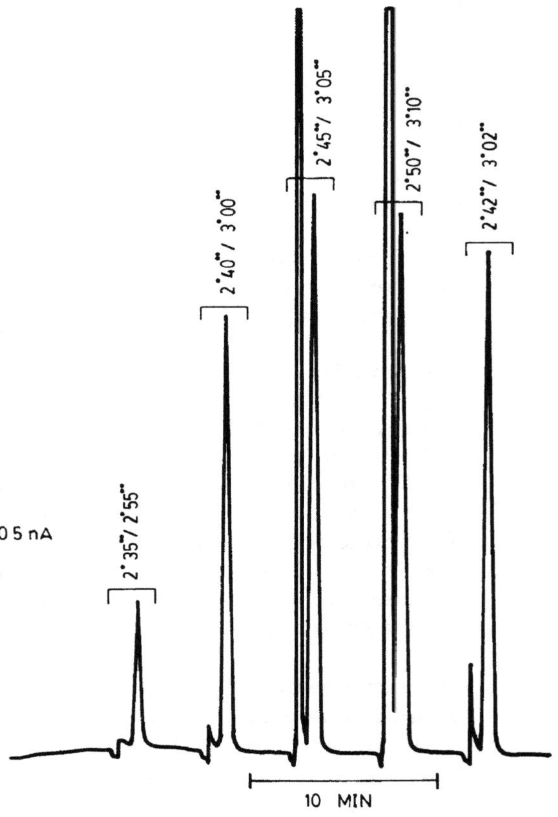

Figure 5. Switching time survey for 5 injections of an NADH/uric acid mixture (4.00×10^{-6} M/2.00×10^{-5} M) into the HPLC column switching system with amperometric detection. Switching times are noted above each peak set.

of a blank human serum - NADH mixture which was injected into the Lichrosorb-DIOL precolumn. The first peak (retention time of 113 s) was primarily composed of IgG and HSA. The second peak (173 s) was NADH; the shoulder which appears at approximately 190 s was uric acid. Therefore, a heart-cut should begin at approximately 150 s (to remove the macromolecular fraction) and end at approximately 185 s (to remove the late-eluting uric acid among other possible interferences).

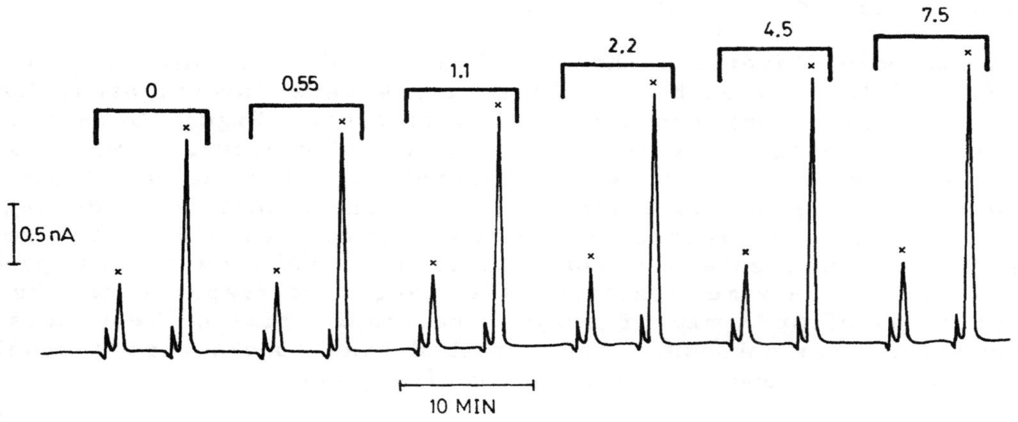

Figure 6. Assay chromatogram for 6 digoxin calibrators (0-7.5 ng/mL). Twelve injections have been made (2 for each calibrator); starred (*) peaks represent enzyme-generated NADH. Enzyme reaction is sampled at 2 min and 7 min; switching time set is 2'42"/3'02".

Switching time surveys were performed to find optimal column switching times i.e. times at which the precolumn is switched in and out of line with the analytical column. The survey involved making repeated injections of an NADH - uric acid solution into the HPLC column switching system. Figure 5 depicts a survey for 5 injections of the mixture. The switching times were variable and are indicated above each set of peaks; the timing interval was held constant at 20 s. The left-hand peak of each switching time set is uric acid; the right-hand peak is NADH. The switching time set of 2'42"/3'02" was chosen for operation in the enzyme immunoassay. This set maximizes the time distance from the HSA retention time. Timing intervals from 10 s to 90 s were evaluated; the 20 s interval produced the least electrode fouling and most precise data. Intervals less than 20 s would have been more desirable from the standpoint of reducing the ballast load on the analytical column, but at short time intervals manual activation of the valves was a limiting factor with respect to precision.

The C-18 analytical column separated NADH from possible coeluting interferences and a disturbance resulting from the switching process. This column switching scheme was effective in removing electroactive and electrode-passivating interferences; no reduction in mean NADH peak height was observed after 44 injections of serum-enzyme reaction mixtures. No deterioration in chromatographic performance was observed after 176 serum injections.

The precision of the HPLC column switching-amperometric system was consistently good. For example 11 injections of a 5.28 x 10^{-6} M NADH solution gave a relative standard deviation of 0.3% (X = 9.65 nA; s = 0.03 nA).

Enzyme Immunoassay

A significant advantage of the homogeneous enzyme immunoassay for digoxin is the large amplification quotient, the ratio of NADH concentration to digoxin concentration. The quotient is 3.7 x 10^4 for digoxin calibrator #1 ([digoxin] = 7.05 x 10^{-10} M) after an enzyme reaction time of 5 min. After a 38-min enzyme reaction time, the same calibrator yielded a quotient of 1.6 x 10^5.

A two-point kinetic method, at 2 and 7 min, was used to determine the rate of NADH production. Other times were investigated, however they were either less convenient or imprecise. Figure 6 is a representative chromatogram of the six digoxin calibrators. Twelve injections were made; two for each calibrator. The number above each bracketed set of peaks is the digoxin calibrator concentration (ng/mL). The starred peaks represent enzymatically-generated NADH. The smaller starred peak represents the concentration of NADH which was produced after 2 min of enzyme reaction; the larger corresponds to the NADH concentration after 7 min of enzyme reaction. Similar peak sets were produced for each unknown. The unmarked small peaks were a result of residual uric acid and the column switching process.

A calibration plot was constructed to evaluate the unknown digoxin concentrations. The difference in peak heights (nA) for each calibrator was computed and plotted versus digoxin concentration on Syva's graph paper. This is a modified log-logit grid which has been customized for spectroscopic measurements and specific immunoassay conditions (Dietzler et al., 1980; Wellington, 1980). When using the paper to plot the amperometric response versus digoxin concentration, the spectroscopic units of the ordinate were ignored. Figure 7 is a calibration plot for five digoxin standards. The ordinate, representing the rate of NADH production, was arbitrarily scaled with respect to the change in current (nA/5 min). The abscissa is logarithmic with respect to the digoxin concentration. Note that the plot is linear (best-fit line by free hand, as prescribed) over the range of 0.55 ng/mL to 7.5 ng/mL.

The within-run precision for this immunoassay using the amperometric HPLC column switching system was good. For the assay of six digoxin calibrator #3 replicates (2.2 ng/mL), the relative standard deviation of the amperometric response was 2.3% (x = 2.15 nA; s = 0.05 nA). Syva reports a relative standard deviation of less than 10% (20 replicate assays of a single serum sample at 2.0 ng/mL digoxin) for their prescribed spectroscopic method of detection. Other investigators who have either used the homogeneous digoxin immunoassay as prescribed (Drost et al., 1977; Linday and Drayer, 1983; Rumley et al., 1980) or modified the assay (Brunk and Malmstadt, 1977; Eriksen and Andersen, 1978; Rosenthal et al., 1976) report relative standard deviations for within-run analysis ranging from 3.3% to 7.1% (prescribed) and 2.7% to 4.8% (modified).

Precision studies of an aqueous mixture of NADH and uric acid were run before and after the digoxin homogeneous enzyme immunoassay to evaluate both the condition of the columns with respect to retention times and the working electrode with respect to mean peak height. No

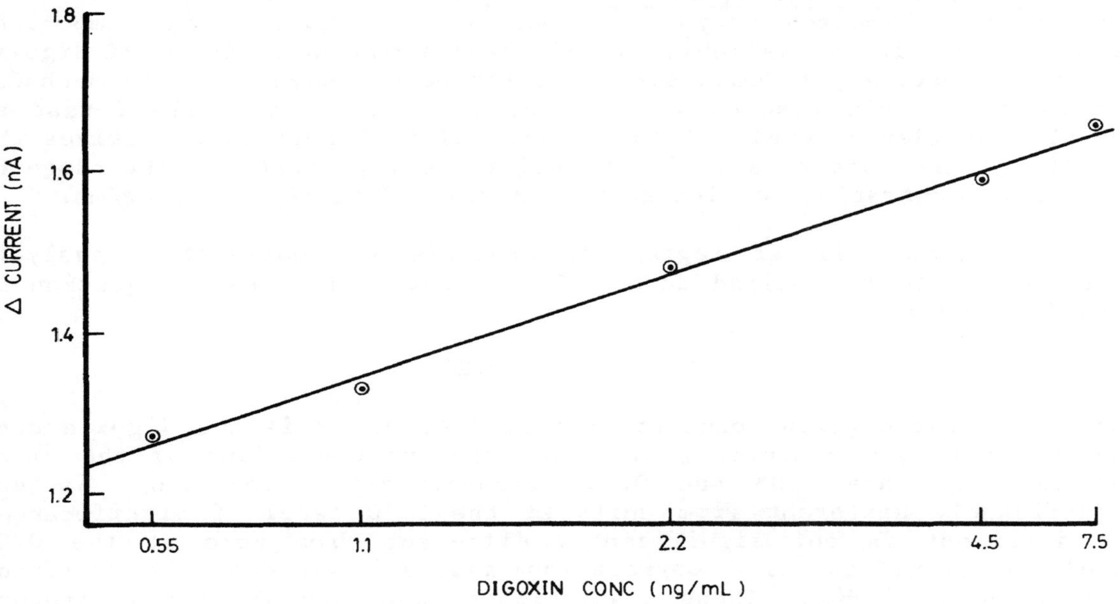

Figure 7. Digoxin log-logit calibration plot.

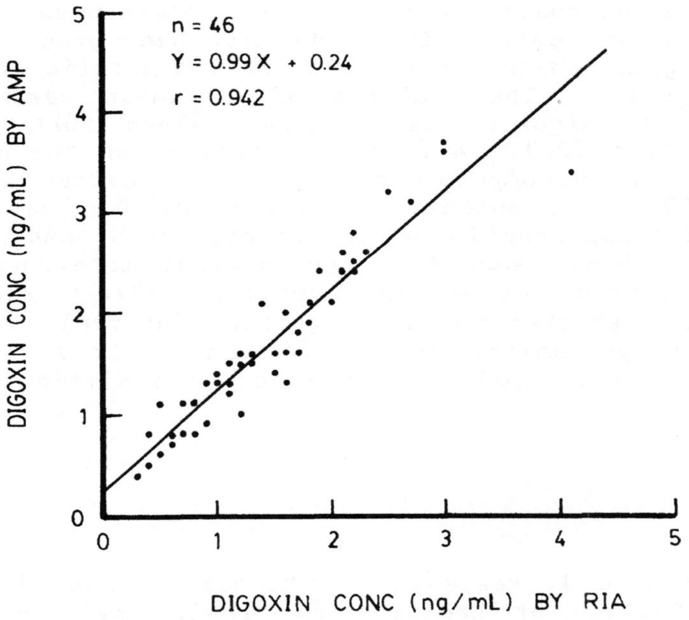

Figure 8. Digoxin correlation plot; 46 human serum samples from patients receiving digoxin therapy analyzed by the amperometric approach and ARIA HT^{R}.

significant changes resulted, and the HPLC column switching system was considered stable and effective in removing interferences.

Comparison With Radioimmunoassay Accuracy of the amperometrically-based enzyme immunoassay was evaluated by comparison with a commercially available digoxin radioimmunoassay (ARIA HT digoxin system). Forty-eight human serum samples were assayed by both methods; two of the sample results were rejected as outliers by the Q-test at the 99% confidence level. A correlation plot, Figure 8, summarizes the results of the comparison; 46 data pairs are represented. The range of digoxin concentrations evaluated by RIA was 0.3 ng/mL to 4.1 ng/mL.

Both simple linear regression analysis and correlation analysis were applied to the paired data. The following regression equation of Y on X resulted:

$$Y = 0.99X + 0.24$$

where X is the digoxin concentration by RIA, and Y is the digoxin concentration by amperometric EIA. The standard deviations of the slope and intercept are 0.05 and 0.09, respectively. The slope is not significantly different from unity at the 0.10 level of significance. The intercept is not significantly different from zero at the 0.01 level of significance. Correlation analysis yielded a correlation coefficient of 0.942. These statistics demonstrate absence of proportional and constant errors, and strong correlation between the methods.

SUMMARY

Results which have been obtained using the amperometric HPLC column switching approach to the digoxin homogeneous EIA are quite good. Linear digoxin calibration plots have been produced over a range of 0.55 - 7.5 ng/mL, which encompasses the therapeutic range (0.9 - 2.0 ng/mL) for digoxin. The analysis of 46 human serum samples from patients receiving digoxin therapy gave values that correlate well (0.942) with values (0.3 - 4.1 ng/mL) obtained on the same samples by RIA. Furthermore, the amperometric approach provided a high level of precision (% RSD = 2.3) which is necessary for good assay performance. HPLC column switching provided a convenient, rapid, and precise on-line sample clean-up which removed electrochemical interferences prior to amperometric detection of enzyme-generated NADH. The amperometric method requires a shorter enzyme incubation interval (5 min) than the prescribed spectrophotometric method (30 min). If a calibration plot has already been registered, a sample can be analyzed in less than 12 min.

ACKNOWLEDGMENT

We thank Amadeo J. Pesce and F. Michael Hassan of the University of Cincinnati College of Medicine for their assistance. Financial support provided by NSF Grant CHE-8217045 is gratefully acknowledged.

REFERENCES

Apffel, J. A., Alfredson, T. V. and Majors, R. E., 1981, Automated on-line multi-dimensional high-performance liquid chromatographic techniques for the clean-up and analysis of water-soluble samples, J. Chromatogr. 206:43-57.

Bastiani, R. and Wilcox-Thole, W. L., 1982, Recent developments in homogeneous enzyme immunoassay, in: "Clinical Laboratory Annual"; Hamburger; Basakis; Eds., Vol. 1, pp 289-338.

Bresnahan, W. T., Moiroux, J., Samec, Z. and Elving, P. J., 1980, Nucleotides and related substances: conformation in solution and at solution electrode interfaces, Bioelectrochem. Bioenerg. 7:125-155.

Brunk, S. D. and Malmstadt, H. V., 1977, Adaptation of the EMIT serum digoxin assay to a mini-disc centrifugal analyzer, Clin Chem. 23:1054-1056.

Buchholz, K., Godelmann, B. and Molnar, I., 1982, High performance liquid chromatography of proteins: analytical applications, J. Chromatogr. 238:193-202.

Dietzler, D. N., Weidner, N., Tieber, V. L., McDonald, J. M., Smith, C. H., Ladenson, J. H. and Leckie, M. P., 1980, Adaptations of the EMIT theophylline assay to kinetic analyzers: the relationship of reaction kinetics to calculation procedures, Clin. Chim. Acta. 101:163-181.

Drost, R. H., Plomp, T. A., Teunissen, A. J., Maas, A. H. J. and Maes, R. A. A., 1977, A comparative study of the homogeneous enzyme immunoassay (EMIT) and two radioimmunoassays (RIA'S) for digoxin, Clin. Chim. Acta. 79:557-567.

Eggers, H. M., Halsall, H. B. and Heineman, W. R., 1982, Enzyme Immunoassay with flow-amperometric detection of NADH, Clin. Chem. 28:1848-1851.

Elving, P. J., Bresnahan, W. T., Moiroux, J. and Samec, Z., 1982, NAD/NADH as a model redox system: mechanism, mediation, modification by the environment, Bioelectrochem. Bioenerg. 9:365-378.

Engvall, E. and Pesce, A. J., Eds., 1978, "Quantitative Enzyme Immunoassay"; Scandinavia Journal of Immunology, Blackwell Scientific: Oxford, Vol. 8, Supplement No. 7, pp 129.

Eriksen, P. B. and Andersen, O., 1978, Homogeneous enzyme immunoassay of serum digoxin with use of a bichromatic analyzer, Clin. Chem. 25:169-171.

Erni, F., Keller, H. P., Morin, C. and Schmitt, M., 1981, Application of column switching in high-performance liquid chromatography to on-line sample preparation for complex separations, J. Chromatogr. 204:65-76.

Jarvis, R. F., 1979, The future outlook for enzyme immunoassays, Antibiotics Chemother. 26:105-117.

Koch-Weser, J., 1972, Serum drug concentrations as therapeutic guides, N. Engl. J. Med. 287:227-231.

Linday, L. and Drayer, D. E., 1983, Cross reactivity of the EMIT digoxin assay with digoxin metabolites, and validation of the method for measurement of urinary digoxin, Clin. Chem. 29:175-177.

Maggio, E. T., Ed.,1980, "Enzyme Immunoassay"; CRC Press: Boca Raton, pp 295.

Monroe, D., 1984, Enzyme immunoassay, Anal. Chem. 56:920A-931A.
Oellerich, M., 1980, Enzyme immunoassays in clinical chemistry: present status and trends, J. Clin. Chem. Clin Biochem. 18:197-208.
Orten, J. M. and Neuhaus, O. W., 1982, "Human Biochemistry", Tenth Ed.; C.V. Mosby: St. Louis, p 437.
Rosenthal, A. F., Vargas, M. G. and Klass, C. S., 1976, Evaluation of enzyme-multiplied immunoassay technique (EMIT) for determination of serum digoxin, Clin. Chem. 22:1899-1902.
Roumeliotis, P. and Unger, K. K., 1979, Preparative separation of proteins and enzymes in the mean molecular-weight range of 10,000-100,000 on lichrosorb DIOL Packing by high-performance size-exclusion chromatography, J. Chromatogr. 185:445-452.
Rumley, A. G., Trope, A., Rowe, D. J. F. and Hainsworth, I. R., 1980, Comparison of digoxin analysis by EMIT with immophase and dac-cel radioimmunoassay, Ann. Clin. Biochem. 17:315-318.
Schmidt, Jr, D. E., Giese, R. W., Conron, D. and Karger, B. L., 1980, High performance liquid chromatography of proteins on a diol-bonded silica gel stationary phase, Anal. Chem. 52:177-182.
Schuurs, A. H. W. M., Van Weemen, B. K., 1977, Enzyme-immunoassay, Clin. Chim. Acta. 81:1-40.
Wellington, D., 1980, Mathematical treatments for the analysis of enzyme-immunoassay data, in "Enzyme Immunoassay"; Maggio, E. T., Ed.; CRC Press: Boca Raton, Chapter 12.
Wisdom, G. B., 1976, Enzyme-immunoassay, Clin. Chem. 22:1243-1255.
Zettner, A. and Duly, P. E., 1974, Principles of competitive binding assays (saturation analyses). II. sequential saturation, Clin Chem. 20:5-14.

ANTIBODY TO AN ENZYME AS THE MODULATOR IN SEPARATION-FREE ENZYME IMMUNOASSAYS WITH ELECTROCHEMICAL SENSOR

T.T. Ngo

Department of Developmental and Cell Biology
University of California
Irvine, California 91717, USA

INTRODUCTION

Separation-free enzyme immunoassays for both haptens and macromolecules have been developed by using enzyme modulator as the label (Ngo and Lenhoff, 1980; Ngo, 1983; Finley et al., 1980; Place et al., 1983; Blecka et al., 1983; Bacquet and Twumasi, 1984; Brontman and Meyerhoff, 1984; Dona, 1985). In this chapter, enzyme modulators are defined as compounds capable of bringing about the modifications of the catalytic activity of an enzyme by inhibiting or enhancing (activating) the enzyme activity. The modification of an enzyme activity by enzyme modulators can be achieved through either non-covalent or covalent interactions between the enzyme and its modulators. Non-covalently interacting modulators are generally reversible modulators, such as reversible low molecular weight enzyme inhibitors (Webb, 1963) or enzyme activators (Wong, 1975), allosteric effectors (Stadtman, 1970), transition state analogs (Wolfenden, 1972; Lienhard, 1973; Ngo and Tunnicliff, 1981) and inhibitory antibodies to an enzyme (Marucci and Mayer, 1955; Visek et al., 1967; Cinader, 1976; Arnon, 1977). Covalently interacting enzyme modulators consist of irreversible modulators such as active-site directed irreversible enzyme inhibitors (Baker,1967; Shaw, 1980), mechanism based inhibitors (Rando, 1974; Abeles, 1978) and enzyme mediated modification of enzyme activity (Krebs, 1972).

Amongst the above-listed enzyme modulators, antibodies to an enzyme appear to be the most useful, versatile and easily adaptable ones for use in developing separation-free enzyme immunoassays. Ngo and Lenhoff (1980) first demonstrated that anti-peroxidase, when covalently-linked with dinitrophenyl groups, can be used as a modulator label for developing a separation-free enzyme immunoassay for dinitriphenyl groups. Subsequently, Dona (1985) using a similar approach developed a separation-free enzyme immunoassay for cortisol in human serum by using a cortisol conjugated Fab fragmant of an anti-(glucose-6-phosphate dehydrogenase) antibody as the modulator and glucose-6-phosphate dehydrogenase as the indicator enzyme.

A separation-free enzyme immunoassay for a macromolecule (human serum albumin) has also been developed by using antibody to adenosine deaminase as the modulator and a conjugate of antigen-adenosine deaminase as the indicator enzyme conjugate (Brontman and Meyerhoff, 1984).

In this chapter I described the strategies for developing separation-free enzyme mediated electrochemical immunoassays by using an ammonium ion-selective electrode as the sensor, ammonia-generating enzymes as the indicator enzymes and the inhibitory anti-enzyme antibody as the modulator.

ANTIBODY TO AN ENZYME AS A MODULATOR IN SEPARATION-FREE ENZYME IMMUNOASSAYS

There are two approaches to using antibody to an enzyme in developing separation-free enzyme immunoassays:

(1) Labeled anti-enzyme antibody approach

Here the antibody to the indicator enzyme (AE) is covalently linked to analyte ligands (L) or derivatives of it to form a stable covalent antibody to the enzyme and ligand conjugate (AE-L). The AE-L has a dual function: (a) as a derivative of the analyte ligand, it competes with the analyte ligand for a fixed and limited amount of antibody (Ab) to the analyte ligand; (b) as modified modulator, it can still modulate the activity of the indicator enzyme. To be useful for the development of separation-free enzyme immunoassays, the AE-L conjugate must, however, completely or greatly lose its modulating effect when it combines with an antibody to the analyte ligand (Ab). The overall process of the assay is depictrd in Fig. 1. It can be seen that in the absence of analyte ligand (L), reaction I would not occur. At the same time, through reaction II, AE-L and Ab (with their concentrations fixed and kept constant) would combine to form Ab:AE-L complexes. The Ab complexed AE-L is no longer able to modulate the activity of the indicator enzyme (E). Only the free, uncomplexed AE-L is functionally able to modulate the enzyme activity. As the concentration of analyte ligand (L) increases, however, it would compete more successfully with AE-L for Ab (i.e. reaction I would dominate) leaving more AE-L free to complex the indicator enzyme (reaction III) thereby increasingly modulating the enzyme activity, and consequently affecting the rate of substrate conversion to product via reaction IV. Depending upon the properties of the modulator, the enzyme activity is icreased by a positive modulator (enzyme activator) or it is decreased by a negative modulator (enzyme inhibitor). When an enzyme inhibitor is used as the modulator, the assay will show a decreasing enzyme activity with increasing concentration of the analyte ligand (see the schematic standard curve in Fig. 1).

(2) Approach using unlabeled antibody to the enzyme

In this approach the analyte ligands are covalently linked to the indicator enzyme molecule (L-E), not to the anti-enzyme antibody molecule (AE). The function of L-E is to serve as an indicator enzyme that can generate electro-active species (e.g. ammonium ion) and also to serve as an analog of the analyte ligand that can compete effectively for the same antibody binding site (Ab). The free, uncomplexed L-E can bind AE and lead to the formation of AE:L-E complexes which posses very much diminished enzyme's specific activity (Reaction III in Fig. 2). Figure 2 showed, in the absence of analyte ligand (L), reaction I would not take place, consequently all of L-E would combine with the antibody to the analyte ligand (Ab) to form Ab:L-E complexes (reaction II) which retained the full enzymatic activity and therefore would be able to enzymatically generate electro-active products (reaction IV). The antibody to the indicator enzyme (AE) would not be able to combine with the E portion of the

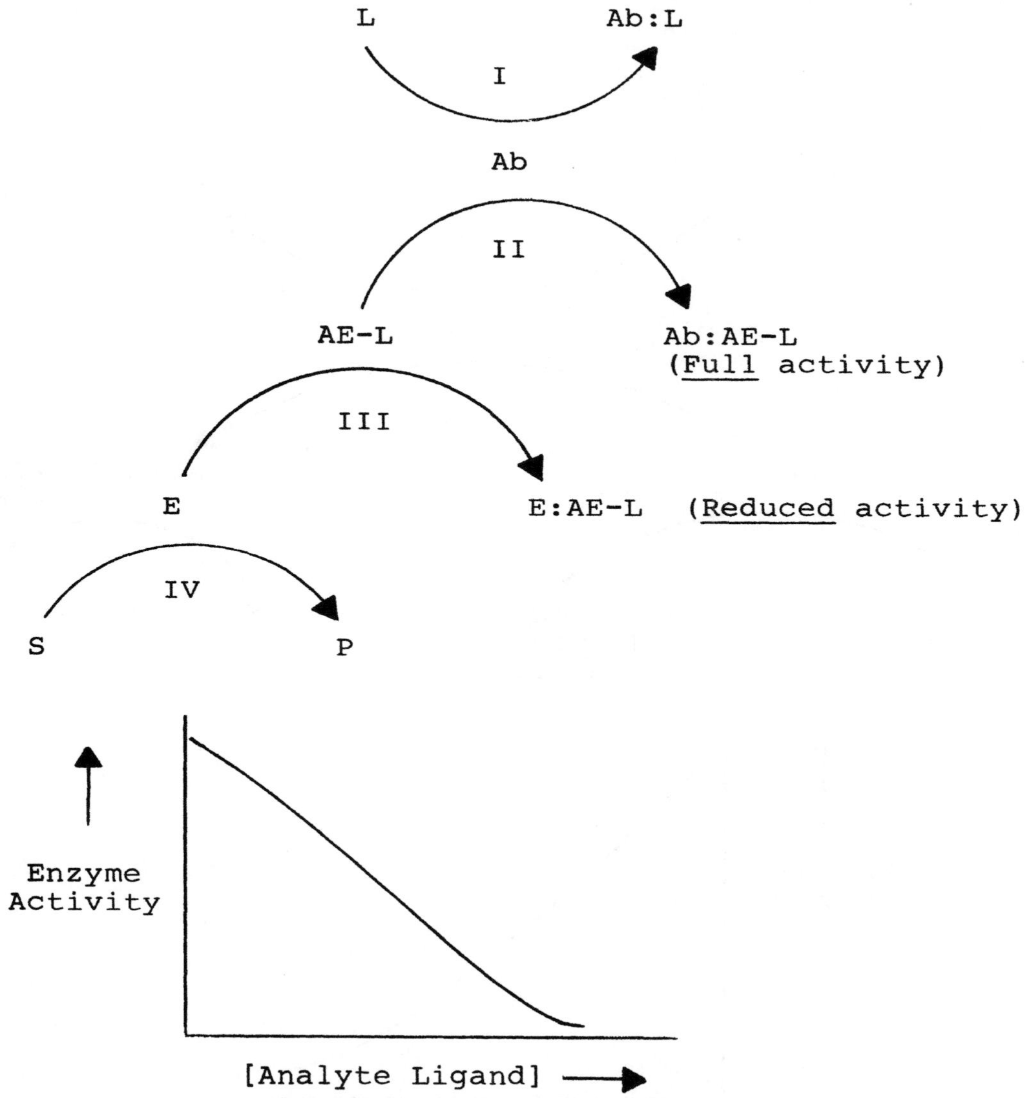

FIG. 1. Principle of the reactions involved in a separation-free enzyme immunoassay based on analyte ligand conjugated antibody to the indicator enzyme (AE-L) and its schematic standard curve. L= Analyte ligand; Ab= Antibody to analyte ligand; AE-L= Analyte ligand conjugated antibody to the indicator enzyme; E= Indicator enzyme; S= Substrate and P= Product of enzymatic reaction.

Ab:L-E complex and therefore would not be able to modulate the enzyme activity of the anlyte ligand-indicator enzyme conjugate. The protective effects of Ab in the Ab:L-E complexes presumably emanated from a physical hindrance created by the bulky AB which prevented the binding of AE to L-E in the form of the complex Ab:L-E. Only the free, uncomplexed L-E can combine with AE (reaction III). Conversely, in the presence of high concentration of L, reaction I would dominate and tie-up most of Ab in Ab:L complexes, leaving most of the L-E unprotected and free to combine with AE to form enzymatically much less active L-E:AE complexes (reaction III). The net result is a diminished overall enzyme activity with increasing analyte ligand concentrations as is shown in the schematic standard curve.

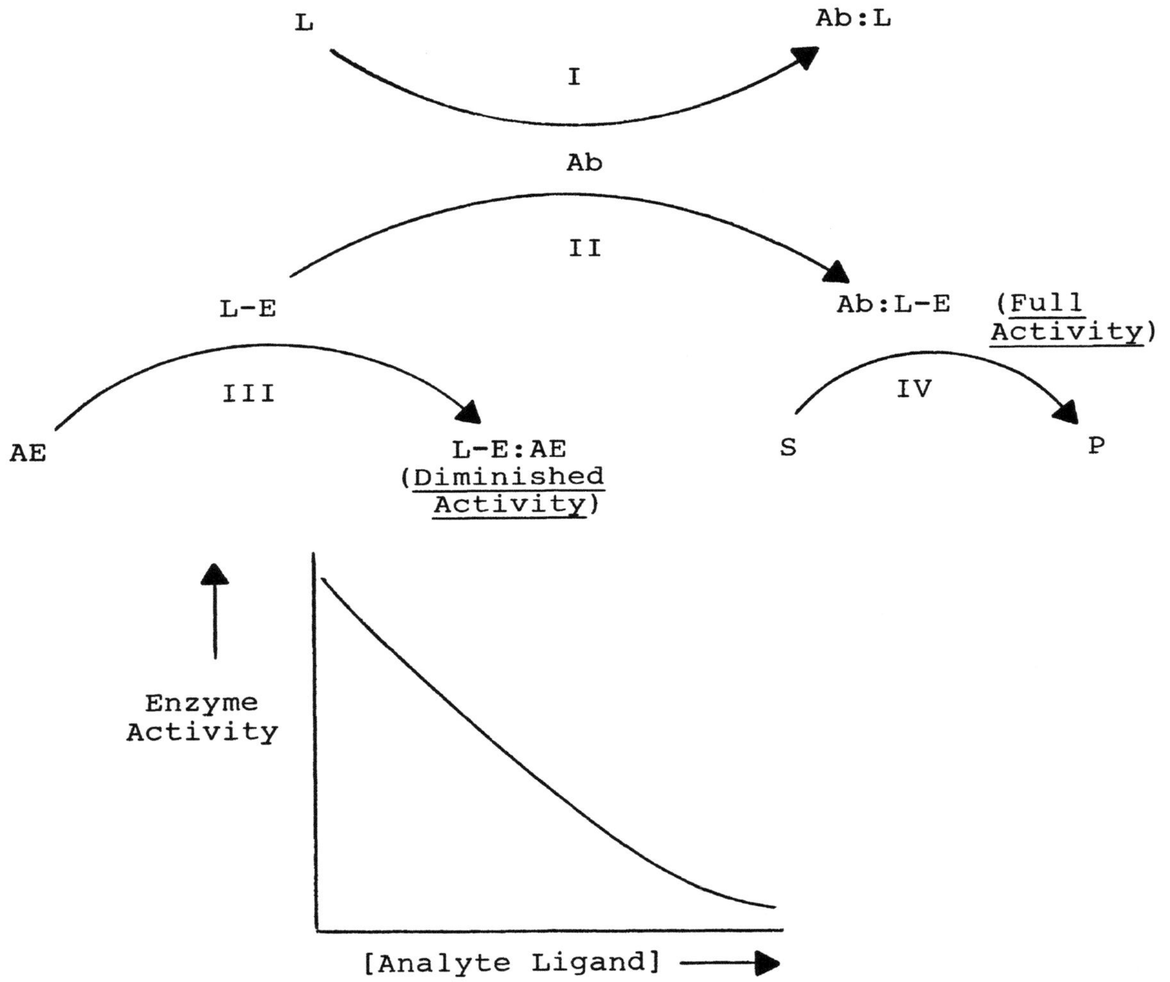

FIG. 2. Principle of the reactions involved in a separation-free enzyme immunoassay using analyte ligand-enzyme (L-E) conjugate and antibody to the enzyme (AE) and its schematic standard curve. L= Analyte ligand; Ab= Antibody to analyte ligand; L-E= Analyte ligand conjugated indicator enzyme and AE= Antibody to indicator enzyme.

MATERIALS AND METHODS

For the purpose of developing a separation-free enzyme mediated immunoassay using an electrochemical sensor, the indicator enzymse should be selected from those that catalyze reactions capable of generating electro-active products from electro-inert substrates and their activity can be modulated by a modulator, e.g. inhibited by antibodies to the enzyme itself. Ideally the enzyme should also meet the following criteria: (a) high turnover rate; (b) inexpensive and readily available in large quantities and in a highly purified form; (c) absence of such enzyme activity in biological fluid under assay conditions and (d) stable under storage and experimental conditions. For potentiometric assays, the enzymes urease and adenosine deaminase are suitable beacause they catalyze reactions that form ammonium ion as the reaction product from electro-inert substrates. Furthermore the antibodies to these enzymes are highly inhibitory (Marrucci and Mayer, 1955; Visek et al., 1967; Brontman and Meyerhoff, 1984). The antibodies to these indicator enzymes have so far been of polyclonal origins. In addition to using the whole antibody molecule (Ngo and Lenhoff, 1980; Brontman and Meyerhoff, 1984), it may, in some instances, be more advantageous to use the proteolytic fragments of immunoglobulins, i.e. the Fab and F(ab')2, in preparing the AE-L conjugate (Dona, 1985). It is anticipated that more specific and inhibitory monoclonal antibodies, and hence are better modulators, can be obtained by using hybridoma technology (Kennett et al., 1980; Sevier et al., 1981; Vora, 1985). Amongst many advantages in using monoclonal antibody for immunoassay systems (Siddle, 1985), the following properties are most relevant and desirable for the development of separation-free potentiometric immunoassays: (a) indefinite supply of antibody with constant characteristics having low non-specific binding ; (b) defined affinity and specificity to a particular epitope and (c) easy purification of the monoclonal antibody.

Several potentiometric sensors that were designed for measuring ammonia gas or ammonium ion have been used to assay deaminating enzymes. For examples urease, arginase, adenosine deaminase; 5'-AMP deaminase and aspariginase have been assayed by using ammonia gas sensing electrode (Meyerhoff and Rechnitz, 1979; Gebauer et al. 1979; Gebauer and Rechnitz, 1982). Urease has also been assayed by using an air-gap ammonia gas sensing microelectrode (Joseph, 1985). Recent developments in the area of chemically sensitive field effect transitor have generated devices that are sensitive to pH, ammonia etc. which are applicable for measuring the activity of deaminating enzymes (Caras and Janata, 1980; Lowe, 1985). Deaminating or ammonia producing enzymes can also be assayed amperometrically. For example, Kirstein et al. (1985) used the pH dependence of electrochemical oxidation of hydrazine in the Tafel region as the basis for monitoring urease activity amperometrically.

EXPERIMENTAL STRATEGIES

After choosing a suitable indicator enzyme and generating either polyclonal or monoclonal inhibitory antibodies to the enzyme, the next critical steps are the preparation of conjugates of analyte ligand to either the antibody to the indicator enzyme (AE-L) or the indicator enzyme (L-E). These conjugates must be chemically stable and functional. That is, in the absence of antibody to the analyte ligand (Ab), the free, uncomplexed AE-L should be able to bind and bring about a substantial inhibition on the activity of the indicator enzyme, and in the case of systems that use L-E, the antibody to the indicator enzyme (AE) should still inhibit the enzyme activity of the L-E conjugate. Furthermore, it is necessary to demonstrate that the free, uncomplexed antibody to the analyte ligand (Ab), upon binding to the L portion of AE-L can nullify the inhibitory action of the AE-L on the indicator enzyme. Similarly it should be demonstrated that the antibody to the analyte ligand (Ab) can combine with the L portion of the analyte ligand-indicator enzyme conjugate (L-E) forming Ab:L-E complexes and effectively proctecting the indicator enzyme conjugate (L-E) from being inhibited by AE.

The conjugation of analyte ligands to either the antibody to the indicator enzyme or the indicator enzyme itself can be achieved by a variety of chemical means. For low molecular weight, monomeric analyte ligands or ligand analogs having groups that are or that can be rendered reactive toward some amino acid residues of the protein molecule, a direct coupling of these ligands to antibodies or enzymes can be used (Ngo and Lenhoff, 1980; Dona; 1985). High molecular weight, polymeric ligands, i.e. antigens and proteins, are generally conjugated to the antibody to the indicator enzymes or to the indicator enzyme (as it is in the preparation of L-E conjugates) by using homo or hetero-bifunctional crosslinkers (Kennedy et al., 1976; Ishikawa et al., 1983).

The sequence of experiments involved in developing a separation-free potentiometric enzyme immunoassay using ligand labeled antibody to the enzyme is shown schematically in Fig. 3. The first series of experiments involved testing the inhibitory properties of the AE-L conjugates toward the indicator enzyme (assuming that the modulator is an enzyme inhibitor). These are usually carried out by adding varying amounts of the AE-L conjugates to solutions each containing a limited and constant amount of the indicator enzyme. The enzyme activity is expected to decrease with increasing concentrations of the AE-L conjugate (Diagram A in Fig. 3). The next series of experiments are critical, because they can demonstrate the feasibility of using the AE-L for developing a separation-free immunoassay. The experiments are designed to test whether the binding of anti-analyte ligand antibodies to AE-L can impair the inhibitory action of AE-L. In such experiments, the concentrations of the indicator enzyme and AE-L are held constant with the amounts of antibody to the analyte ligand being varied. Should the experiments proceed as predicted, the enzyme activity would be minimal in the absence of antibody to the analyte ligand. Conversely, the enzyme activity would be maximal with high concentration of the antibody to the analyte ligand (Diagram B, Fig. 3). Following the second series of experiments, the standard curves for the determination of the analyte ligand can generally be established from experiments in which the concentrations of the indicator enzyme, AE-L and the antibody to the analyte ligand are kept constant. The only variable is the concentration of the analyte ligand (Diagram C, Fig. 3).

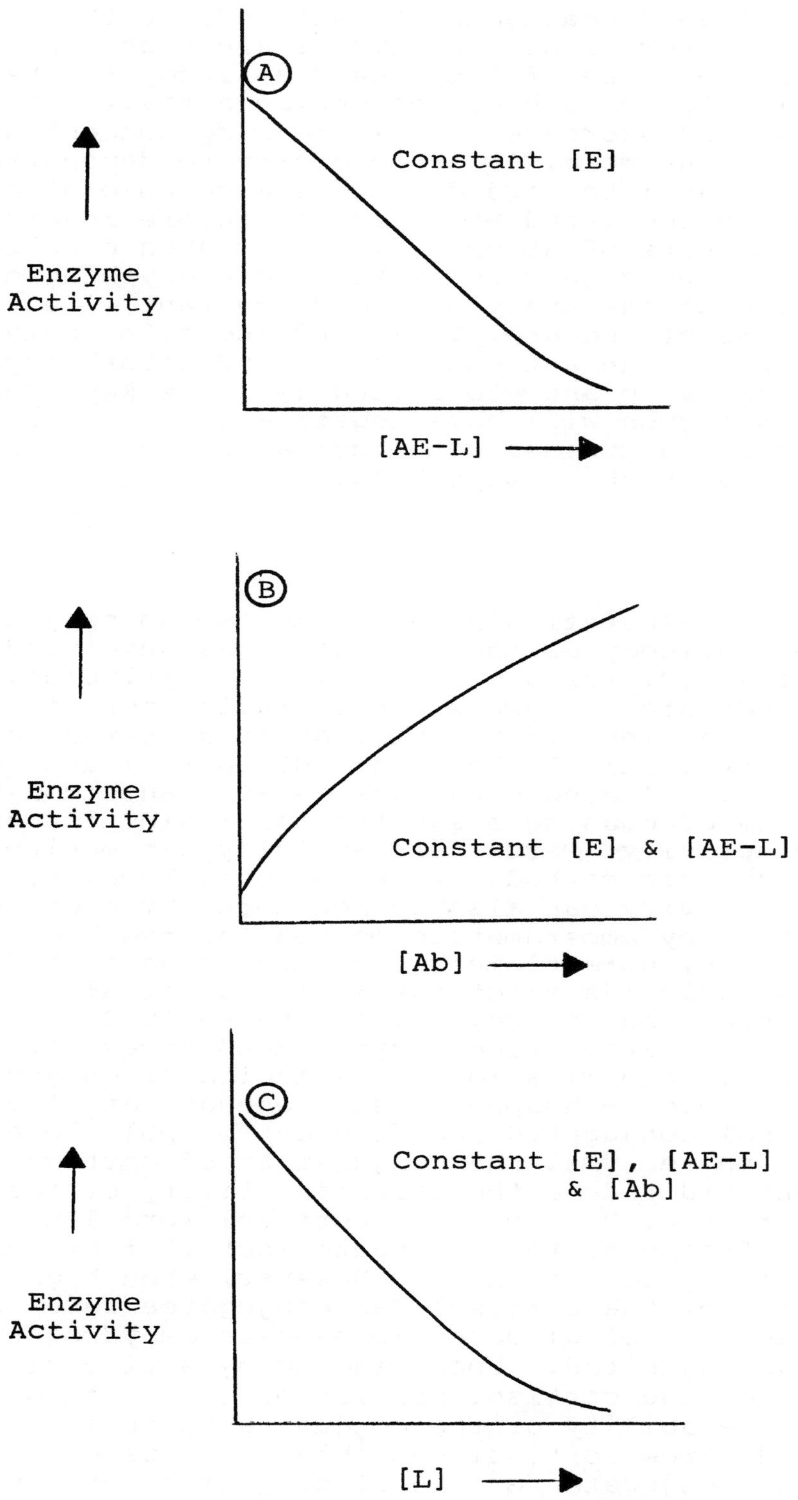

FIG. 3. Sequence of experiments in developing a separation-free enzyme immunoassay using antibody to an enzyme covalently-linked with analyte ligand (AE-L).

In the case where the indicator enzyme, not the antibody to the enzyme, is labeled with analyte ligands, we would also obtain results similar to those described for systems that use analyte ligand conjugated to antibody to the enzyme (Fig. 3). The first series of experiments are those designed to evaluate the response of L-E toward antibody to the indicator enzyme (AE). To be usable, the enzyme activity of L-E conjugate should decrease with increasing concentrations of the (AE). Furthermore, it is necessary to demonstrate that antibodies to analyte ligands (Ab) are capable of protecting L-E from being complexed by AE and therefore reversing the inhibitory effects of AE on L-E. The standard calibration curve can then be generated by keeping everything except the concentration of the analyte ligand constant, i.e. the concentrations of the analyte ligand-indicator enzyme conjugate (L-E), the antibody to the indicator enzyme (AE) and the antibody to analyte ligand (Ab) are kept constant. The standard curves will have negative slopes, i.e. the enzyme activity decreases with increasing concentrations of analyte ligand in the test sample.

RESULTS

The first separation-free enzyme immunoassay using an anti-enzyme antibody as the modulator was developed by Ngo and Lenhoff (1980) for a model analyte, DNP-lysine. In this assay the indicator enzyme was horseradish peroxidase. The modulator was an inhibitory antibody to horseradish peroxidase covalently linked with DNP to form anti-enzyme antibody-analyte ligand conjugate (AE-L). Such conjugate was still capable of cuasing a substantial inhibition of the peroxidase activity. Peroxidase activity was monitored by spectrophotometric method. It is obvious, however, that the peroxidase activity can also be monitored by electrochemical means, such as by amperometric method for hydrogen peroxide measurment or by potentiometric method using an iodide or fluoride ion selctive electrode (Boitieux et al., 1978 & 1979; Alexander and Maltra, 1982). Recently Dona (1985) developed a separation-free enzyme immunoassay for human serum cortisol determination. The indicator enzyme was a bacterial glucose 6-phosphate dehydrogenase and the modulator was a cortisol conjugated Fab fragment of polyclonal IgG antibody to the enzyme. The conjugation of cortisol to the Fab fragment did reduce the immunoreactivity of the fragment. When an average of 9.5 moles of cortisol were linked to one mole of Fab fragment, the conjugate lost 45 % of its native immunoreactivity to the enzyme. However, when high enough concentration of the cortisol-Fab conjugates were used, greater than 80 % of glucose-6-phosphate dehydrogenase activity was inhibited. Upon binding by anti-cortisol antibodies to the cortisol portion of the cortisol-Fab conjugate, the ability of the conjugate to inhibit the enzyme was impaired. Free cortisol was able to compete with the cortisol-Fab conjugate for the binding sites of antibody to cortisol. The activity of glucose 6-phosphate dehydrogenase was monitored colorimetrically for the production of NADH via an auxiliary enzyme system (diaphorase) which converted a tetrazolium salt to a colored formazan. The assay has a working range of 20 to 640 μg/l of cortisol and utilizes 50 ul of serum. The absorbance at 580 nm due to the formation formazan was 1.4 at zero cortisol level and decreased to approximately 0.85 at 640 µg/l cortisol which represented a 40 % inhibition of the enzyme activity. A minimum of about 9-10 cortosol molecules per molecule of Fab is required for a functional cortisol-Fab conjugate. The concentration of NADH produced from the enzymatic reaction can also be measured by amperometric monitoring of the rate of oxygen depletion due to a peroxidase-manganese(II) catalyzed oxidation of NADH by molecular oxigen by using a Clark oxygen electrode (Cheng and Christian, 1977; Cheng and Christian, 1979; Christian, 1982; Nikolelis et al., 1979). The activity of glucose 6-phosphate dehydrogenase can also be monitored by amperometric measurement of the rate of NADH oxidation at a platinum electrode, i.e. Clark hydrogen peroxide electrode (Broyles and Rechnitz, 1986).

Brontman and Meyerhoff (1984) used the antibody to an indicator enzyme to develop a different type of separation-free enzyme immunoassay. They selected an ammonium ion producing enzyme, adenosine deaminase as the indicator enzyme. The enzymic reaction was monitored by using an ammonium ion selective electrode. Rabbit anti-adenosine deaminase antibodies were shown to be inhibitory to the enzyme. Brontman and Meyerhoff covalently linked human serum albumin to the indicator enzyme rather than to the antibody to the indicator enzyme. The enzyme activity of human serum albumin-adenosine deaminase conjugate can be inhibited by the anti-adenosine deaminase antibody up to 70 % of the original activity. However, after binding with antibody to human serum albumin to the albumin portion of the conjugate of human serum albumin-adenosine deaminase, the enzyme activity of the conjugate is protected from being inhibited by the anti-adenosine deaminase antibody. The working range of the assay was 0.1 to 10 µg of the protein.

DISCUSSION

Antibodies to an enzyme that are capable of inhibiting the enzyme activity have been shown to be excellent modulators for developing separation-free enzyme immunoassays for both low and high molecular weight analytes (Ngo and Lenhoff, 1980; Ngo, 1983; Ngo, 1985; Brontman and Meyerhoff, 1984; Dona, 1985). In addition to using the whole antibody molecule, the proteolytic fragments of IgG, such as Fab can also be used (Dona, 1985). The advantages in using antibody to an indicator enzyme as a modulator in developing separation-free enzyme immunoassay are: (1) the ease of obtaining anti-enzyme antibody, because enzymes are generally highly antigenic (Cinader, 1976; Arnon, 1977); (2) the ability to obtain highly specific and potent inhibitory monoclonal antibody by hybridoma technology (Kennett et al., 1980; Sevier et al., 1981; Vora, 1985); (3) the ease of covalent cojugating analyte ligand to the antibody molecule (Kennedy et al., 1976; Ishikawa et al., 1983); (4) maintenance of the immunological specificity of the antibody after chemical linking with analyte molecules; (5) good stability of antibody molecules under normal storage and assay conditions and (6) the ease of preparing Fab or F(ab)'2 from the whole antibody by proteolytic fragmentations.

The selection of a proper indicator enzyme is crucial for the development of a sensitive and convenient assay system. The enzyme should have (1) a high turn-over rate so that a small amount of it can be ready and easily measured; (2) no inteference from the sample matrix; (3) an antibody that can rapidly and preferably completely inhibit the enzyme activity and (4) a good stability.

The size of the analyte ligand plays an important role in deciding the way the assay should be developed. Assays for low molecular weight analytes should preferably be developed by using anti-enzyme antibody conjugated with analyte ligands as those shown by Ngo and Lenhoff (1980) and Dona (1985). This is because of the binding of the anti-analyte antibody to the analyte ligand portion of the ligand-modulator conjugate can effectively prevent the conjugate from modulating (inhibiting, in the case where the modulator is an enzyme inhibitor) the enzyme presumably due to a steric or a physical hindrance created as a result of the binding. Should high molecular weight anlayte ligands be linked to an anti-enzyme antibody, the inhibitory effects of such a conjugate may not be altered upon binding of the conjugate by the anti-analyte ligand antibody. This is likely due to the large size of the analyte ligand that the binding of the ligand portion of the ligand-modulator conjugate by the antibody to the ligand may not be sufficient to creat a steric hindrance capable of altering the activity of the conjugate. In such a case, the approach that uses the native, unmodified antibody to the indicator enzyme and indicator enzyme conjugated with analyte ligsnds should be adopted (Brontman and Meyerhoff, 1984).

In order to develop separation-free enzyme mediated potentiometric immunoassays, it is necessary to use an enzymes that can be conveniently assayed potentiometrically. Adenosine deaminase is one such enzymes. It catalyses the production of ammonium ion by deaminating adenosine (Brontman and Meyerhoff, 1984). Another enzyme that shows great potentials is jack bean urease. This enzyme has previously been used as the enzyme label in several heterogeneous enzyme immunoassay systems (Meyerhoff and Rechnitz, 1979; Chandler et al., 1982). As a label for enzyme immunoassay, urease offers several advantages over other commonly used enzymes (peroxidase, alkaline phosphatase, B-galactosidase): (1) higher turnover rate; (2) non-toxic and stable substrate; (3) linear reaction time course; (4) stable ligand-enzyme conjugates; (5) mild reaction stopping conditions; (6) absence of urease in mammalian systems and (7) enzyme activity can be monitored spectrophotometrically or electrochemically. Urease has been conjugated to both low molecular weight ligand such as cyclic AMP and high molecular weight antigen such as bovine serum albumin and imminoglobulins (Meyerhoff and Rechnitz, 1979; Chandler et al. 1982). Futhermore, highly specific and inhibitory antibody to urease can be obtained by immunizing animals (Marucci and Mayer, 1955; Visek et al., 1967).

CONCLUSION

Novel separation-free enzyme amplified immunoassays for both haptens and antigens can be developed by a combined use of antibody to an indicator enzyme and a suitable electrochemical sensor. Suitable indicator enzymes included adenosine deaminase, horseradish peroxidase, glucose 6-phosphate dehydrogenase and urease. Antibodies to these enzymes have been obtained by injecting animal with the respective enzyme as the immunogen. It is likely that monoclonal antibodies to these enzymes can also be generated by using the well established hibridoma technique. Several gas sensing membrane electrodes, ammonium ion selective electrodes and other electrochemical sensors can be used as the detector.

REFERENCES

Abeles, R. H., 1978, Suicide enzyme inactivators, in: "Enzyme -Activated Irreversible Inhibitors", N. Seiler, M.J. Jung and J. Koch-Weser, editors, pp. 1-12, Elsevier/North-Holland Biomedical Press, Amsterdam.

Alexander, P.W. and Maltra, C., 1982, Enzyme-linked immunoassay of human immunoglobulin G with the fluoride ion selective electrode, Anal. Chem. 54:68-71.

Arnon, R., 1977, Immunochemistry of lysozyme, in: "Immunochemistry of Enzyme and Their Antibodies", M.R.J. Galton, editor, pp. 1-28, Wiley, New York.

Bacquet, C. and Twumasi, D.Y., 1984, A homogeneous enzyme immunoassay with avidin-ligand conjugate as the enzyme-modulator. Anal. Biochem., 136:387-490.

Baker, B.R., 1967, "Design of Active-Site Directed Irreversible Enzyme Inhibitors", Wiley, New York.

Blecka, L.J., Shaffar, M. and Dworschack, R., 1983, Inhibitor Enzyme Immunoassays for quantitation of various haptens: A review, in: "Immunoenzymatic Techniques", S. Avrameas, P. Druet, R. Masseyeff and G. Feldman, editors, pp. 207-214, Elsevier Science Publishers, Amsterdam.

Boitieux, J.L., Desmet, G. and Thomas, D., 1978, Potentiometric determination of hepatitis B surface antigen in biological fluids, Clin. Chim. Acta, 88:329-336.

Boitieux, J.L., Desmet, G. and Thomas, D., 1979, An antibody electrode, preliminary report on a new approach in enzyme immunoassay, 25:318-321.

Brontman, S.B. and Meyerhoff, M.E. 1984, Homogeneous enzyme-linked assays mediated by enzyme antibody; a new approach to electrode-based immunoassays, Anal. Chim. Acta, 162:363-367.

Broyles, C.A. and Rechnitz, G.A.,1986, Drug antibody measurement by homogeneous enzyme immunoassay with amperometric detection, Anal. Chem., 58:1242-1245.

Caras, S. and Janata, J., 1980, Field effect transitor sensitive to penicillin, Anal. Chem., 52:1935-1937.

Chandler, H.M., Cox, J.C., Healy, K., MacGregor, A., Premier, R.R. and Hurrell, J.G.R., 1982, An investigation of the use of urease-antibody conjugates in enzyme immunoassays, J. Immunol. Meth., 53:187-194.

Cheng, F.S. and Christian, G.D., 1977, Amperometric measurement of enzyme reactions with an oxygen electrode using oxidation of reduced nicotinamide adenine dinucleotide, Anal. Chem., 49:1785-1788.

Cheng, F.S. and Christian, G.D., 1979, A coupled enzymatic method to measure blood lactate by amperometric monitoring of the rate of oxygen depletion with a Clark oxygen electrode, Anal. Chim. Acta, 91:295-301.
Cinader, B., 1976, Enzyme-antibody interaction, in: "Methods of Immunology and immunochemistry", M.Chase and C. Williams, editor, pp. 313-375, Academic Press, New York.
Dona, V., 1985, Homogeneous colorimetric enzyme inhibition immunoassay for cortisol in human serum with Fab anti-glucose 6-phosphate dehydrogenase as a label modulator, J. Immunol. Meth., 82:65-75.
Finley, P.R., Williams, R.J. and Lichti, D.A., 1980, Evaluation of a new homogeneous enzyme inhibitor immunoassay of thyroxine with use of a bichromatic analyzer, Clin. Chem., 26:1723-1726.
Gebauer, C.R., Meyerhoff, M.E. and Rechnitz, G.A.,1979, Enzyme electrode-based kinetic assays of enzyme activities, Anal. Biochem., 95:479-482.

Gebauer, C.R. and Rechnitz, G.A., 1982, Deaminating enzyme labels for potentiometric enzyme immunoassay, Anal. Biochem., 124:338-348.
Ishikawa , E., Imagawa, M., Hashida, S., Yoshitake, S., Hamaguchi, Y. and Ueno, T., 1983, Enzyme-labeling of antibodies and their fragments for enzyme immunoassay and immunohistochemical staining, J. Imminoassay, 4:209-327.
Joseph, J.P., 1985, An enzyme microsensor for urea on an ammonia gas electrode, Anal. Chim. Acta, 169:149-156.
Kennett, R.H., McKearn, J.J. and bechtol, K.B. editors, 1980, "Monoclonal Antibodies, Hybridoma: A New Dimension in Biological Analysis", Plenum, Press, New York.
Kennedy, J.H. Kricka, L.J. and Wilding, P., 1976, Protein-protein coupling reactions and the applications of protein conjugates, Clin. Chim. Acta, 70:1-31.
Kirstein, D., Kirstein, L. and Scheller, F. 1985, Enayme electrode for urea with amperometric indication: Part 1 - Basic principle, Biosensors, 1:117-130.
Krebs, E.G., 1972, Protein Kinases, in: "Curent Topics in Cellular Regulation", B.L. Horecker and E.R. Stadtman, editors, pp. 99-133, Academic Press, New York.
Lienhard, G.E., 1973, Enzymatic catalysis and transition state theory, Science, 180:149-154.
Lowe, C.R., 1985, An introduction to the concept and technology of Biosensors, Biosensors, 1:3-16.
Marucci, A.A. and Mayer, M.M., 1955, Quatitative studies on the inhibition of crystalline urease by rabbit anti-urease Arch. Biochem. Biophys., 54:330-340.
Meyerhoff, M.E. and Rechnitz, G.A., 1979, Electrode based enzyme immunoassay using urease conjugates, Anal. Biochem., 95:483-493.
Ngo, T.T., 1983, Enzyme modulator mediated immunoassay (EMMIA), Int. J. Biochem., 15:583-590.
Ngo, T.T., 1985, Enzyme modulator as label in separation-free immunoassays: Enzyme modulator mediated immunoassay (EMMIA), in: "Enzyme-Mediated Immunoassay", T.T. Ngo and H.M. Lenhoff, editors, pp. 52-72, Plenum Press, New York.

Ngo, T.T. and Lenhoff, H.M., 1980, Enzyme modulators as tools for the development of homogeneous enzyme immunoassays, FEBS Letters, 116:285-288.
Ngo, T.T. and Tunnicliff, G., 1981, Inhibition of enzymic reactions by transition state analogs: An approach for drug design, Gen. Pharmacol, 12:129-138.
Nikolelis, D.P., Painton, C-D. D. and Mottola, H.A., 1979, The peroxidase-catalysed oxidation of NADH as an indicator reaction for repetitive determinations by by sample injection in closed flow-through systems: The determination of LDH in blood serum, Anal. Biochem., 97:255-263.
Place, M.A., Carrico, R.J., Yeager, F.M., Albarella, J.P. and Boguslaski, R.C., 1983, A colorimetric immunoassay based on enzyme inhibitor method, J. Immunol. Meth., 61:209-216.
Rando. R.R., 1974, Chemistry and enzymology of k cat inhibitors, Science, 185:320-324.
Sevier, D.E., David, G.S., Martinis, J., Desmond, W.J., Bartholomew, R.M. and Wang, R., 1981, Monoclonal antibodies in clinical immunology, Clin. Chem., 27:1979-1806.
Shaw, E.N., 1980, Design of irreversible inhibitors, in: "Enzyme Inhibitors as Drugs", M. Sandler, editor, pp. 24-42, University Park Press, Baltimore.
Siddle, K., 1985, Properties and applications of monoclonal antibodies, in: "Alternative Immunoassays", W.P. Collins, editor, pp. 13-37, Wiley, New York.
Stadtman, E.R., 1970, Mechanism of enzyme regulation in metabolim, in: "The Enzyme", P.D. Boyer, editor, Vol. 1, pp. 397-459, Academic Press, New York.
Visek, W.J., Iwert, M.E., Nelson, N.S. and Rust, J.H., 1967, Some immunological properties of jack bean urease and its antibody, Arch. Biochem. Biophys., 122:95-104.
Vora, J.L., 1985, Monoclonal antibodies in enzyme research: Present and potential applications, Anal. Biochem., 144:307-318.
Webb, J.L., 1963, Enzyme and Metabolic Inhibitors, Vols. 1-3, Academic Press, New York.
Wolfenden, R., 1972, Analog approaches to the structure of the transition state in enzyme reactions, Acc. Chem. Res., 5:10-18.
Wong, J.T.F., 1975, "Kinetics of Enzyme Mechanism", pp. 39-72, Academic Press, New York.

IMMUNOCHEMICAL DETERMINATION OF CREATINE KINASE ISOENZYME-MB

Shia S. Kuan* and George G. Guilbault

Department of Chemistry
University of New Orleans
New Orleans, Louisiana 70148

INTRODUCTION

In recent years, emphasis on the treatment of myocardial infarction is gradually shifting away from observation and documentation toward prompt management procedures. Early detection and confirmation of coronary occlusion by physical or chemical tests can provide the doctors with reliable information in planning an early operation in order to avoid future tissue damage. Most of the tests used in the past are for documentation. In general, EKG's are not useful for early detection of myocardial infarcts since the electrocardiographic changes usually occur after onset. Many chemical tests are neither specific nor sensitive in detecting myocardial distress. Among the chemical tests, the measurement of CK-MB, the cardiac specific isoenzyme, appears to be the most specific and reliable procedure to obtain confirmative information for early diagnosis of acute myocardial infarction.

Clinical procedures commonly used for the separation of CK-MB are electrophoresis, chromatography and immunochemical methods. Electrophoretic separation of CK-MB in gel is tedious, laborious, subject to non-specific band interference and lacks sensitivity. More simplified ion exchange procedures suffer from incomplete separation and poor recovery of each isoenzyme. Thus, these methods are less suitable for fast detection of CK-MB at low levels. Generally speaking, immunochemical procedures are rapid, more specific, reliable and less cumbersome.

Immunochemical methods recently developed can be classified into two categories, namely the immunoprecipitation test and the immunoinhibition test. The former can more accurately measure the activity of pure subunit isoenzymes, for example, CK-MM or CK-BB using anti-CK-BB or anti-CK-MM. But this test requires a second antibody and centrifugation for separation of antibody-bound from unbound fraction; therefore it is more tedious and time-consuming. In the immunoinhibition test, the active site of CK-M subunit is blocked by goat-anti-human CK-M-IgG (termed Inh-CK-M-antibody) and the activity

*Food & Drug Administration, 4298 Elysian Fields Avenue, New Orleans, La. 70122

of CK-B remaining is assayed. This test does not require a separation step and hence is more rapid, simple, and easy to manipulate and automate. However, the measurement of CK-MB using Inh-CK-M has a prerequisite for usefulness that the CK-BB activity be never or only very rarely present in serum. Practically, CK-BB activity is found in serum only in exceptional cases (Galen, 1975; Konttingen, et al., 1973; Roberts, et al., 1975), and so, the test is mostly valid. Another assay, which combines the two tests, was commercially introduced by Roche Diagnostics to overcome the nonspecificity of the single antibody step. In this test, the activity of CK-MB is first measured by the inhibition test and is followed by the precipitation test in order to eliminate or subtract out interfering activities (e.g. CK-BB, adenylate kinase, or atypical CK) from the activity of CK-MB by the first test.

The method most commonly employed for the assay of creatine kinase after a separation step is the use of a CK-HK-NADP coupled reaction (Scheme I). The NADPH produced can be measured directly by observing the UV change or by measuring the color or fluorescent intensity through the coupling of NAD/NADH to a chromophor or fluorophor.

Two methods have been developed in our laboratory using 2-p-iodophenyl-3-p-nitrophenyl-5-phenyl tetrazolium chloride and resazurin as hydrogen acceptors, respectively, for colorimetric and fluorometric determination of CK-MB. Both methods give good sensitivity and selectivity but require additional reagents and time for incubation and measurement. These subsequent steps make the procedures less attractive for the assay of CK-MB.

Because of many inherent advantages, such as simplicity, rapidity, adaptability to interfacing with other instruments for automation and freedom of interferences from tubidity, electrochemical methods have recently become useful analytical tools for the assay of biologically active materials and metabolites in the living organism. Electrochemical methods for the determination of NADH or NADPH (reduced form of nicotinamide adenine dinucleotide phosphate) have been well developed. Methods include direct amperometric measurement of NADH or NADPH at solid electrodes (Blaedel and Jenkins, 1975; Thomas and Christian, 1975) and coupling NADH or NADPH to different redox agents, such as 2,6-dichlorophenol-indophenol (Smith and Olson, 1975), Bindschedler's Green (Smith and Olson, 1974), and ferricyanide (Thomas and Christian, 1976) for indirect quantitation. The use of an oxygen electrode to monitor the oxygen consumption by NADH in the presence of horseradish peroxidase (Cheng and Christian, 1977, 1978) has been documented as well. Other electrochemical methods based on potentiometric electrodes and various enzyme electrodes have also been discussed (Guilbault, 1978, 1985). The electrochemistry of the $Fe(CN)_6^{3-}$ complex ion in biological media has been well illustrated. Both NADPH and ferricyanide are stable in the presence of oxygen (Peel, 1972), and the voltametric half-wave potential of the $Fe(CN)_6^{3-}/Fe(CN)_6^{4-}$ system is moderately positive, whereas amperometric residual currents in the presence of serum are low. On the basis of the information described above, a novel immunochemical system was then designed and developed in this laboratory for fast, simple and reliable assay of CK-MB. In this chapter, the use of different approaches for the separation and quantitation of CK-MB is described and discussed.

I. Amperometric Measurement of CK-MB After Inhibition by Soluble Antibody

In this system, goat-anti-human CK-M (GAHCK-M) is used to inhibit all CK-M subunits in serum and the residual CK-B activity is detected with a platinum electrode by coupling the NADH to the ferricyanide-diaphorase reaction. Apparatus setup, reagents needed and conditions used are described as follows.

A. MATERIALS AND METHODS

1. Apparatus

The rate of current change at a fixed potential was monitored with a polarographic analyzer (PAR 174A, Princeton Applied Research Corp., Princeton, NJ) using a platinum electrode poised at +0.36 V vs SCE (Saturated Calomel Electrode) and an auxiliary platinum foil electrode. A linear chart recorder (Cole-Parmer Instrument Co., Chicago, IL) was used to record the rates of reaction.

An EC Motomatic motor control stirrer (Model E550M, Electro Chart Corp., Hopkins, MN) and a ministirrer (2 x 8 mm) were used to control the stirring speed during assay.

A circulating water bath with a laboratory immersion heater (Blue M Electric Co., Blue Island, IL) and a circulating pump (Brinkman IC-2, Brinkman, Westbury, NY) were used to maintain a constant temperature within $\pm 0.1^{o}C$.

pH adjustments were performed with a Beckman Research pH meter (Beckman Instrument, Inc., Fullerton, CA).

Autopipets (Finnipipette KY., Helsinki, Finland) of 5-50, 50-200, and 200-1000 μL capacities were used to deliver sera, substrates, and reagents.

2. Reagents and Solutions

A conventional Oliver-Rosalki CK reagent solutoin modified by Dade was prepared as described below by use of chemicals obtained from Sigma Chemical Co. (St. Louis, MO).

Cardiozyme kit is as follows: triethanolamine buffer (pH 7.0), 100 mM; creatine phosphate, 35 mM; glucose, 20 mM; magnesium acetate, 10 mM; adenosine diphosphate (ADP), 1 mM; nicotinamide adenine dinucleotide phosphate, 0.6 mM; adenosine monophosphate, 10 mM; glucose 6-phosphate dehydrogenase, 2.4 U/assay; hexokinase, 2.4 U/assay; reduced glutathione, 0.3 mM.

CK-M inhibiting antibodies (from goat) were obtained from E. Merck (Darmstadt, Germany).

Potassium ferricyanide crystals [$K_4Fe(CN)_6 \cdot 3H_2O$] were obtained from Matheson Coleman and Bell (Norwood, OH). Potassium chloride was puchased from Fisher Scientific Co. (Fair Lawn, NJ).

3. Procedure

Immerse the three electrodes into 2 mL of Cardiozyme kit solution in a 10-mL minibeaker maintained at $37^{o}C$ using a thermal cell circulated with water from a constant-termperature water bath. Pipet 0.1 mL of 0.1 M potassium chloride solution, 0.1 mL of 5.4 U/assay of diaphorase, and 0.1 mL of creatine kinase MB isoenzyme control of serum and preincubate 2 min at 37° C. Stir the assay solution with a ministirrer at a constant rate of 800 rpm. Fix the potential of the polarographic analyzer at +0.36 V vs. SCE, push the control button, and initiate the electrochemical reaction by introducting 0.1 mL of freshly prepared 3.4×10^{3} M ferricyanide solution into the assay solution. Record the rate of current change with time and calculate

SCHEME 1

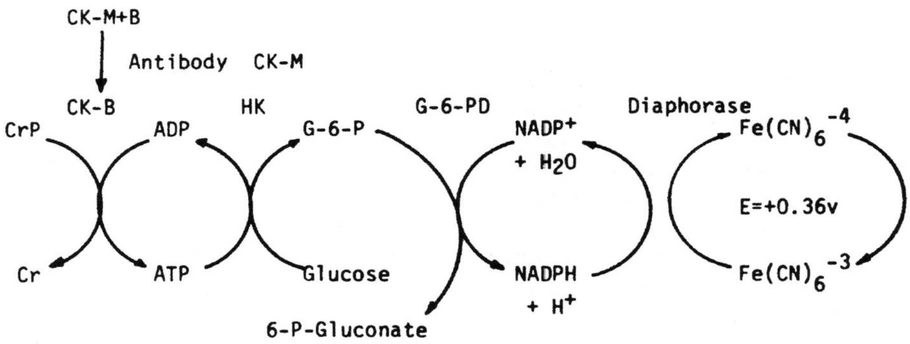

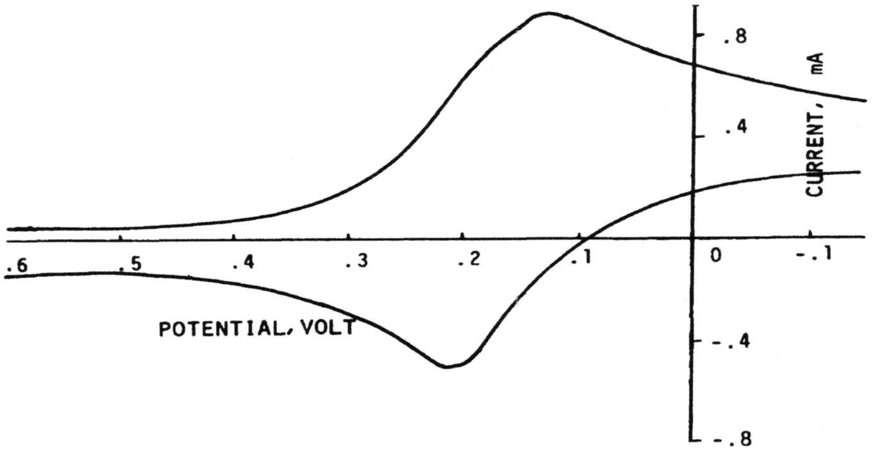

Fig. 1. The Cyclic Voltammogram of Ferricyanide-Ferrocyanide Species in TEA buffer(0.1M, pH 7). Scan rate = 5 mV/s.

the activity of creatine kinase MB after immunoinhibition from a calibration curve of $\Delta i/\Delta t$ vs. activity. Rinse the electrode with dilute nitric acid between each assay to ensure that the electrode surfaces are clean and the results are reproducible.

The calibration plot of $\Delta i/\Delta t$ vs. CK-MB activity is linear up to 875 U/L.

B. RESULTS

1. Preliminary Study

The current-voltage curve of a 3.4×10^{-3} M solution of ferricyanide in 0.1 M. pH 7, triethanolamine bufer, with 0.1 M potassium chloride as a supporting electrolyte, was recorded by use of the three-electrode cell with a platinum working electrode (Figure 1). The potential was scanned from 0.6 to -0.15 V vs. SCE in an unstirred solution at a scan rate of 5 mV/s. The curve illustrates the feasibility of using ferricyanide in TEA buffer system (30-32). In order to ensure that the applied potential is high enough to effect the anodic oxidation

$$Fe(CN)_6^{4-} \rightleftharpoons Fe(CN)_6^{3-} + e^- \quad (1)$$

we applied a potential of +0.36 V in the assay.

The electron transfer reactions

$$Fe(CN)_6^{3-} + NADPH \xrightleftharpoons{diaphorase} Fe(CN)_6^{4-} + NADP^+ \quad (2)$$

and

$$Fe(CN)_6^{4-} \rightleftharpoons Fe(CN)_6^{3-} + e^- \quad (1)$$

were studied at a fixed potential of +0.36 V and a constant rate of stirring. The current level was proportionally changed by varying the concentrations of ferricyanide, reduced nicotinamide adenine dinucleotide phosphate (NADPH), or diaphorase. The electrode transfer in these two coupling reactions is so fast that an instantaneous jump of the current was seen (instead of a slope of current change).

2. Studies on Optimum Conditions

The maximal activity of CK-MB in the electrochemical assay was found to occur at 37°C in 0.1 M triethanolamine hydrochloride buffer pH 7.0, containing 0.1 M potassium chloride, 5.4 U/diaphorase, and 3.4 mM ferricyanide. Preincubation is required for obtaining a stable base line before initiating the electrochemical reaction; a 2-min period was found to be optimum.

3. Effect of Stirring

The assay solution was stirred by a 2 x 8 mm ministirrer at a constant rate of stirring. The stirring action increases the transport rate of the assay species at the electrode, thus increasing the sensitivity. This mode of mass transport is called convection or hydrodynamic transport. Without stirring of the assay solution, the reaction rate is hardly detectable, hence the stirring rate should remain constant during the assay. In general, the reaction rate increases with an increase in stirring speed, but the noise level increases significantly and a whirlpool is formed when the stirring speed is 1000 rpm or higher. Therefore, a speed of 800 rpm was used, and satisfactory results were obtained.

4. **Effect of Activators**

The activator for the creatine kinase, reduced glutathione, is an electroactive species itself (E^o = -0.25 V) (Rao, et al, 1977). At a concentration of 9 x 10^{-3} M (Dade kit formula), reduced glutathione reacts with the electron acceptor ferricyanide, thus interfering with the assay. Several other activators, such as cysteine, thioglycerol, 2-mercaptoethanol, n-acetylcysteine (NAC), dithiothreitol (DTT), dithioerythritol (DTE), and 2-aminoethylisothiouronium bromide (AET) (Rao, et al., 1977) were tried as substitutes. Unfortunately, similar interferences were obtained. But if the concentration of the activator is reduced to 3 x 10^{-4} M, and a preincubation of the CK-MB sample for 2 min is effected, reproducible rates using this electrochemical approach are obtained. Peck et al. (1965) successfully used the same approach (i.e., reduction of activator concentration) for the assay of adenosine 5'-phosphosulfate (APS) reductase in the presence of ferricyanide.

5. **Effect of Other Trapping Agents for NADPH**

2,6-Dichlorophenolindophenol (DCPIP) was investigated as a NADPH trapping agent (instead of using ferricyanide) for the determination of CK-MB in this study. An applied potential of +0.217 V was used. The reproducibility of the results was not as good as that of ferricyanide, probably due to the ill-defined redox peak curves in the cyclic voltammogram.

6. **Precision**

Run-to-run and day-to-day precision studies were performed to evaluate the new method for CK-MB assay. Two levels of pooled blood sera (normal, 8.1 U/L, and abnormal, 203 U/L) were analyzed for three consecutive days. Table I shows the results obtained from the proposed method. Excellent precision was observed.

7. **Recovery Study**

Cardiozyme controls (Dade) were added to the pooled serum to bring the CK-MB activity from 23.0 U/L to values ranging from 35.5 to 313.0 U/L. The average recovery was found to be 97.8%.

8. **Specificity**

Interferences from adenylate kinase (Szasz, et al., 1976a, 1976b) and endogenous polyvalent cations (Szasz, et al., 1977) acting upon serum creatine kinase have been reported. Several other common interfering substances present in serum were added to the specimen to study their effect on the accuracy of the assay. The interference effect of all diverse substrate is negligible (Table II).

9. **Comparison Study**

Fifteen fresh sera ranging from 8 to 285 CK-MB U/L were assayed by the proposed electrochemical method and compared with those of the Helena electrophoresis method (from Hotel Dieu Hospital). The linear regression analysis of CK-MB gave a coefficient correlation of 0.999.

Table I. Study of Precision of Electrochemical Assay

	Normal value	Normal value
	Run to Run	
av	8.1 U/L[a]	203.1 U/L[a,b]
N[c]	6 runs	6 runs
std. dev.	0.26 U/L	2.6%
	Day to Day	
av	8.07 U/L	204.6 U/L
N	3 days	3 days
std. dev.	0.25 U/L	7.8 U/L
CV	3.1%	3.8%

[a] Value of each run was the average of three assays.
[b] U/L = units/liter. [c] N = Number of runs.

Table II. Interference Effects of Species in Electrochemical Assay

Species	Concn. (mg/dL)[a]	Result (U/L)[b]	Discrepancy (U/L)[b]
Control serum	34.0 U/L	34.0	0
Creatine	6.5	32.4	-1.6
Ascorbic acid	6.2	30.0	-4.6
Uric acid	44.0	42.3	+8.3
Bilirubin	1.8	32.6	-1.4
Hemoglobin	100.0	39.0	+5.0

[a] dL = deciliter. [b] U/L = units/liter.

10. Electrochemical Measurement of CK-MB After Immunoseparation by Immobilized GAHCK-M

In addition to its high cost, precision becomes the problem of concern in using soluble antibody for CK-MB assay because the affinity of GAHCK-M toward creatine kinase is sensitive to the change of temperature and pH during assay. Besides, soluble antibody is more susceptible to heat denaturation and microbial attack. Furthermore, the titer of polyclonal antisera varies from batch to batch. All of these may cause irreproducible results. However, the above problems can be eliminated or minimized through immobilization.

A useful immobilized antibody is one which is physically confined or localized in some way with retention of its immunological activity, and which can be used repeatedly and continuously. There are four principal methods for the immobilization of antibodies: physical adsorption, entrapment, protein cross-linking and covalent attachment. Of these, covalent coupling appears to be best because it provides strong linkages which hold the antibodies in a stable conformation. The reusability of the covalently bound immunoadsorbents is another advantage. The following experiments carried out in this laboratory have demonstrated the feasibility of using immobilized antibody for the separation of CK-MB, which is in turn quantified amperometrically.

II. USE OF IMMOBILIZED REAGENTS - THE IMMUNO-STIRRER

A. Materials and Methods

1. Apparatus

The kinetic measurements of creatine kinase (CK) and the residual CK (B subunit) were performed with a Cary 17 spectrophotometer, using the Oliver-Rosalki enzyme system (see below). A PAR 174H polarographic analyzer (Princeton Applied Research Corp., Princeton, NJ) was used for measurements with a three-electrode system.

An E.C. Motormatic motor control (Electro-Craft, Inc., Hopkins, MN) was used to maintain a steady stirring speed during assay. A Hall-effect digital sensor was employed to sense the magnetic field, and the trigger was amplified and displayed on a Heathkit frequency counter (Model CB-101 with modifications, circuit diagram available on request).

A Brinkman IC-2 circulator (Brinkman, Westbury, NY) was used to heat and pump water into a home-made cell to keep the temperature constant during assay; a laboratory immersion heater (Blue M Electric Co., Blue Island, IL) was occasionally used as an auxiliary heater.

Autopipettes (Finnpipette, Helsinki, Finland) of 5-50, 50-200, and 200-1000- µl capacities were used to deliver sera, substrates and reagents. A home-made rotating porous cell with removable lid was designed to facilitate easy loadinng and unloading of immobilized proteins . The body was machined from polyethylene and at the bottom of the cell was drilled a hole into which a small cylindrical magnet was inserted and sealed by epoxy cement to keep the solution from reacting with the metal. The upper chamber, which accommodates the immobilized antibodies, had six holes drilled and a removable, tight-fitting cover. Nylon net (37- µm pore size, Tetko, Inc., NY) was glued onto the lid and around the wall inside the chamber using plastic rubber (Woodhill Chemical Sales Corp., Cleveland, OH) to prevent the immobilized antibodies from leaching out of the stirrer.

2. Reagents and Solutions

Modified Oliver-Rosalki CK reagents (prepared according to the Dade formula; Dade Div. of American Hospital, Miami, FL) were: triethanolamine buffer (pH 7.0), 100 mM; creatine phosphate, 70 mM; glucose, 40 mM; magnesium acetate, 20 mM; adenosine diphosphate (ADP), 2 mM; $NADP^+$. 0.6 mM; adenosine monophosphate, 20 mM; glucose-6-phosphate dehydrogenase, 4.8 U/assay; hexokinase, 4.8 U/assay; reduce glutathione 18 mM (spectrophotometric assay) or 0.6 mM (electrochemical assay), obtained from Sigma Chemical Co., St. Louis, MO.

CK-M inhibiting antibodies were used (anti-human, from goats, E. Merck, Darmstadt, Germany). Each vial can be constituted to 2 ml, of which 0.1 ml can inhibit 800 U l^{-1} of CK-M by 99%.

The immobilization medium, sodium phosphate buffer, 0.1 M, pH 7.0, was prepared from $NaH_2PO_4.H_2O$ (Mallinckrodt). The charaterization medium, triethanolamine buffer, 0.1 M, pH 7.0, was prepared from triethanolammonium chloride (Sigma).

A 2-ml portion of 50% dimethylsulfoxide and 2 ml of 0.05 M glycine - HCl buffer (pH 2.3) were used as the regeneration medium for the immobilized antibodies. Phosphate-buffered saline was used to wash away the organic solvent and acidic solution, and to restore the biological activity of the immobilized antibodies. It was prepared by adding 0.9% sodium chloride solution to 0.1 M, pH 7.0 sodium phosphate buffer. Borate-buffered saline, used to store the immobilized antibodies during refrigeration, was prepared by adding 0.9% sodium chloride solution to 0.01 M, pH 8 sodium borate buffer.

3. Coupling Reagents and Carriers

Alkylamine porous glass beads (40-80 mesh) were supplied by Corning Glass Works (Corning, NY). Cellulose beads (65 mesh) were a gift from Dr. L. F. Chen, Department of Chemical Engineering, Purdue University, West Lafayette, IN. Glutaraldehyde (Eastman Kodak) and cyanogen bromide (Sigma) were used as coupling and activating reagents.

B. Immobilization Procedures for Inh-CK-M Antibodies

1. Immobilization on Alkylamine Glass Beads

Wash 0.25 g of alkylamine porous glass beads three times (5 ml each time) with double-distilled deionized (D^3) water in a 20-ml vial, then degas with 7 ml of D^3 water under vacuum in a desiccator for 10 min, until no more gas bubbles evolve from the glass beads, indicating that all the pores in the beads are filled with water. To these washed and drained beads, add 10 ml of 2.5% glutaraldehyde solution (25% glutaraldehyde diluted 1:10 with 0.1 M sodium phosphate buffer, pH 7), sufficient to cover the glass beads. Place the reaction mixture in a desiccator attached to the aspirator for 1 h to remove air and gas bubbles from the particles. Remove and filter the activated beads on a medium-porosity Gooch funnel and wash three times with distilled water (10 ml each time). Make sure that all the unbound glutaraldehyde is washed off.

Mix 1 ml (one half of the contents of the Merck vial) of antibody in sodium phosphate buffer (0.1 M, pH 7) with the alkylamine derivative prepared above. Allow to react in a cold room for 24 h, with gentle shaking. Wash the immobilized antibodies at least three times with the sodium phosphate buffer (pH 7) to remove unbound antibodies.

2. Immobilization on Cellulose Beads with Cyanogen Bromide

Wash 0.3 g of 65-mesh cellulose beads three times (5 ml each time) with D^3 water in a 20-ml vial. Transfer the beads to a beaker containing 15 ml of acetonitrile and 10 ml of 0.1 M phosphate buffer in which 3 g of cyanogen bromide has been dissolved, and adjust the apparent pH of the solution to 11.0 $\pm$0.2. Activate the carrier at room temperature with shaking for 2 h, filter and wash successively with acetone (20 ml) and ethanol (20 ml).

Mix the activated beads with 1 ml of 0.1 M sodium phosphate buffer (pH 7) containing 1/2 of the contents of a vial of Inh-CK-M antibodies, as above. Carry out the coupling reaction with gentle shaking at 4°C. After 24 h, wash the product with the phosphate buffer (pH 7) to remove any unbound antibodies.

C. Assay Procedures

1. Pretreatment before assay (stirring incubation)

Mix 0.3 ml of CK-MB isoenzyme control or serum with 3 ml of 0.1 M tri-ethanolamine buffer (pH 7) in a 10-ml beaker. Place the immuno-stirrer in the solution, which is held at 37°C with the use of the water-circulating cell. Stir at 200 rpm for 30 min.

2. Procedure

Place 1.1 ml of the CK solution obtained after pretreatment in a 1-cm spectrophotometric quartz cell containing 1.0 ml of the modified reagent mixture. Initiate the reaction by the addition of 0.1 ml of $NADP^+$ solution (10 mg ml^{-1}), and record the rate continuously for 7 min at 500 nm. Measure the slope (A Δt^{-1}), to give a measure of the CK-MB activity. Alternatively, measure the residual activity amperometrically as described previously.

3. Regeneration

Put the immuno-stirrer, which has been used for 5 pretreatments, into the dimethyl sulfoxide regeneration medium. Stir at 200 rpm at 37°C for 20 min. The dimethyl sulfoxide lowers the surface tension of the liquid medium and breaks the Van der Waals' forces holding antibody to antigen in the complex (the pH 2.3 glycine-HCl bufffer dissociates the complex by the change in pH). Next, put the immuno-stirrer into the phosphate-buffered saline regeneration medium, and stir at 200 rpm and 37°C for 10 min to wash away the organic solvent and acidic solution. At the same time, this restores the biological capacity of the immobilized antibodies.

When not in use, keep the immuno-stirrer in a borate buffered saline solution at 0-4°C to prevent microbial attack.

D. RESULTS

1. Optimization of the Assay Procedure

The inhibitory efficiency (I.E.) is defined as:

I.E. (%) = 100(T.A.- E.A.)/(T.A - R.A.)

where T.A. is total CK activity, E.A. is the experimental activity after immuno-stirring treatment, and R.A. is the residual activity (i.e., the B-subunit activity). Judging from the profile of I.E. vs. incubation time, maximum I.E. (100%) occurs after 25 min incubation at 37°C at 200 rpm (Fig. 2). To insure complete inhibiton during the pretreatment, 30 min was used; this is believed enough to cover the clinically useful range.

A stirring rate of 200 rpm was found to give the greatest inhibition. The I.E. decreases at a higher stirring speed (300 rpm) because the fast revolution caused a vortex on top of the stirrer which resulted in poor transport and bad contact between antigen and immobilized antibodies. It is hard to attain antigen-antibody equilibrium with fast rotation, and vigorous rotation distorts the conformation of the macromolecules (i.e. antigen and/or antibody). A temperature of 37°C was chosen for incubation because the physiological feasibility of maximum formation of antigen-antibody complex is greatest at this temperature.

2. Studies on the Regeneration of the Immuno-Stirrer

The regeneration procedure consists of two stages, the dissociation of the antigen-antibody complex, and the restoration of the inhibition capacity of the immobilized antibody. Regeneration was carried out at 37°C and 200 rpm. It is not necessary to regenerate

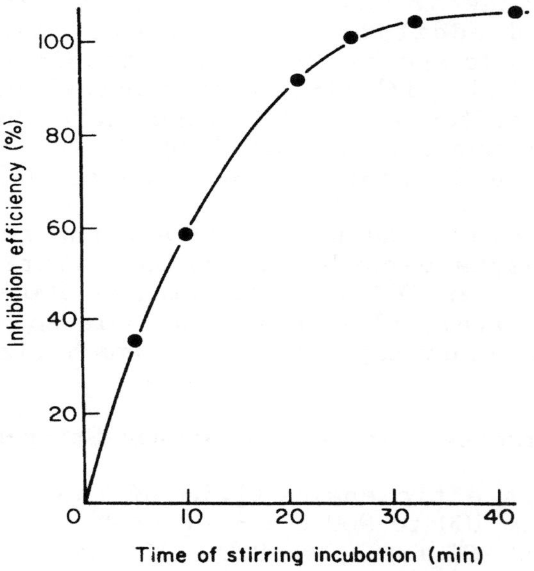

Fig. 2. The Effect of Time of Stirring Incubation on the inhibition efficiency.

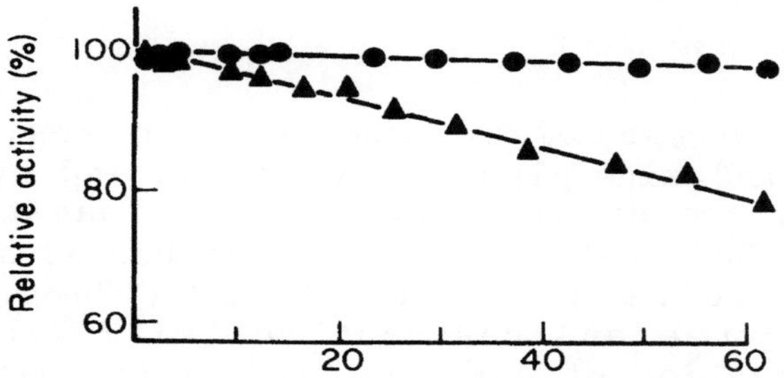

Fig. 3. The Long-term Operational Stability of the of the Immuno-stirrers. (o) Glutaraldehyde-alkylamine glass beads; (▲) cyanogen bromide cellulose beads.

the immuno-stirrer after each run. Since the inhibition capacity of the immuno-stirrer was sufficient to bind many more antigen molecules, the immuno-stirrer could be re-used five times without regeneration in routine clinical analysis.

The antibody molecules bind to the antigen through a variety of forces, which may include electrostatic, hydrophobic, Van der Waals' forces and hydrogen bonding. The dissociation of the complex can be achieved by a change in one or more parameters, such as ionic strength, pH, and addition of a competitive hapten, protein denaturants, chaotropic agents or surface tension- lowering agents. In the proposed method a 1:1 mixture of dimethyl sulfoxide and 0.05 M pH 2.3 glycine-HCl buffer was used to dissociate the complex. A 20-min incubation at 37°C and 200 rpm was found to be sufficient for this purpose. Prolonged incubation would cause denaturation of the immobilized antibodies.

The second medium used to regenerate the immobilized immuno-stirrer was a phosphate-buffered saline solution (0.1 M, pH 7). Incubation for 10 min at 37°C with stirring at 200 rpm was found to be sufficient to clean away all traces of organic solvent and to restore the biological inhibitory capacity of the immobilized antibodies.

3. Performance Characteristics of the Immuno-Stirrer

The inhibition efficiencies (I.E.) of the stirrer for a set of samples ranging from 100 to 900 U l^{-1} of CK-M were investigated. The efficiencies were 92-99% for up to 800 U l^{-1} of CK-M. At 900 U l^{-1} the I.E. dropped to 80%. Therefore the total inhibition capacity of an immuno-stirrer was estimated to be 800 U l^{-1}; the average inhibition efficiency was 97.8%.

A within-day precision study was done to evaluate the reliability of the immuno-stirrer for CK-MB determinations. Five determinations of normal (8.0 U l^{-1}) and of abnormal (187 U l^{-1}) samples were done; 5.5% and 4.6% were obtained as the coefficients of variation, respectively. Ten serum samples were pooled as the matrix base for a recovery study. Cardiozyme controls (of known value) were added to the pooled sera to give CK-MB values of 33-168 U l^{-1}. The results (Table III) show that the average recovery was 98.1%.

4. Stability Study

Two sets of immobilized immuno-stirrers, prepared from glutaraldehyde- alkylamine glass (g.a.g.) beads and cyanogen bromide-cellulose (c.b.c.) beads were used for a long-term operational stability study. Each set of stirrers was run 52 times over a two-month span. The results are shown in Fig. 3. The g.a.g. beads showed a good response curve and constant activity with a day-to- day coefficient of variation of 4.0%; no loss of activity was evident by the end of the test. The c.b.c. beads gave a decrease in activity after 10 assays over three days, with a coefficient of variation of 8.4%.

III. THE IMMUNOREACTOR

In order to improve the precision, inhibitory capacity and shorten the time required per assay, an immunoreactor was prepared and tried to replace the immunostirrer described above.

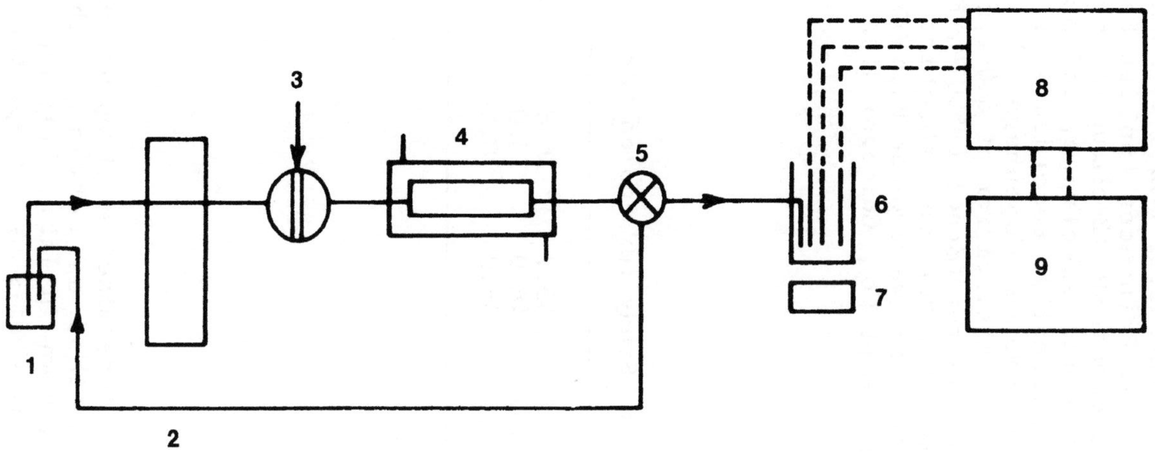

Fig. 4. Semi-On-line Measurement of CK-MB with electrochemical detection: 1, buffer solution, 0.1M TEA (pH 7.0); 2, peristaltic pump; 3, sample injection; 4, immunoreactor with circulating warm water; 5, three-way valve; 6, reaction cell; 7, magnetic stirrer; 8, polarographic analyzer, PAR 174 A; 9, chart recorder.

A. Materials and Methods

1. Apparatus

The system (Figure 4) contains a peristaltic pump (Model 403, Scientific Industries, Bohemia, NY), a Valco injector (Valco Instrumental Co., Houston, TX), an immunoreactor, a three-way valve (home made), Teflon tubing (0.038 in. i.d. and 1/16 in. o.d.), and an electrochemical detector. The activity of CK is measured from the rate of current change at a fixed potential (+0.36 V) maintained by a polarographic analyzer (PAR 174A, Princeton Applied Research Co., Princeton, NJ) using a platinum electrode (Beckman Instruments, Inc., Fullerton, CA), a saturated calomel electrode, and an auxiliary platinum foil electrode. The current is recorded on a strip chart recorder (Servograph REC 61, Radiometer, Copenhagen, Denmark). The solution in the cell was mixed by a ministirrer at 600 rpm controlled by an EC Motomatic Motor Control Stirrer (Model E 550 M, Electrochart Co., Hopkins, MN). The immunoreactor and reaction cell were maintained at 37 $\pm$ 1°C by a circulating warm water bath (Cenco Central Science Co. and Cole-Parmer Instrument Co., Chicago, IL).

Table III. Recovery Study for Pooled Sera

Found (U l^{-1})	18.0	34.2	45.7	76.8	160.0
Expected (U l^{-1})	18.0	33.0	48.0	78.0	168.0
Recovery (%)	–[a]	103.6	95.2	98.5	95.2

[a]Calibration standard.

2. Reagents and Solutions

The activity of CK-MB was determined spectrophotometrically by the modified Oliver-Rosalki method as described above. All ingredients and concentrations (shown in Table IV)) were recommended by the German Society for Clinical Chemistry, except for glutathione which interfered in the electrochemical assay - its concentration was reduced significantly. All chemicals were obtained from Sigma Chemical Co., St. Louis, MO, and were prepared daily prior to assay. The buffer solution was adjusted with 0.1 M triethanolamine (TEA)- HCl to pH 7.0. Potassium ferricyanide was obtained from Matheson, Colman and Bell, Norwood, OH; sodium m-periodate, benzamidine, sodium borohydride ($NaBH_4$), and diaphorase (from pig heart, activity 2 units per mg) were obtained from Sigma Chemical Co. CK-M inhibiting antibody was gift from D. H. Lang, E. Merck & Co., Darmstadt, Germany.

CK controls, of CK-MM and CK-MB, were obtained from Roche Diagnostics (Nutley, NJ).

3. Preparation of the Immunoreactor

CK-M inhibiting antibody was immobilized onto Glycophase-CPG 200 (pore diameter, 200 Å; particle size, 37-74 μm; Pierce Chemical Co., Rockford, IL). A 0.25-g portion of Glycophase-CPG was washed with double distilled water under vacuum in a desiccator to ensure the penetration of water into the pores. After washing, the Glycophase-CPG was placed in a vial containing 10 mL of $NaIO_4$ (6 mM) in 0.1 M acetate buffer, pH 5, and was gently shaken for 1 h at 25°C. The oxidized CPG support was washed on a Gooch funnel with distilled water. Antibody, 25 mg in 2 mL of 0.1 M borate buffer (sodium borate and HCl, pH 8.5) containing 1 mM of benzamidine, was mixed with CPG support and was shaken gently for 24 h at 5°C. After 40 min, 5 mg of $NaBH_4$ was added to reduce the Schiff base formed between the aldehyde groups on the support and the amino group of the antibody. The glycophase bound antibody was washed several times with 0.1 M phosphate buffer (pH 7.0) and then loaded onto a small Teflon tube (1/4 in. o.d.) by gentle suction. The reactor was stored in a borate buffer, saline (1%) solution (0.01 M, pH 8.0) at 5°C until use.

Table IV. Reagent Concentrations Used in the CK-MB Determination

Chemicals	Reagents Used/Assay, mM
Triethanolamine buffer (pH 7.0)	100
Creatine phosphate	70
Glucose	40
Glutathione	0.6
Magnesium acetate	20
Adenosine diphosphate (ADP)	2
Adenosine monophosphate (AMP)	20
Nicotinamide adenine Dinucleotide phosphate (NADP)	1.2
Glucose 6-phosphate dehydrogenase	4.8 IU
Hexokinase	4.8 IU

B. Procedure

Pump 3 mL of TEA bufer into the immunoreactor and inject 0.1 mL of either CK isoenzyme control (Roche Diagnostics, Nutley, NY) or sample serum into the reactor by a syringe through the three-way valve. Then recycle the mixture for 10 min at a flow rate of 1.5 mL/min. After completion of antigen-antibody binding, withdraw the solution and place into the three electrode cell compartment containing 1 mL of CK reagent (described in Table I), 0.3 mL of KCl (1.67 M), and 0.1 mL of diaphorase (5 mg/mL). Mix the solution with a ministirrer for 3 min at 37°C, add 0.1 mL of $K_3Fe(CN)_6$ solution (3.4×10^{-3} M) to initiate the electrochemical reaction, and measure the rate of current change at a fixed potential (+0.36) vs. SCE. Read the activity of CK-MB from a calibration curve of $\Delta i/\Delta t$ vs. activity of CK-MB (which is linear from 20 to 250 units/L).

C. Results

In this study, we used both Corning alkylamine glass (Corning Glass, NY) and Glycophase-CPG for antibody immobilization. The former gives better antibody stability, but nonspecific adsorption of proteins in the sample matrix occurs which interferes with the accuracy of the assay. Therefore, Glycophase-CPG is a better support for this purpose.

A comparison study on the inhibition efficiency of the immunoreactor and the immunostirrer was conducted. The immunoreactor permits a 100% binding of antigen in only 5 min compared to over 30 min using an immunostirrer (Figure 5).

Twelve serum samples, six normal and six abnormal, were assayed for within day and day to day precision by this method. Both showed the excellent precision of this new method. The higher coefficient of variation (CV) found in each case for the normal serum was due to the impairment of sensitivity of this method.

Table V. Recovery Study Using the Immunoreactor

Amt. found, U/L	Amt. expected, U/L	Recovery, %
20.0	20.0	
32.5	31.2	104.2
56.0	53.6	104.4
80.5	76.0	105.9
95.0	92.4	102.8
125.0	120.8	103.5
		104.2[a]

[a] Mean.

Linearity of the calibration plot of $\Delta i/\Delta t$ vs. CK-MB activity extendes from 20 to 250 units/L.

A serum control (CK-MB = 132 U/L) was added in varying amounts to a pooled serum containing the activity of CK-MB, 22 U/L. The average recovery was 104.2% (Table V). Similar recoveries were obtained upon addition of serum control (CK-MM, 100 U/L, and CK-MB, 132 U/L) to this same CK-MB pooled serum. This showed that CK-MB can be distinguished from CK-MM.

The immunoreactor was evaluated for stability in long period operation. Fifty-six assays over a 40-day period could be performed without significant change in binding functionality (Figure 6). Equivalent results are obtained if 60 assays are performed over 2-5 days. After this period, the CK-MB activity increases due to small inhibitory capacity of the reactor. The decrease in capability of the reactor to bind CK-MB is due mostly to the decrease of free binding sites of the immobilized CK-M inhibiting antibody. Attempts were made to regenerate the reactor using the procedure reported in a previous paper. However, the reactor did not give the same reactivity as when it is new.

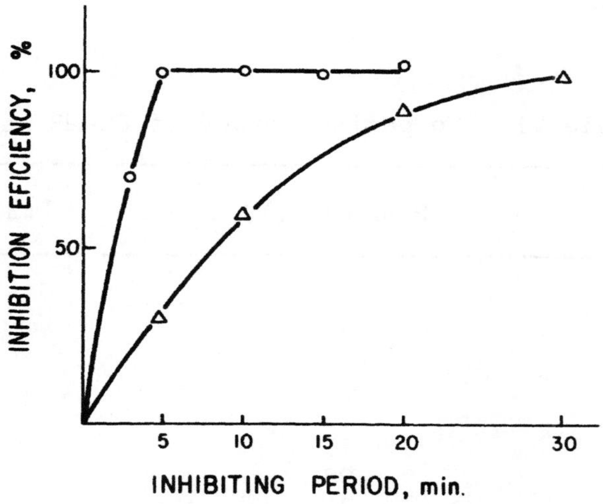

Fig. 5. Comparison of the Inhibition Efficiency of the Two Immobilized Antibody Systems: (O) Immunoreceptor; (Δ) Immonoreceptor.

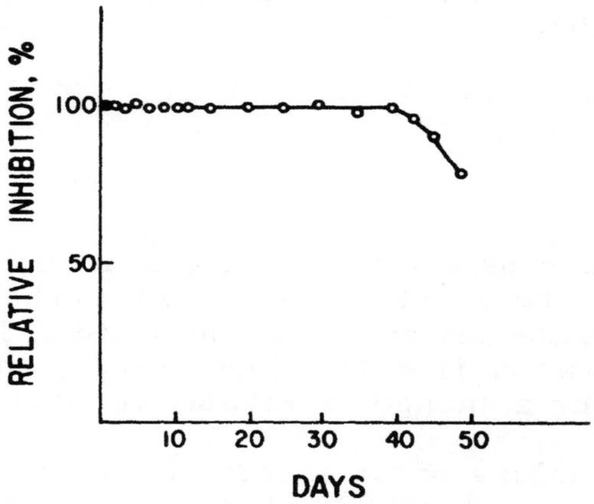

Fig. 6. Stability of the Immunoreactor.

Seventeen fresh serum samples, with known CK-MB values determined by a UV-electrophoresis method were obtained from Methodist Hospital, Houston, TX, and were analyzed for the activity of CK-MB by the method developed. The resulting linear regression curve obtained was: Y (hospital) = 1.06X (reactor) + 3.60, with a correlation coefficient of 0.997 (Table VI).

Table VI. Comparison Study of CK-MB Assay[b]

Sample No.	Hospital ($\bar{Y}$)	Immunoreactor ($\bar{X}$)[a]
1	80	75
2	58	53
3	267	245
4	107	99
5	59	50
6	70	62
7	98	98
8	71	65
9	23	20
10	102	98
11	46	40
12	25	20
13	53	45
14	75	68
15	57	52
16	44	40
17	25	21

[a] Average of three-five runs. [b] Linear regression analysis: $\bar{Y} = 1.06\ \bar{X} + 3.60$, r - 0.997.

IV. CONCLUSIONS

There has been considerable progress in clinical pathology for health care and treatment of disease. Nevertheless, the important question is how precise and accurate can a medical decision be made on the basis of information from the laboratory. And also how fast and easily can we detect a change of metabolic conditions caused by the onset of disease.

Through the coupling of an electroactive pair $Fe(CN)_6^{-3}/Fe(CN)_6^{-4}$ to NADH produced from the HK-GPD system, the first procedure can be employed for the determination of creatine kinase with the use of soluble GAHCK-M and a platinum electrode. The method is rapid, simple, sensitive and specific; the linear range is extended to 875 U/L; and positive myocardial infarction cases can be easily determined in 10 min. This is very helpful to physicians for timely diagnosis of patients with myocardial infarction. Even though this procedure appears to meet all requirements claimed previously, an excess amount of GAHCK-M must be added into the assay system in order to assure complete inhibition of possible high levels of CK-MB in serum, thus greatly increasing the cost per assay. Nevertheless, this inhibition

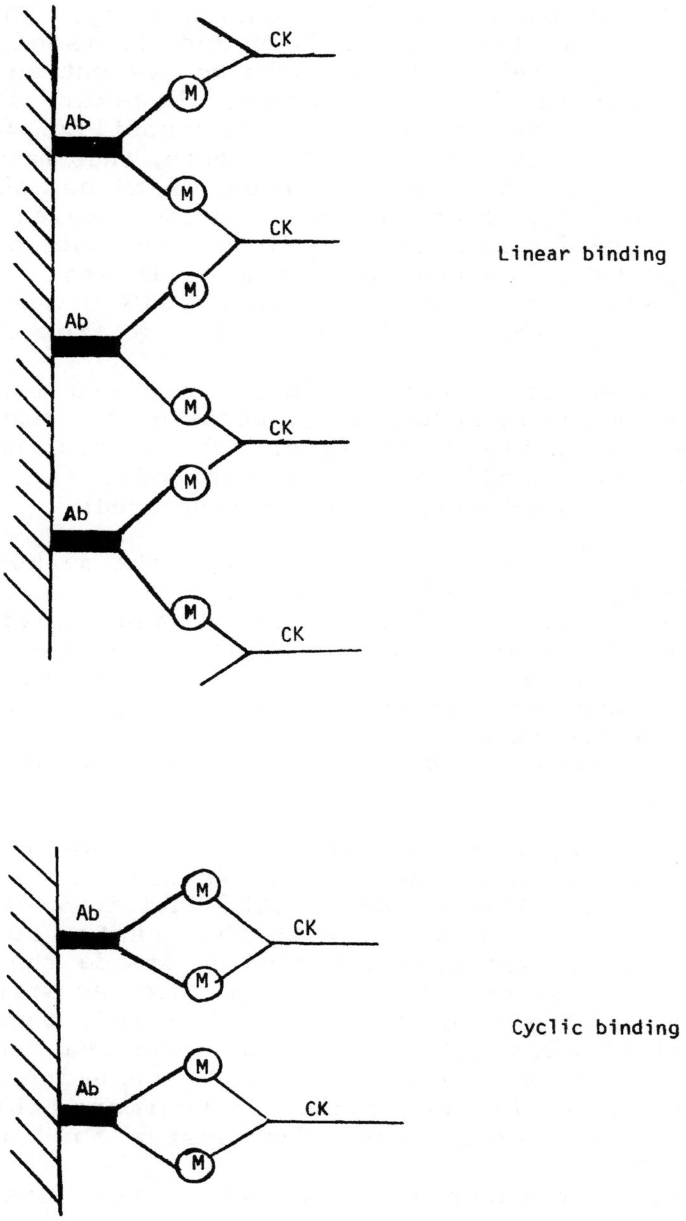

Ab- Antibody ; CK- Creatine Kinase

Fig. 7. Binding Schemes for CK-Antibody.

procedure will be very useful in the future, if an inexpensive GAHCK-M monoclonal antibody can be produced through the use of either hybridoma or gene splicing technology.

Immobilized GAHCK-M was tried in this laboratory in an attempt to reduce the cost per assay and improve the precision of the first method developed. It was found in these studies that immobilized GAHCK-M appeared to bind CK-MM tightly, but not CK-MB, and hence can be used for the separation of CK-MB during assay. The mechanism and type of interaction between GAHCK-M and CK isoenzymes were not investigated. It is believed that both monovalent and bivalent type interaction occur in the assay system. However, it appears most likely that CK-MM binds bivalently to the immobilized GAHCK-M, to form a linear or a cyclic complex (Fig. 7) or both, thus being held tightly onto the immobilized antibody. Whereas, CK-MB has only one subunit available for binding, the monovalent binding complex between the CK-MB and immobilized GAHCK-M tends to dissociate under assay conditions. This is due probably to the bulky B subunit which creates steric hinderance between the rigid immobilized GAHCK-M and the M subunit, thereby forcing the whole CK-MB molecule out from the immobilized GAHCK-M matrix.

We have also demonstrated that the CK-MB isoenzyme can be determined easily and reproducibly by means of an immuno-stirrer, and that problems previously encountered, such as cost and precision of the assay and the stability of the antibody, can be eliminated. However, the following drawbacks are encountered:

1. The reaction rate and precision of the method depend on how well the stirring speed can be regulated.

2. The capacity to load immobilized antibody into an immunostirrer is limited and frequent change or renewal of immunostirrer is necessary if hundreds of assay needed to be done continuously. Each immunostirrer, freshly prepared or renewed, must be calibrated before use.

3. The time required to complete an assay, 30-40 minutes, is too long.

An immunoreactor, with a semi-on-line setup, was next used in order to overcome the above drawbacks. As shown in the experimental section, the time required to obtain 100% inhibition has been reduced from 30 min to 5 min. This is due to the greater surface area and more reactive sites available for binding inside the reactor. With semi-on-line setup, several tedious steps, such as stirring, solution transferring, and long assay time are eliminated, thus improving the precision. It is also anticipated that the reaction time can be shortened and sensitivity increased, by reducing the volume of circulating solution in the system. Furthermore, this setup can be easily adapted to a continuous flow system for fully automated analysis of CK-MB.

New physical methods of diagnosing heart disease, such as Echocardiograms - ultrasonic images of heart and blood vessels and nuclear magnetic resonance imaging technique, are now on the horizon. Although these techniques may become very useful in assessing myocardial infarction in the future, intensive use of chemical test for CK-MB is anticipated, since the immunochemical test is rapid, simple, sensitive and specific for early detection of myocardial infarction. It is predicted that the measurement of CK-MB with immunoseparation and electrochemical detection will become a useful and important test, when an inexpensive monoclonal GAHCK-M becomes available and the procedure is fully automated.

REFERENCES

Blaedel, W. A., and Jenkins, R. A., 1975, Studies of the electrochemical oxidation of reduced nicotinamide adenine dinucleotide, Anal. Chem., 47:1337-140.

Cheng, F. S., and Christian, G. D., 1977, Amperometric measurement of enzyme reactions with an oxygen electrode using air oxidation of reduced nicotinamide adenine dinucleotide, Anal. Chem., 49:1785-1788.

Cheng, F. S., and Christian, G. D., 1978, Enzymatic determination of blood ethanol with amperometric measurement of rate of oxygen depletion, Clin. Chem., 24:621-626.

Galen, R. S., 1975, The enzyme diagnosis of myocardial infarction, Prog. Hum. pathol. 6:141-145.

Guilbault, G. G., 1978, "Theory, design and biomedical applications of solid state chemical sensors", P. W. Cheng et al, Ed. CRC Press, Inc., p. 193, Florida.

Guilbault, G. G., 1985, Analytical uses of immobilized enzymes, Marcel Dekker, Inc., New York.

Konttinen, A. and Somer, H., 1973, Specificity of serum creatine kinase isoenzymes in diagnosis of acute myocardial infarction, Br. J. Med., i:386-392.

Peck, H. D. Jr., Deacon, T. E., and Davidson, J. T., 1965, Studies on adenosine 5'-Phosphosulfate reductase from desulvibrio desulficans and thiobacillus thioparus. I. Assay and Purification, Biochim. Biophys. Acta, 96:429-435.

Peel, J. L., 1972, in J. R. Norris and D. W. Ribbons (Eds.), Methods in Microbiology Academic Press, New York.

Rao, P. S., Evans, R. G., and Mueller, H. S., 1977, Experimental determination of activation potentials of creatine kinase isoenzymes in human serum and their significance, Biochem. & Biophys. Research Comm., 78:648-652.

Roberts, R., Gowda, K. S., Ludbrook, P. A. and Sobel, B. E., 1975, Specificity of elevated serum MB creatine phosphokinase activity in the diagnosis of acute myocardial infarction, Am. J. Cardiol, 36:433-440.

Smith, M. D., and Olson, C. L., 1974, Differential amperometric measurement of serum lactate dehydrogenase activity using bindschedler's green, Anal. Chem., 46:1544-1548.

Smith, M. D., and Olson, C. L., 1975, Differential amperometric determination of alcohol in blood or urine using alcohol dehydrogenase, Anal. Chem., 47:1074-1077.

Szasz, G., Gruber, W., and Bernt, E., 1976a, Creatine kinase in serum. 1. Determination of optimum reaction conditions, Clin. Chem. 22:650-655.

Szasz, G., Gerhardt, W., Gruber, W., and Bernt, E., 1976b, Creatine kinase in serum. 2. Interference of adenylate kinase with the assay, Clin. Chem. 22:1806-1810.

Szasz, G., Gerhardt, W., and Gruber, W., 1977, Creatine kinase in serum. 3., Further study of adenylate kinase inhibitors, Clin. Chem. 23:1888-1891.

Thomas, L. C., and Christian, G. D., 1975, Voltammetric measurement of reduced nicotinamide-adenine nucleotides and application to amperometric measurement of enzyme reaction, Anal. Chim. Acta, 78:271-275.

Thomas, L. C., and Christian, G. D.,1976, Amperometric measurement of hexacyanoferrate (III)-coupled dehydrogenase reactions, Anal. Chim. Acta, 82:265-270.

IMMUNOSENSOR BASED ON LABELED LIPOSOME ENTRAPPED-ENZYMES

Makoto Haga

Faculty of Pharmaceutical Sciences
Science University of Tokyo
12, Ichigaya Funagawara-machi, Shinjuku-ku, Tokyo 162, Japan

INTRODUCTION

There has been great interest recently in the development of new, simple, sensitive, specific immunoassays for the quantitative determination of small molecules in biological fluids. Liposome immunoassay, a new method which utilizes either complement-dependent immune lysis (Uemura and Kinsky, 1972), cytolysin-mediated lysis (Freytag and Litchfield, 1984; Litchfield et al., 1984) or lysis induced by lateral phase separation (Ho and Huang, 1985), has been used for detecting specific antibodies and small molecules. Detection of the release of an entrapped aqueous maker is used to monitor the lysis of liposomes. A number of markers have been utilized in this assay, such as glucose (Kinsky et al., 1969), spin labels (Humphries and McConnell, 1974; Wei et al., 1975; Hsia and Tan, 1978; Chan et al., 1978), fluoresent probes (Smolarsky et al., 1977; Geiger and Smolarsky, 1977) and electroactive compounds (Shiba et al., 1980). We encapsulated an enzyme as a release marker and combined the liposome immunoassay with an enzyme immunosensor (Haga et al., 1980, 1981).

The enzyme immunosensor, developed by Aizawa et al. (1976), offers an instantaneous assay method and a simple, easy to use system. However, its sensitivity was insufficient to measure drugs at the nanogram level. In this report, I have described a new simple and sensitive liposome immunosensor, developed by combining the advantages of liposome immunoassays and enzyme immunosensor.

The method is based on the liposome lysis induced by a specific antiserum and complement which is monitored by the release of entrapped enzymes. The principle of the liposome immunosensor is shown schematically in Fig.1. An unknown concentration of antigen, a known quantity of an antibody, a complement and a sensitized liposome are incubated. Competition occurs between the free antigen and the hapten which is bound on the liposome (Fig.1a), as each tries to bind to the antibody. The complement activated by the antigen-antibody reaction induces the lysis of the liposome and entrapped enzyme (HPO) is then released (Fig.1b). The enzymatic activity is directly proportional to the immune lysis of liposome, which relates inversely to the concentration of free antigen. When a pulse of substrate is injected, the enzyme catalyzes the following reaction:

$$2NADH + O_2 + 2H^+ \longrightarrow 2NAD^+ + 2H_2O$$

Oxygen in the solution is consume during the reaction (Cheng and Christian,

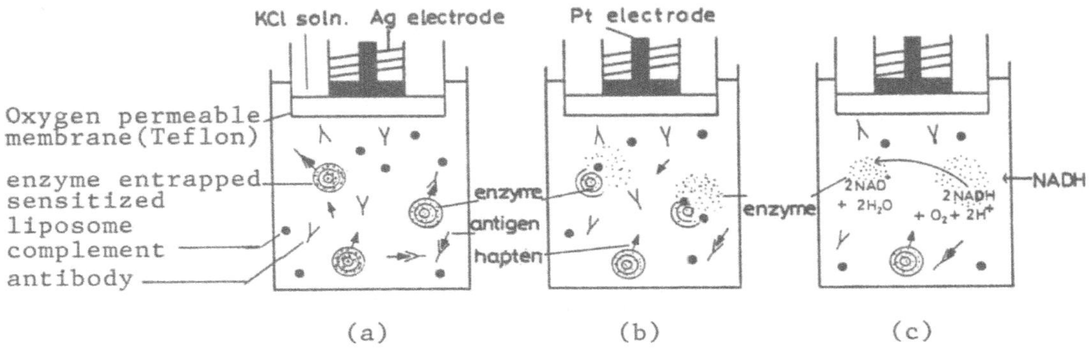

Fig. 1. Principle of liposome immunosensor. (a)immunological reaction, (b)lysis by complement and (c)enzymatic reaction.

1977, 1979). The depletion of oxygen can be detected by a Clark-type oxygen electrode (Fig.1c) while the amperometric current can be registered on a chart recorder. The effect of oxygen back-diffusion from the atmosphere on the initial rate is negligible due to the rapid reaction that takes place.

The important feature of the liposome immunosensor is that the antigen-antibody complexing reaction is amplified twice by both the liposome lysis and the enzymatic reaction. This method also has the advantage of eliminating the necessity of using a radiolabeled antigen. Theophylline, widely used in the treatment of acute and chronic respiratory diseases and recently suggested for the prevention of apnea in infants, was chosen as the model molecule for the assays. In a system using theophylline as a lipid hapten, 1 ng/ml of theophylline was detected.

MATERIALS AND METHODS

Reagents

5,6-Diamino-1,3-dimethyluracil was purchased from Aldrich Chemical Company (Milaukee, Wisc.). Additional reagents were obtained from the following sources: bovine serum albumin (BSA), L-α-phosphatidylethanolanine (PE), cholesterol (Chol), dicetylphosphate (DCP), β-nicotinamide adenine dinucleotide reduced (NADH) and horseradish peroxidase (HPO) from Sigma Chemical Company (St. Louis, Mo.); guinea pig complement from Miles Laboratories, Inc. (Elkhart, Ind.); theophylline and theobromine from Nakarai Chemicals, Ltd. (Kyoto, Japan); caffeine from Kanto Chemicals Company, Ltd. (Tokyo); hypoxanthine and uric acid from Tokyo Kasei Kogyo Company, Ltd. (Tokyo); and [8-^{3}H] theophylline from the Radiochemical Centre (Amersham, England). Complete Freund's adjuvant was obtained from Iatron Laboratories (Tokyo). Unisil was a product of Clarkson Chemical Company, Inc. (Williamsport, Penn.). Other chemicals used were commercial products of high purity, and all operations utilized freshly redistilled water.

Preparation of Antigen

8-(3-Carboxypropyl)-1,3-dimethylxanthine was synthesized according to the method described by Cook et al. (1976). Briefly, 5,6-diamino-1,3-dimethyluracil (15 g, 88.2 mmol) and glutaric anhydride (20.1 g, 177 mmol) were refluxed in 150 ml of N,N-dimethylaniline under nitrogen for 2.5 h under a Dean-Stark trap. Another 50 ml of N,N-dimethylaniline was added and refluxed for another 30 min. The mixture was cooled and filtered. The resulting solid was washed with benzene and recrystallized from water to yield crystals (10.54 g, 44.9%). The N-hydroxysuccinimide ester method (Anderson et al., 1964) was used for the coupling reaction of hapten to BSA. The hapten (144.7 mg) was dissolved in 1.5 ml of dimethylformamide and N-hydroxysuccinimide (62.6 mg) and dicyclohexylcarbodiimide were added. The mixture was then stirred for 1 h and a freshly prepared solution of BSA (200 mg) in 10 ml of phosphate buffer (pH 7.0) was added. The solution was maintained at 25°C for 4 h and then dialyzed. The theophylline-BSA conjugate was isolated and lyophilized. Incorporation of hapten was determined by differential ultraviolet spectrophotometry (Erlanger et al., 1975) at 274 nm. The number of moles of theophylline per mole of BSA was about 18.

Immunization

Male New Zealand white rabbits were immunized with antigen (2 mg), which was dissolved in 0.9% NaCl solution (1 ml) and emulsified with an equal volume of Freund's complete adjuvant. The immunogen was given subcutaneously every 2 weeks and blood was collected 4 months after the initial immunization. Immunological potencies were determined by measuring the binding of [8-^{3}H]theophylline. A mixture of diluted serum solution (100 μl), [8-^{3}H]theophylline solution (50 μl), normal rabbit serum (100 μl) and 0.2 M Tris-HCl buffer pH 7.6 (250 μl) was incubated at 4°C for 24 h, then free ligand and antibody-bound ligand were separated by the addition of saturated ammonium sulfate solution (500 μl). After centrifugation at 1500 x g for 30 min at 4°C, the precipitate was dissolved in 500 μl of distilled water. The solution was dissolved in 6 ml of scintillation fluid (6 g of DPO, 0.5 g of POPOP, 333 ml of Triton X-100, 667 ml of toluene) and its radioactivity was measured in a liquid sintillation counter (Beckman LS 7000). The final dilution ratios of antibody, which bound 50% of the ^{3}H-theophylline added, ranged from 1 : 800 to 1 : 1300.

Synthesis of Sensitizer

A theophylline-PE composite was prepared according to the method described by Six et al. (1973) by acylation of phosphatidylethanolamine with the acid chloride of theophylline hapten. Theophylline hapten and thionyl chloride were dissolved in redistilled benzene and the solution was refluxed for 1 h and the reaction mixture was dried under reduced pressure to remove excess thionylchloride. PE (dissolved in redistilled benzene containing triethylamine) was then added to the dried acid chloride of theophylline hapten and the mixture was stirred at room temperature under a nitrogen atmosphere for 20 h. Theophylline-PE was isolated by chromatography with preparative silica gel plates after initial extraction in which most of the excess acid chloride of theophylline hapten was converted to a compound soluble in methanol-water. Silica gel plates were developed in a system of chloroform-methanol-water (70:30:5). Only one of the five bands (R_f of about 0.7) was transfered to a Unisil chromatography column. The column (dimensions: 1.5 x 25 cm) was developed by using 350 ml of an 8 : 1 chloroform-methanol mixture. The fractions containing theophylline-PE composite (checked by UV spectrophotometry at 278 nm) were collected and dried under reduced pressure.

LIS Solution

The liposome immunosensor (LIS) solution was 0.1 M Tris-HCl (pH 8.0) which contained 0.05 M NaCl, 1 mM $MgCl_2$, 0.15 mM $CaCl_2$ and 0.13 mM $MnSO_2$. Mn^{2+} is necessary to perform the aerobic oxidation of NADH by HPO and Ca^{2+} and Mg^{2+} are necessary for the immune lysis of liposomes by complement.

Preparation of Sensitized Liposomes

Actively sensitized multilamellar liposomes were prepared using an appropriate amount of sensitizer and a basic lipid mixture containing PC, Chol and DCP (molar ratio, 1 : 0.75 : 0.1). An aliquot of the lipid mixture, containing 20 μmol of PC, was dissolved in chloroform and added to a small conical flask. The solvent was removed by rotary evaporation and the dried lipid film inside the flask was then dispersed in 2 ml of LIS solution containing HPO. For removal of the untrapped enzymes, the liposomes were washed twice by ultracentrifugation (100000 x g for 60 min). The pellet was resuspended in 1 ml of LIS solution and stored in a nitrogen atmosphere at -20°C. From one batch of liposomes, approximately 10 assays could be performed. The transition temperature of these liposomes was not measured, but it was assumed to be below 0°C (Brendzel et al., 1980).

Assay Procedure

The sample chamber of the apparatus used in this work was sealed to prevent atmospheric oxygen from entering the system and was thermostated at 30°C with a Luada Model K2RD temperature controller. A homemade Clark-type oxygen electrode with a Beckman oxygen-permeable Teflon membrane was used to monitor the oxygen concentration. The sample solution (400 μl) containing theophylline solution (40 μl), liposomal suspension (100 μl), anti-theophylline antiserum (40 μl), guinea pig serum (as a source of complement, 40 μl) and LIS solution (180 μl) was incubated at 30°C for 30 min. The sample solution was then stirred with a magnetic stirrer at 300 rpm. After the output current reached a steady state, a pulse of the substrate (10 μl of 200 mM NADH) was introduced through a rubber plug. The output current, reflecting the changed concentration of oxygen, was amplified and registered on a Hitachi Model 056 chart recorder. From the initial rate of current decrease (-di/dt), the concentration of theophylline was determined.

RESULTS

Sensitizer-concentration Dependence of Immune Lysis

Theophylline-PE can actively sensitize liposomes prepared with PC, Chol, and DCP. It should be noted that the anti-theophylline antibodies used in the present study seem to be exclusively of the IgG class. Accordingly, a minimum of two lipid antigen-antibody complexes in close proximity on the membrane surface is required to activate the complement sequence. Fig. 2 shows the effect of the sensitizer concentration relative to PC (S, mol %) on the immune lysis of liposome. Saturation was achieved with 0.6% theophylline-PE, i.e., 1 molecule of antigen for every 167 molecules of PC. This ratio is fairly consistent with the value derived for the active sensitization of liposomes by ε-dinitrophenylated aminocaproyl phosphatidylethanolamine (Chan et al., 1978), where maximum sensitization was obtained at S=0.4. The 50% lysis was attained at S=0.4.

Effect of NADH and HPO Concentration on Sensitivity

Under the optimal conditions for immune lysis of liposomes, sensitivity depends strongly on the amount of NADH added to the reaction mixture and the amount of HPO incorporated in the liposomes. The maximal concentration of NADH in the Tris-HCl buffer was about 200 mM at pH 8.0 and 30°C. The dissolved oxygen did not noticeably influence the output current when 10 μl of NADH was injected into the sample solution. We therefore, examined the effect of HPO concentration on the maximum rate of oxidation in the presence of 10 μl of 200 mM NADH. The decreasing rate of the current leveled off at a saturation point of about 30 IU of HPO, at which point the reaction became nearly zero order with respect to the enzyme. In the following experiment we used 15 IU of HPO, taking into consideration both the sensitivity and the consumption of the enzyme.

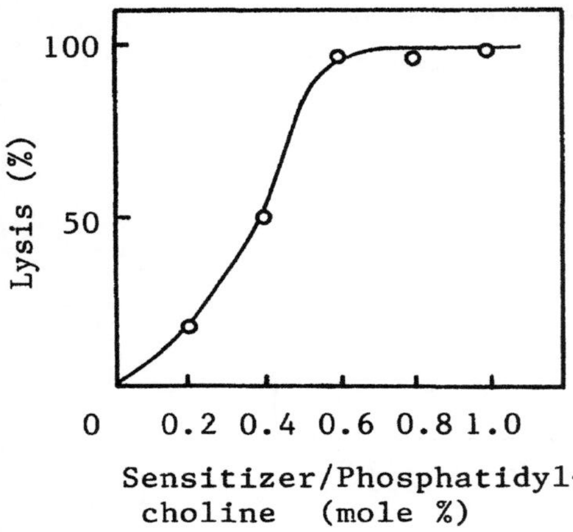

Fig. 2

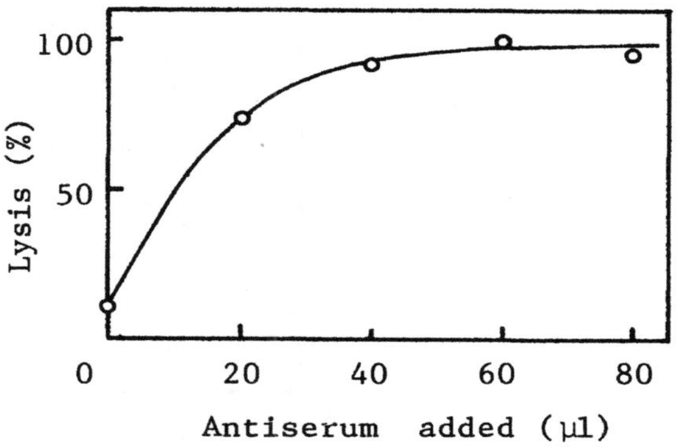

Fig. 3

Fig. 2. Dependence of immune lysis of liposomes on sensitizer concentration. Liposomes were prepared with various amounts of theophylline-PE. Liposomal suspension (100 µl), 40 µl of anti-theophylline antiserum, 40 µl of guinea pig serum, and LIS solution were incubated at 30°C, pH 8.0 for 30 min. The degree of immune lysis was measured from the initial rate of current decrease after addition of 10 µl of 200 mM NADH.

Fig. 3. Effect of specific antiserum concentration of immune lysis of liposomes. Liposomal suspension (100 µl, S=0.6), 40 µl of guinea pig serum, LIS solution, and varying amounts of antiserum as indicated on the abscissa were incubated at 30°C, pH 8.0 for 30 min. The degree of immune lysis was measured after addition of 10 µl of 200 mM NADH.

<u>Effect of Antiserum and Guinea Pig Serum Concentration on Immune Lysis</u>

In the presence of complement components in native serum, immune lysis of sensitized liposomes was induced by a sera containing a specific anti-thophylline antibody. Because of the presence of complement components in native antiserum, the sera was then heated for 30 min at 56°C to inactivate the cytolytic complement activity. Without the presence of guinea pig complements, heated anti-theophylline antiserum did not cause lysis of liposomes. The maximal range of specific immune lysis of liposomes was determined by comparing the HPO activity, in the case of immune lysis, with that of the case of complete lysis by freezing and thawing. The specific immune lysis of liposomes released about 80% of the total entrapped HPO. This behavior can probably be ascribed to the multicompartment nature of the liposomes.

To represent the true end point of specific immune lysis, we define this maximum as 100% lysis. As shown in Fig.3, approximately 40 µl of the antiserum was required to produce 100% lysis. This result suggests that a higher antiserum concentration would be favorable for the formation of antigen-antibody complexes. Fig.4 shows the effect of guinea pig serum as a source of complement on the immune lysis of the liposomes. Approximately 40 µl of guinea pig serum was required to produce 100% lysis. Accordingly, higher concentration of guinea pig serum would favor the activation of C1 components.

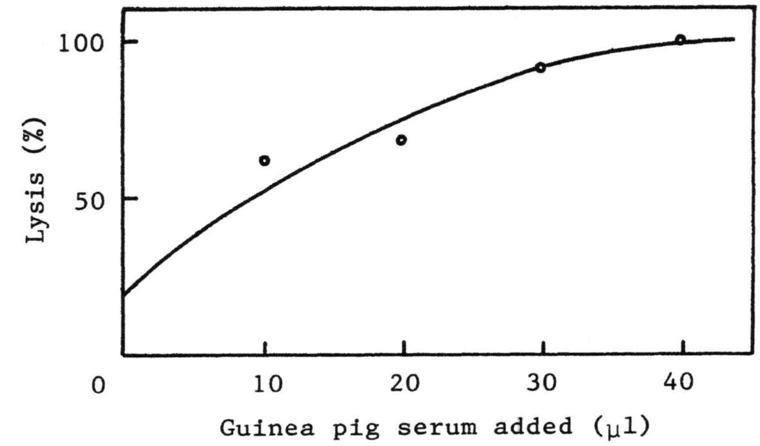

Fig. 4. Effect of guinea pig serum (complement) concentration on immune lysis of liposomes. Liposomal suspension (100 μl; S=0.6), 40 μl of anti-theophylline antiserum, LIS solution, and an increasing volume of guinea pig serum were incubated at 30°C, pH 8.0 for 30 min. The degree of immune lysis was measured after addition of 10 μl of 200 mM NADH.

Effect of Incubation Time, Temperature and pH

The effect of incubation time on immune lysis is illustrated in Fig.5. Under optimal conditions, the immune lysis of theophylline-PE sensitized liposomes was induced almost within 15 min, which is consistent with the results obtained in the investigation of kinetics of glucose release from actively sensitized liposomes by Six et al. (1973), Uemura et al. (1972). Our results clearly indicate that by allowing a short incubation time,

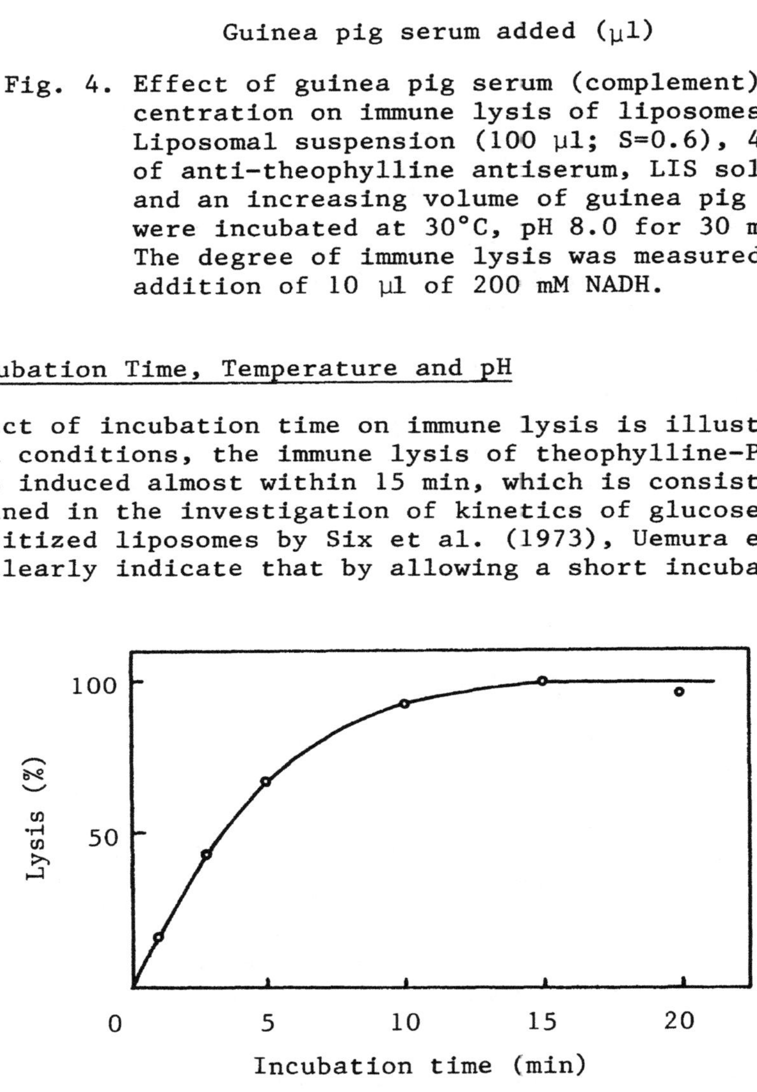

Fig. 5. Incubation time dependence of immune lysis of liposomes. Liposomal suspension (100 μl ; S=0.6), 40 μl of anti-theophylline antiserum, 40 μl of guinea pig serum and LIS solution were incubated at 30°C, pH 8.0 for various lengths of time. The degree of immune lysis was measured after addition of 10 μl of 200 mM NADH.

liposome immunosensor can be applied to the rapid assay of theophylline. To obtain high sensitivity and reproducible results, the selection of pH and incubation temperature is important. The maximum lysis observed at about 30°C provides the basis for optimal assay condition. The optimal pH, about 8.0 in the assay solution is consistent with the value obtained for NADH oxidation by HPO in Tris-succinate buffer (Cheng et al., 1977).

Sensitivity of Liposome Immunosensor

From the results described above, the optimal assay conditions were determined to be as follows: 100 µl of liposomal suspension (S=0.6), 40 µl of heated antiserum, 40 µl of guinea pig serum, 180 µl of LIS solution and theophylline standard or an unknown sample in a total volume of 400 µl. The mixture was then incubated for 30 min at 30°C. After incubation, 10 µl of 200 mM NADH was introduced and the output current was recorded at 30°C. The sensitivity of the liposome immunosensor is demonstrated in the calibration curve developed by using standard solution of theophylline (Fig.6). The initial rate of the current decrease, which proportionally increased with the degree of immune lysis, was inversely related to the concentration of theophylline. In the range of 4×10^{-9} to 2×10^{-8} M, the rate of current decrease was proportional to the logarithmic concentration of theophylline. The calibration curve shows that the method can detect concentration of theophylline as low as 0.7 ng/ml.

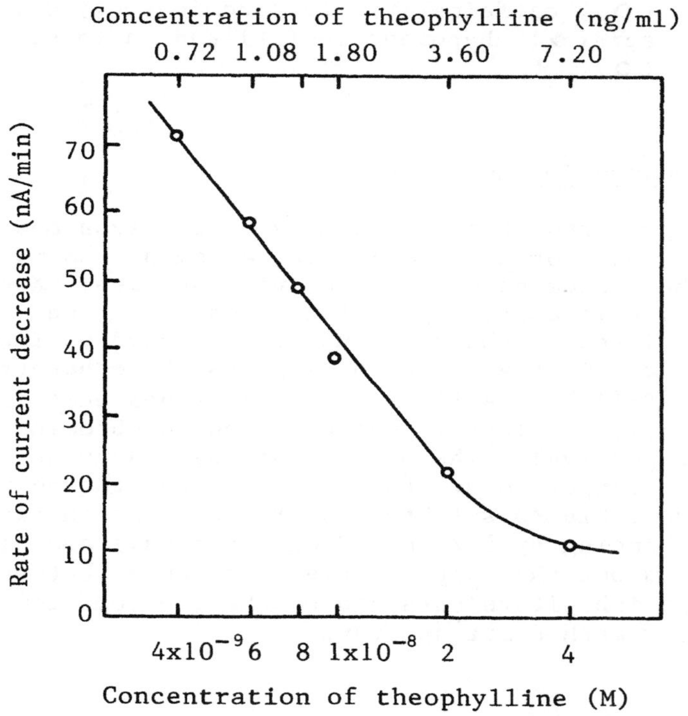

Fig. 6. Standard curve for assay of theophylline. Theophylline calibrators (40 µl) were mixed with 100 µl of liposomal suspension (S=0.6), 40 µl of guinea pig serum, 40 µl of anti-theophylline antiserum, and LIS solution. The mixture was incubated at 30°C, pH 8.0 for 30 min and then 10 µl of 200 mM NADH was added.

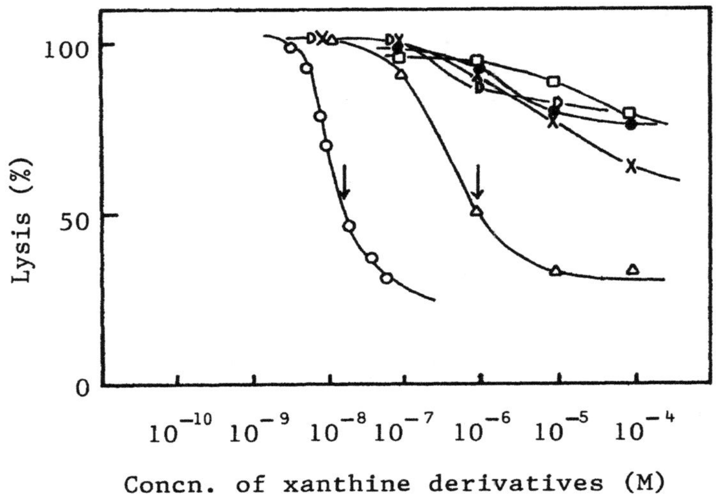

Fig. 7. Cross-reactivity of xanthine derivatives in liposome immunosensor. The procedure is identical to that described in the legend of Fig.6 except that 40 μl aliquots of varying concentration of inhibitors were added instead of theophylline calibrators. The vertical arrows indicate the 50% inhibition of immune lysis. The symbols used are as follows: theophylline (○), caffeine(△), theobromine(×), uric acid(●), hypoxanthine(□) and xanthine (D).

Specificity of Liposome Immunosensor

The specific property of the immune lysis of liposomes sensitized with theophylline-PE, as monitored by the liposome immunosensor, was further demonstrated by the following cross-reactivity studies. Xanthine derivatives, theobromine, uric acid, hypoxanthine, caffeine and xanthine, which are structurally related to theophylline, were tested as inhibitors. The results are summarized in Fig.7, which displays the concentration required to inhibit immune lysis by 50% (IC_{50}). It was found that caffeine was the most effective inhibitor. IC_{50} of caffeine and theobromine was 1×10^{-6} and 1×10^{-4} M, respectively. The other available xanthine derivatives were even less efficient competitors. For a comparison, the degree of cross-reactivity of these inhibitors (obtained from the concentration required to inhibit immune binding by 50% in radioimmunoassay) are listed in Table I. It should be pointed out that appreciable differences between the cross-reactivity of these inhibitors obtained by the liposome immunosensor and the radioimmunoassay method are apparent.

DISCUSSION

The liposome immunosensor described here combines the advantages of the spin membrane immunoassay and the enzyme immunosensor. Because of the amplification of the antigen-antibody complexing reaction by liposome lysis and enzymatic reaction, a more sensitive system was developed. In our study using the same anti-theophylline antiserum, the sensitivity range of liposome immunosensor was $\sim 4 \times 10^{-9}$ M compared with $\sim 10^{-4}$ M for both the spin membrane immunoassay and the radioimmunoassay. Accordingly, the

Table I. Specificity of Antisera to Theophylline

Inhibitor	Percentage cross-reactivity[a)]
Theophylline	100
Caffeine	8.5
Theobromine	0.1
Xanthine	< 0.0001
Hypoxanthine	< 0.0001
Uric acid	< 0.0001

[a)] Values were determined from the concentrations required to inhibit immune binding by 50% in radioimmunoassay.

sensitivity of the liposome immunosensor may be limited by its affinity rather than the electrode system. The specificity of liposome immunosensor was confirmed by the measurement of cross-reactivities of structurally related haptens. Comparing the cross-reactivities of caffeine and theobromine (obtained from the values of 50% lysis in liposome immunosensor) with those of 50% binding in the radioimmunoassay, the cross-reactivities were greatly improved with the liposome immunosensor. The reasons, although unclear, may be attributed to the apparent decrease in the free inhibitor concentration in the presence of liposomes.

A significant advantage of the liposome immunosensor is the elimination of the separation procedure required in radioimmunoassays and enzyme immunosensors. Furthermore, the liposome immunosensor is easier to manage and removes the need to use expensive equipment and the use of radioisotopes. Incubation takes about 30 min, measurement is virtually instantaneous and analytical results can be displayed directly in electric signal output. For practical use, the amount of sample used is an important factor. The liposome immunosensor system requires a smaller amount of sample (total 400 µl) than enzyme immunosensor (∿2 ml).

A disadvantage of the liposome immunosensor was the difficulty in preparing uniform batches of sensitized liposomes. If we use small unilamellar vesicles instead of multilamellar vesicles, a more uniform size distribution and complete release of enzyme can be obtained. However, its sensitivity may be decreased because the small unilamellar vesicles entrap fewer enzymes than multilamellar vesicles. These problems have been solved in several subsequent studies by using large unilamellar vesicles (Freytag, 1984; O'Connell, 1985).

The sensitivity of the liposome immunosensor may be improved by trapping enzymes that have a high turnover number. Furthermore, with use of a monoclonal antibody, a higher sensitivity may be obtained. With appropriate modifications, the liposome immunosensor offers a practical, simple, rapid and sensitive method for measuring the serum concentration of drugs.

SUMMARY

A new, simple and sensitive liposome immunosensor was developed by combining the advantages of the spin membrane immunoassay and the enzyme immunosensor. Sensitive detection was possible because of the double amplification of the antigen-antibody complexing reaction by the liposome lysis and the enzymatic reaction. The concept of double chemical amplification that we have introduced with the liposome immunosensor should be useful in measuring various antigens in biological fluids.

ACKNOWLEDGEMENTS

The author is indebted to Dr. Hiroshi Itagaki and Mr. Shinya Sugawara who contributed to the investigation described here.

REFERENCES

Aizawa, M., Morioka, A., Matsuoka, H., Suzuki, S., Nagamura, Y., Shinohara, R. and Ishiguro, I., 1976, An enzyme immunosensor for IgG, J. Solid Phase Biochem., 1:319-328.

Anderson, G.W., Zimmerman, J.E. and Callahan, F.M., 1964, The use of esters of N-hydroxysuccinimide in peptide synthesis, J. Amer. Chem. Soc., 86:1839-1842.

Brendzel, A.M. and Miller, I., 1980, Effects of lipid-soluble substances on the thermotropic properties of liposome filtration, Biochim. Biophys. Acta, 601:260-270.

Chan, S.W., Tan, C.T. and Hsia, J.C., 1978, Spin membrane immunoassay: Simplicity and specificity, J. Immunol. Methods, 21:185-195.

Cheng, F.S. and Christian, G.D., 1977, Amperometric measurement of enzyme reactions with an oxygen electrode using air oxidation of reduced nicotinamide adenine dinucleotide, Anal, Chem., 49:1785-1788.

Cheng, F.S. and Christian, G.D., 1979, A coupled enzymatic method to measure blood lactate by amperometric monitoring of the rate of oxygen depletion with a Clark oxygen electrode, Clin. Chim. Acta, 91:295-301.

Cook, C.E., Twine, M.E., Myers, M., Ameson, E., Kepler, J.A. and Taylor, G.F., 1976, Theophylline radioimmunoassay: Synthesis of antigen and characterization of antiserum, Res. Commun. Chem. Pathol. Pharmacol., 13:497-505.

Erlanger, B.G., Borek, F., Berser, S.M. and Lieberman, S., 1975, Steroid-protein conjugates I. Preparation and characterization of conjugates of bovine serum albumin with testosterone and with cortisone, J. Biol. Chem., 328:713-727.

Freytag, J.W. and Litchfield, W.J., 1984, Liposome-mediated immunoassays for small haptens (Digoxin) independent of complement, J. Immunol. Method, 70:133-140.

Geiger, B. and Smolarsky, M., 1977, Immunochemical determination of ganglioside GM_2 by inhibition complement-dependent liposome lysis, J. Immunol. Methods, 17:7-9.

Haga, M., Itagaki, H., Sugawara, S. and Okano, T., 1980, Liposome immunosensor for theophylline, Biochem. Biophys. Res. Commun., 95:187-192.

Haga, M., Sugawara, S. and Itagaki, H., 1981, Drug sensor: Liposome immunosensor for theophylline, Anal. Biochem., 118:286-293.

Ho, R.J.Y. and Huang, L., 1985, Interactions of antigen-sensitized liposomes with immobilized antibody: A homogeneous solid-phase immunoliposome assay, J. Immunol., 134:4035-4040.

Hsia, J.C. and Tan, C.T., 1978, Membrane immunoassay: Principle and applications of spin membrane immunoassay, Ann. N.Y. Acad. Sci., 308:139-148.

Humphries, G.K. and McConnell, H.M., 1974, Immune lysis of liposomes and erythrocyte ghosts loaded with spin label, Proc. Nat. Acad. Sci. U.S.A., 71:1961-1964.

Kinsky, S.C., Haxby, J.A., Zopf, D.A., Alving, C.R. and Kinsky, C.B., 1969, Complement-dependent damage to liposomes prepared from pure lipids and Forssman hapten, Biochemistry, 8:4149-4158.

Litchfield, W.J., Freytag, J.W. and Adamich, M., 1984, Highly sensitive immunoassays based on use of liposomes without complement, Clin. Chem.,30:1441-1445.

O'Connell, J.P., Campbell, R.L., Fleming, B.M., Mercolino, T.J., Johnson, M.D. and McLaurin, D.A., 1985, A high sensitive immunoassay system involving antibody-coated tubes and liposome-entrapped dye, Clin. Chem., 31:1424-1426.

Shiba, K., Umezawa, Y., Watanabe, T., Ogawa, S. and Fujiwara, S., 1980, Thin-layer potentiometric analysis of lipid antigen-antibody reaction by tetrapentylammonium (TPA^+) ion loaded liposomes and TPA^+ selective electrode, Anal. Chem., 52:1610-1613.
Six, H., Uemura, K. and Kinsky, S., 1973, Effect of immunoglobulin class and affinity on the initiation of complement-dependent damage to liposomal model membranes sensitized with dinitrophenylated phospholipids, Biochemistry, 12:4003-4011.
Smolarsky, M., Teitlbaum, D., Sela, M. and Gitler, C., 1977, A simple fluorescent method to determine complement-mediated liposome immune lysis, J. Immunol. Methods, 15:255-265.
Uemura, K., Kinsky, S.C., 1972, Active vs. passive sensitization of liposomes toward antibody and complement by dinitrophenylated derivatives of phosphatidyl ethanolamine, Biochemistry, 11:4085-4094.
Wei, R., Alving, C.A., Richards, R.L. and Copeland, E.S., 1975, Liposome spin immunoassay: A new sensitive method for detecting lipid substances in aqueous media, J. Immunol, Methods, 9:165-170.

MEASUREMENT OF COMPLEMENT FIXATION WITH ION SELECTIVE MEMBRANE ELECTRODES

Paul D'Orazio

Ciba-Corning Diagnostics Corp.
63 North Street
Medfield, MA 02052

INTRODUCTION

The immunochemical action of Complement (C) toward biological membranes provides a host with an extremely sensitive defense mechanism against invading organisms. The proteins that collectively comprise Complement represent 4-5% of the total serum proteins in humans (Alper,1974). These proteins have enzymatic activity in an inactive native state until rapidly and systematically triggered into action by the invasion of organisms foreign to the living system. Complement produces irreversible lytic damage to a sensitized cell membrane, that is, one which has bound to it a specific antibody. The lytic action of C is assayed based on its ability to produce lysis of sensitized red blood cells and is expressed in CH_{50} units, defined as the amount of Complement activity required to produce lysis of 50% of the cells in a sensitized sheep erythrocyte suspension. The degree of lysis is a function of the hemolytic Complement activity present and may be measured quantitatively by monitoring the release of intracellular oxyhemoglobin colorimetrically at 541 nm (Kabat and Mayer, 1971).

The clinical importance of the Complement lysis phenomenon is its ability to be used as an indicator system in Complement fixation studies. Complement, as well as producing lysis of sensitized red blood cells, may be "fixed" or inactivated by any soluble or insoluble antigen-antibody complex. Thus, a detection system for immune reactions is possible even in the absence of precipitation or agglutination by measuring the remaining Complement activity following addition of a known Complement level to a sample thought to contain the antigen-antibody complex. Likewise, the system may be made quantitative with respect to antibody or antigen in the presence of optimum amounts of the other components.

The flow diagram of Figure 1 demonstrates the concept. Complement, typically from guinea pig serum, along with a specific antigen, are added to a test serum thought to contain the antibody to that antigen. If the antibody is present, an antigen-antibody complex will form and will fix some portion of the Complement. Sensitized sheep red blood cells (SRBC) are then added to the reaction mixture. The term "sensitized" implies that the cells have bound to their surface a rabbit antibody toward a substance found in the cell membrane. The unfixed Complement will lyse a quantity of the red blood cells inversely proportional to the amount of antibody present in the original sample. A maximum amount of cell lysis takes place in the absence

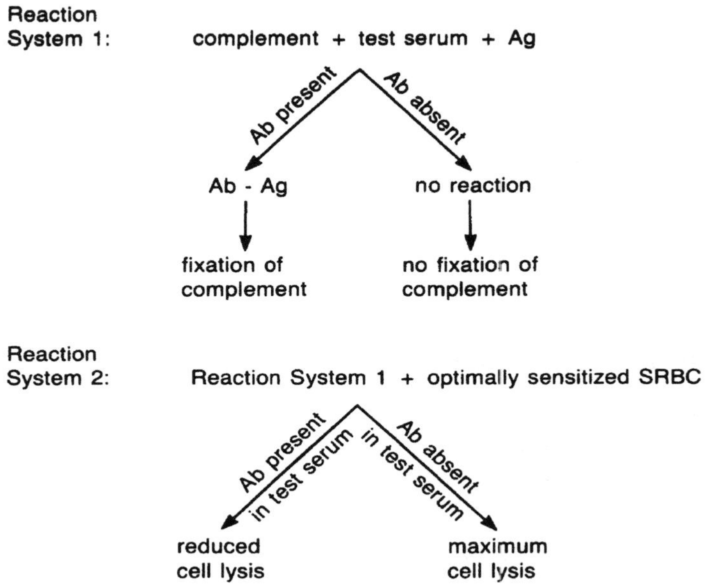

Figure 1. The Complement Fixation Test.

of antibody. This system may be made quantitative toward antibody or antigen by addition of optimum amounts of the other components. A spectrophotometric determination of the oxyhemoglobin released from the SRBC's requires a separation step to remove cells, cell residues and turbid serum components from the reaction mixture.

We describe below an electrochemical method for following the Complement lysis phenomenon based on the release of an electroactive marker from sheep red blood cell ghosts. The marker, the trimethylphenylammonium cation (TMPA+) is loaded into sheep red blood cells by the formation of resealed cell ghosts, explained in detail below. A potentiometric membrane electrode, selective for TMPA+, is then used to monitor the release of the marker ion from these vesicles.

We describe optimization of this system with respect to: i) the indicator system, consisting of the marker-loaded SRBC ghosts, the rabbit antibody to these cells (rabbit anti-sheep hemolysin), guinea pig Complement, and the electrochemical measurement cell, and ii) quantitation of a model antibody (anti-bovine serum albumin) in serum utilizing the Complement fixation phenomenon and the novel indicator system.

MATERIALS AND METHODS

Equipment

The poly(vinyl chloride) based electrode for trimethylphenylammonium (TMPA+) is of the type described by Moody,et.al. (1970) and was constructed by the following method. The tetraphenylborate salt of TMPA+ was precipitated by mixing 0.1 molar solutions of trimethylphenylammonium chloride and sodium tetraphenylborate. The precipitate was gravity filtered, washed with water and air dried overnight. A 5% w/w slurry of this salt in dioctyl phthalate was made and to this was added 33% w/w poly(vinyl chloride) powder. The slurry was thoroughly mixed and thinly spread on a glass slide. A glass tube (2 mm i.d.) was rapidly dipped into this film so that some of the slurry was drawn up into the tube by capillary action. The PVC slurry forms a membrane following heat polymerization at 200 degrees Centigrade for five minutes.

The electrode was fitted with an internal Ag/AgCl wire and a 10 millimolar TMPA-chloride fill solution. The small size of this electrode permits its use, together with a miniature saturated calomel reference electrode, in small sample volumes. A sample size of 750 µl was used in all work described below. The indicator electrode was stored in air when not in use and was briefly soaked in the working buffer prior to use. Under these conditions, the electrode could be used for a period of one month with only negligible deterioration in its response characteristics.

All potentiometric measurements were made at 37 degrees Centigrade in a 15 ml glass thermostatted cell. A Corning Model 12 pH meter and a Heath-Schlumberger SR-204 strip chart recorder were used to monitor and record electrode potential outputs.

Reagents

Triethanolamine buffered saline (TBS) containing 5.0×10^{-4} M Mg^{++} and 1.5×10^{-4} M Ca^{++} at pH 7.4 and an ionic strength of 0.150 M was used as the standard diluent for this work based on the reported suitability of this buffer for Complement fixation studies (Rose,1980). Sodium chloride was replaced with lithium chloride in this diluent owing to the slight interference of high sodium levels at the $TMPA^+$ electrode.

Guinea pig Complement and rabbit anti-sheep hemolysin, both lyophilized reagents, were obtained from Grand Island Biological Co., Grand Island, NY. The guinea pig Complement is serum from normal, healthy, adult guinea pigs and was found to have titers of no less than 1:130 by the U.S. Department of Health, Education, and Welfare standard method (1965). The rabbit anti-sheep hemolysin is produced by injection of rabbits with boiled sheep erythrocyte stromata. The antiserum was titered at no less than 1:1500 by the standard method. The antiserum was pretreated by heating to 56 degrees Centigrade for thirty minutes to remove any Complement activity.

Rabbit antisera to bovine serum albumin were from Miles Laboratories Inc., Elkhart, IN. The lyophilized product was reconstituted with distilled, deionized water before use and heated to 56 degrees for thirty minutes to remove any Complement activity. Antibody titers in milligrams per milliliter were supplied with the product. The antigen, bovine serum albumin (lyophilized, Cohn Fraction V) was from Sigma Chemical Co., St. Louis, MO.

Sheep Red Blood Cell Ghosts Red blood cells that are lysed in a hypotonic medium may be resealed by restoring the lysate to isotonicity. The solutes added to the medium to reach isotonicity will be trapped inside the resealed ghosts. Work with resealed RBC ghosts (Schwoch and Passow,1973) has shown that cells lysed under mildly hypotonic conditions (40 milliosmolar or above) and subsequently resealed regain original membrane structure and function, including cation impermeability. The importance of maintaining membrane structure in the resealed ghosts for purposes of the present work is obvious since the antigenic properties of the cell must be preserved. The molecule toward which rabbit antibodies to sheep cells are formed is a membrane glycolipid known as the Forssman antigen. This substance has antigenic sites which protrude from the surface of the cell membrane and are readily accessible to complex with antibody (Kinsky,1972).

Sheep whole blood was obtained from the Department of Animal Science and Agricultural Biochemistry, University of Delaware, Newark, DE. as an 87% suspension in acid-citrate-dextrose anticoagulant. A volume of blood could be stored refrigerated and remained usable for our purposes for approximately one month.

SRBC ghosts were prepared by the method of DeLoach and Ihler (1977) with some modifications. Cells were washed three times in TBS and resuspended to a final concentration of 50% in TBS. The cells were then dialyzed for 1.5 to 2 hours against a solution of $4X10^{-3}$ M $MgSO_4$ and $1X10^{-5}$ M $CaCl_2$ at pH 6.5 and 4 degrees Centigrade to produce osmotic lysis. Resealing was accomplished by addition of solid TMPA chloride to the lysate to yield a final TMPA+ concentration of 0.138 M, followed by incubation at 37 degrees for one hour. The reformed ghosts were then sensitized in bulk with a 1:750 anti-sheep hemolysin dilution. A ten minute incubation at room temperature is considered sufficient for sensitization (Kent and Fife,1963). The suspension was separated from untrapped marker by dialysis overnight against a solution of 0.154 M LiCl, $1.5X10^{-4}$ M $CaCl_2$ and $5.0X10^{-4}$ M $MgCl_2$ at pH 7.4 and 4 degrees C. The total amount of trapped marker was determined by dilution of a 200 µl aliquot of the ghost suspension to 0.750 ml in TBS. A second aliquot was sonicated in an ultrasonic bath for 1.5 hours and likewise diluted to 0.750 ml in TBS. Measurement of free TMPA+ in both samples allows calculation of the total amount of marker trapped in a given aliquot of the ghost suspension.

We have found that storage of ghosts in the sensitized state greatly reduces the background leakage of marker ion from the ghosts. The method of storing unsensitized ghosts and mixing only the desired aliquot with the hemolysin antibody prior to use resulted in such high rates of leakage that cells were not usable after two or three days. Under the new condition, ghosts could be stored and remained viable for periods as long as two weeks.

Procedures

Complement and Hemolysin Titrations Dilutions of rabbit anti-sheep hemolysin in TBS were made in the following concentrations: 1:10000, 1:6000, 1:3000, 1:1500, 1:750 and frozen until ready for use. Ghosts used for hemolysin titrations were not sensitized in bulk after resealing. A 0.2 ml aliquot of the ghost suspension was mixed with 0.2 ml of the various hemolysin dilutions and incubated for ten minutes at room temperature. A further dilution with 310 µl of buffer was made. The indicator and reference electrodes were immersed into the sample at this point and a stable potential was reached in less than one minute. Addition of 5.6 CH_{50} units of Complement activity, an experimentally determined excess in the presence of the highest antibody concentration, increased the final sample volume to 750 µl. The EMF immediately upon addition of Complement was recorded as the baseline potential (El). A fifteen minute reaction time was required for the cell lysis process to reach completion in the presence of the most concentrated hemolysin dilution. This reaction time was therefore used for all points on the hemolysin and Complement titration curves. The potential change at T=15 minutes relative to the baseline reading was designated ΔE and plotted as a function of relative hemolysin concentration and stock antiserum dilution.

Dilutions of guinea pig Complement in TBS were made in the following ratios: 1:28, 1:15, 1:10, 1:8, 1:7, 1:6.2, and 1:5. Complement titrations were made using ghosts which were sensitized in bulk using the 1:750 hemolysin dilution. Four hundred microliters of this suspension were diluted to 710 µl with TBS. Addition of 40 µl of the above guinea pig Complement dilutions represented 0.2, 0.4, 0.55, 0.7, 0.8, 0.9, and 1.1 CH_{50} units, respectively. ΔE values were plotted as a function of Complement units and as a function of stock Complement dilution.

Serum Antibody Measurements Using the Complement Fixation Test. Quantitation of a model antibody (anti-bovine serum albumin) in serum was used to demonstrate the applicability of the above indicator system to the Complement Fixation Test. The rabbit antiserum to BSA, reconstituted as described above, was diluted in TBS in the following concentrations: 0.04, 0.17, 0.33, 0.50, 0.67, 0.83, and 1.0 mg antibody/ml. Constant aliquots of these dilutions, representing 1 to 24 µg of rabbit antibody, were mixed with an optimum amount of BSA antigen and 11.2 CH_{50} units of Complement activity in a total sample volume of 160 µl followed by incubation at 2-4 degrees C for twenty hours. The samples were then placed on ice or frozen until ready for use.

Experiments to determine free or unfixed Complement in each of the samples were run similarly to the Complement titration in the preceding section. Four hundred microliters of a sensitized SRBC ghost suspension were diluted to 710 µl in TBS. Forty microliters of the test sample were added and the ΔE values at 15 minutes recorded. A series of ΔE values resulting from free Complement activity in the samples was plotted as a function of the amount of antibody in a 25% aliquot of the initial sample.

Complement titration curves were run immediately following the antibody response curves to obtain a quantitative estimate of the number of Complement units fixed at each point on the antibody curve. Complement and hemolysin dilutions were prepared fresh prior to use.

RESULTS

The calibration curve of the TMPA+ ion selective electrode is shown in Figure 2. The electrode is linear to a lower limit of almost $1X10^{-5}$ M TMPA+ in triethanolamine buffered saline at pH 7.4. The selectivity coefficients of this electrode toward alkali metal ions, expressed as $K^{TMPA+,M+}$, are as follows: K+=690, Na+=5200, Li=14000 (Moody and Thomas, 1971). NaCl was therefore replaced with LiCl at 0.128 M as the background electrolyte in the working buffer, permitting a lower limit of detection toward TMPA+. It should be noted, however, that this substitution is not necessary for routine analytical work using the proposed procedure.

Sheep erythrocyte ghosts demonstrate the ability to trap the trimethylphenylammonium cation with slow leakage. The background level of this marker ion may be reduced to $1X10^{-5}$ M by dialysis of the ghost suspension. Complete lysis by sonication of a 200 µl aliquot of the ghosts, prepared as described above, followed by a potentiometric measurement of the TMPA+ cation in a total volume of 750 µl shows a ten-fold increase in the concentration of the marker ion in solution. This increase corresponds to $6.8X10^{-5}$ millimoles of TMPA+ released from 200 µl of the ghost suspension. Leakage of marker as a function of time is slow and negligible in comparison to trapped marker levels and is not measureable during the time course of the experiments described below.

Figure 3 demonstrates a typical hemolysin titration curve in the presence of 5.6 CH_{50} units. The small ΔE value obtained even in the absence of rabbit hemolysin is due to the background action of other guinea pig serum components in sensitizing the SRBC ghosts. If necessary, this effect could be eliminated by pretreatment of the serum (Kabat and Mayer,1971).

The results of Figure 3 show that in the presence of excess Complement and using a 200 µl aliquot of the ghost suspension, a 1:750 dilution of the rabbit hemolysin shows a saturation effect of the indicator system toward

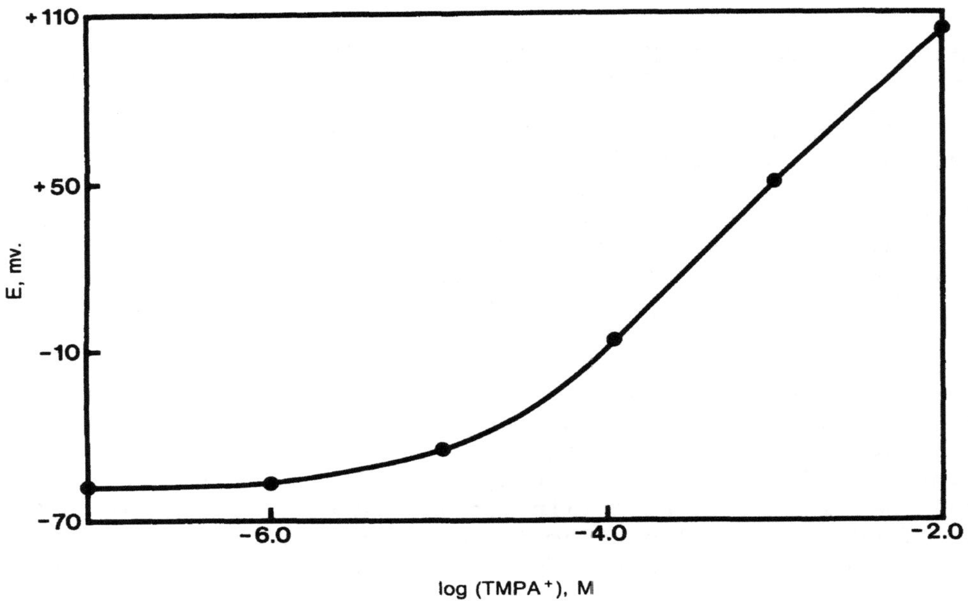

Figure 2. Calibration Curve of the TMPA+ Electrode vs. SCE in pH 7.4 Triethanolamine Buffered Saline.

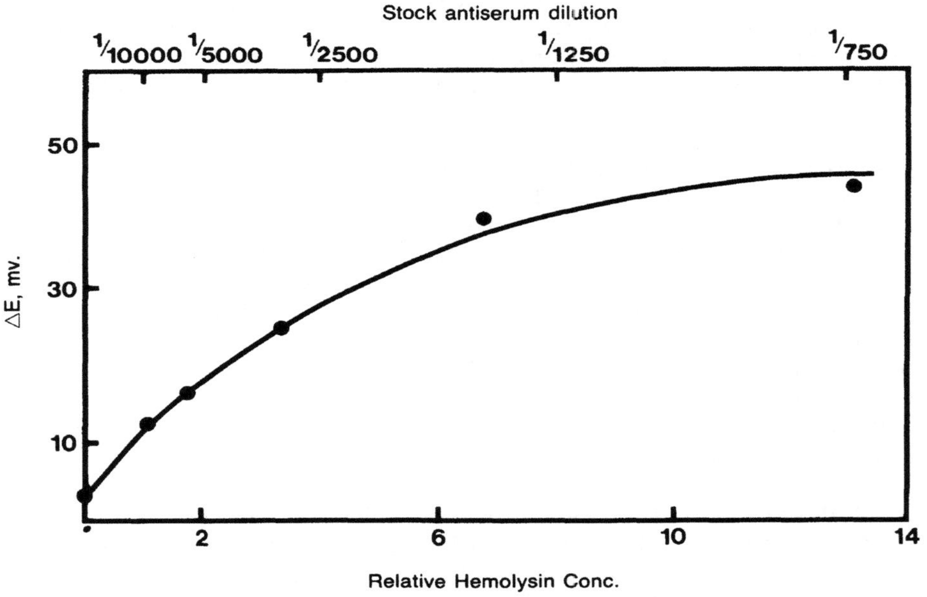

Figure 3. Potentiometric Response Curve to Rabbit Anti-Sheep Hemolysin in the Presence of 5.6 CH_{50} Units.

added antibody. This dilution was therefore used to sensitize the red cell ghosts for the purpose of Complement titration. Figure 4 shows such a titration curve with relative potential output plotted as a function of both stock guinea pig Complement dilution and CH_{50} units. The good sensitivity of the system toward added Complement as well as the early saturation effect with added CH_{50} units is seen.

The above experiments lay the ground work for the development of immunoassays using potentiometric membrane electrodes. By using the Complement lysis of SRBC ghosts as the indicator reaction, it is possible to extend the analytical concept to antibody or antigen determinations. We have chosen the model system:

bovine serum albumin (BSA) + rabbit anti-BSA + Complement

to demonstrate this application. The amount of unfixed Complement in solution following a constant reaction time is inversely proportional to the amount of immune complex present. Moreover, the measured activity of Complement may be related to antibody to BSA or BSA concentration when the other components are present in constant, optimum amounts. A level of added Complement was chosen which lyses close to 100% of a ghost suspension while avoiding a large excess. Based on the available data (Table 1), a starting level of 2.8 Complement units was chosen. Greater accuracy in pipetting is achieved by initially mixing 11.2 CH_{50} units with the antibody and antigen followed by sampling a 25% aliquot of this mixture. This method resembles that of Stein and Van Ngu (1950) since excess Complement is not employed and the reaction mixture is added directly without dilution to the sensitized cells.

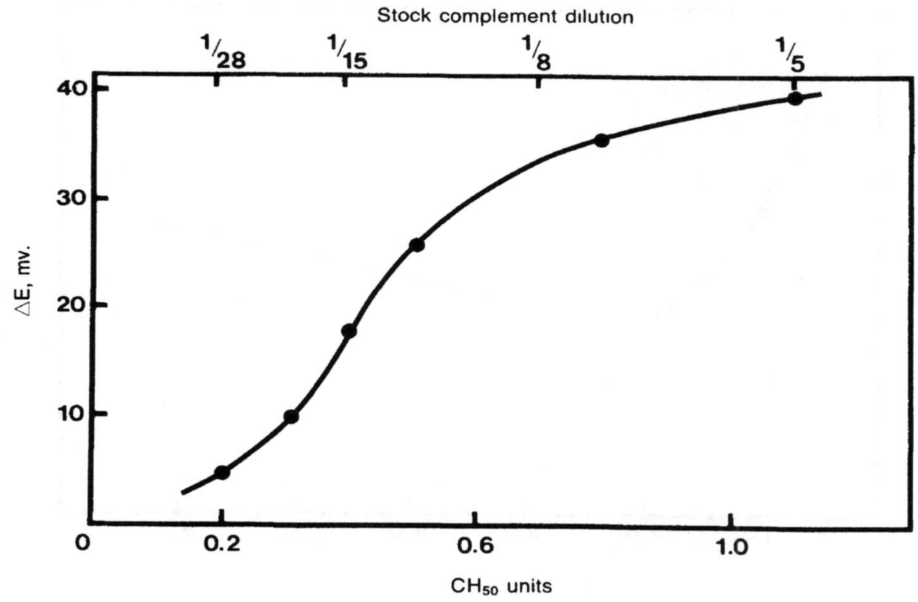

Figure 4. Potentiometric Response Curve to Guinea Pig Complement.

Table 1. Determination of Complement Excess for Hemolysin Titrations and Quantitative Complement Fixation Studies[a]

Guinea Pig Serum Dilution	CH_{50} Units	ΔE, mv
1:5	1.1	46.3
1:2	2.8	58.2
1:1.5	3.7	55.7
undiluted	5.6	57.6

[a]SRBC Ghosts were Sensitized with 1:750 Hemolysin

The upper limits of BSA and anti-BSA which could be employed were determined by the Complement fixing ability of impurities present in the matrices containing these compounds. It was found that an amount of anti-serum corresponding to 24 µg of antibody could be used in the presence of 11.2 CH_{50} units with no fixation of Complement. Likewise, 24 µg of BSA antigen showed no Complement fixing ability. This corresponds to 6 µg of antigen and 6 µg of antibody in the presence of 2.8 Complement units under the conditions of the assay.

Construction of a response curve to anti-BSA required optimization of the amount of antigen to be used with 24 µg of anti-BSA, the highest antibody value employed. The optimum value should produce the maximum fixation of Complement as a function of the antibody to antigen ratio. The curve of Figure 5 results from an experiment in which antibody to antigen ratio is varied from 1:1 to 10:1 by weight in the presence of a constant 24 µg of antibody. The minimum ΔE value, that is, the point of maximum fixation of

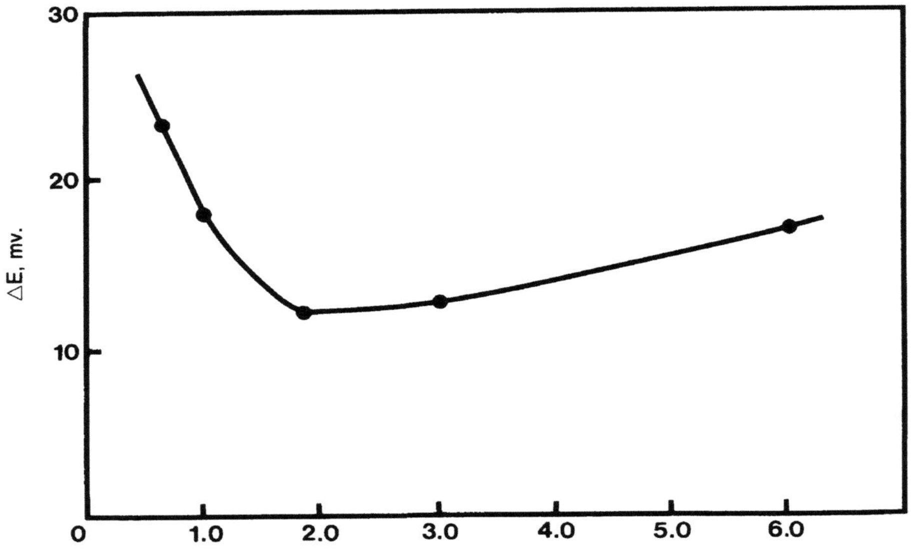

Figure 5. Determination of the Amount of BSA Producing Maximum Fixation of Complement in the Presence of 24 ug of Anti-BSA (6 ug per run).

Complement occurs at an antibody to antigen ratio slightly greater than 3:1. This point corresponds to the equivalence point as defined by precipitin analysis where the maximum amount of immune complex is formed resulting in greater fixation of Complement than in the zones of antibody or antigen excess. A precipitin curve, prepared by the standard quantitative micro-precipitation method (Keleti and Lederer,1974), is shown in Figure 6 and indicates an antibody to antigen ratio at the equivalence point between 3:1 and 4:1, in good agreement with the potentiometric Complement Fixation method.

Based on the results of Figure 5, the amount of BSA antigen chosen for purposes of antibody quantitation was 7.2 µg per total reaction mixture. One to 24 µg of antibody were incubated with this quantity of antigen in the presence of 11.2 CH_{50} units. Figure 7 is a plot of the ΔE values as a function of the amount of antibody in a one-fourth aliquot of the original mixture. It is possible to obtain a quantitative estimate of the Complement units fixed at the various points on this curve by constructing a Complement titration curve similar to Figure 4 immediately following the antibody response curve. In this way, the characteristics of the system, amount of marker trapped within the cell ghosts, concentration of background ion, etc., remain constant. The ΔE values of Figure 7 are read on the curve of Figure 4 and the resulting values subtracted from 2.8 CH_{50} units, the starting Complement level present in a 25% aliquot of the original mixture. These results are shown in Table 2.

Accuracy and precision of the proposed method for the determination of antibody were evaluated using two lots of commercial antisera other than that used as the standard and reading these as unknowns on the curve of Figure 7. The data shown in Table 3 are the result of triplicate determinations of each lot and comparison with the label value of the material from precipitin analysis. The method demonstrates the accuracy and precision necessary to permit quantitative determinations.

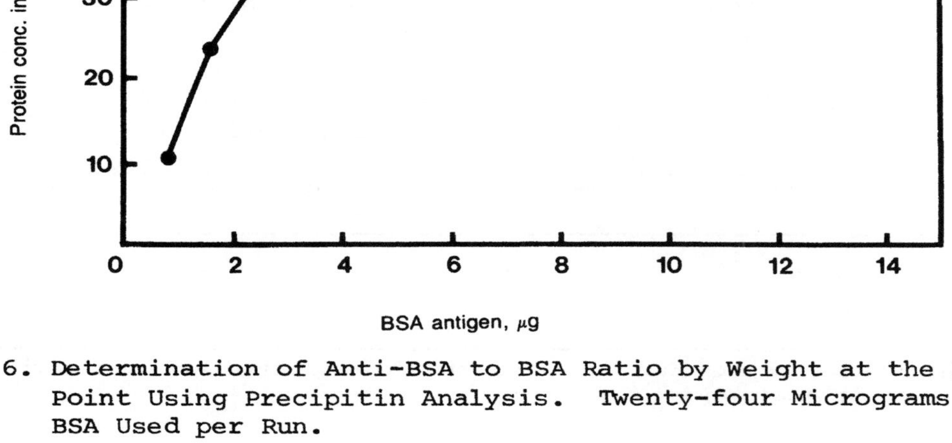

Figure 6. Determination of Anti-BSA to BSA Ratio by Weight at the Equivalence Point Using Precipitin Analysis. Twenty-four Micrograms of Anti-BSA Used per Run.

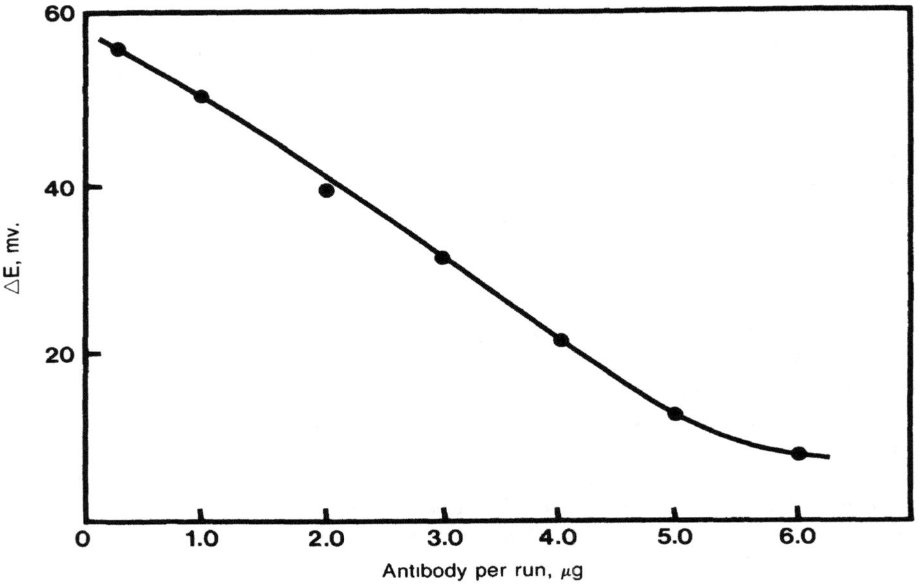

Figure 7. Anti-BSA Calibration Curve Using the Potentiometric Complement Fixation Method.

Table 2. Complement Units Fixed by Varying Amounts of Anti-BSA in the Presence of 1.8 μg of BSA

Anti-BSA, μg	CH_{50} Units Fixed[a]
0.25	not determinable
1.0	1.8
2.0	2.1
3.0	2.2
4.0	2.3
5.0	2.4
6.0	2.4

[a]Out of a total of 2.8 CH_{50} units. Each value represents the mean of three determinations with a standard deviation of ±0.1 units.

Table 3. Potentiometric Electrode Measurement of Anti-BSA in Commercial Antisera

Antiserum Lot	Known Value, mg/ml	Measured Value, mg/ml[a]
R270	2.5	2.6
R148	2.4	2.5

[a]Each value represents the mean of three determinations with a standard deviation of ± 0.2.

DISCUSSION

The application of ion selective electrodes to quantitative Complement fixation studies has been illustrated above for a simple model system. An important implication of this work is that it lays the foundation for the development of immunoassays using the Complement fixation and lysis phenomena in conjunction with electrochemical sensors. Potentiometric membrane electrodes are particularly attractive for this purpose owing to their high selectivities toward potentially interfering species in a complicated matrix such as blood serum. This advantage is clearly illustrated above since the membrane electrode for the marker cation TMPA+ is 690 times more selective for TMPA+ over K+, the blood serum cation for which it shows the least selectivity. Moreover, in comparison to spectrophotometry, membrane electrodes offer several other advantages, discussed below.

The instrumentation required for potentiometric measurements is inexpensive and easy to maintain. Electrodes may be prepared in large numbers and at low cost and may be disposed following deterioration of their response characteristics. This deterioration is typically observed as a decrease in electrode slope or an unstable signal upon continued exposure to blood serum. The mechanism is a loss of ion exchanger (tetraphenylborate) and plasticizer (dioctyl phthalate) from the membrane resulting from the affinity of these species for the lipophilic components of blood serum (Oesch,et.al,1985). As stated above, the lifetime of a TMPA+ sensor, when used in the type of studies described, is about one month. Electrode maintenance, which may be occasionally required, is a wash in a proteolytic enzyme solution to remove adsorbed protein from the surface of the membrane.

Optical measurements of Complement lysis of erythrocytes require a separation step to remove turbid components from the sample. Ion selective electrodes may be used in these solutions without interference and, therefore, the need for separation steps is eliminated. Cell lysis may be measured directly and in real time with the elimination of potential sample handling errors. The above work also presents for the first time a method for studying the kinetics of the Complement lysis phenomenon. In a recent report, Shiba, et.al. (1980) used an electrode selective for tetrapentylammonium (TPA+) to monitor the release of this ion from phospholipid liposomes in response to Complement lysis. They reported negligible short term leakage of TPA+ from the liposomes during their experiments at room temperature. The configuration of their potentiometric cell, while allowing use of a small sample volume, did not permit continuous monitoring of the Complement lysis phenomenon and eliminated the possibility of studying the kinetics of the process, an important advantage of the electrochemical method (Mayer,1978).

Prior to using marker loaded SRBC ghosts to measure hemolytic Complement activity, we attempted loading marker ions such as cadmium and rubidium into antigen-labelled phospholipid liposomes for purposes of measuring Complement lysis. This work was largely unsuccessful owing to the high ion leakage rates from the liposomes at 37 degrees Centigrade. Figure 8 illustrates that the majority of this leakage takes place within ten minutes of incubation at 37 degrees. The particular liposomes used in this case contained 85 mole% phosphatidyl choline, 12 mole% cholesterol, and 3 mole% cardiolipin. The same experiments, repeated at room temperature, demonstrate greatly reduced leakage of ion, particularly over the first five minutes of incubation where it is indistinguishable from the background ion concentration. The enhanced leakage at the elevated temperature did not permit the use of liposomes as the ion carrier in these studies. While room temperature measurements of Complement mediated lysis may have been potentially useful, this temperature represents a further departure from the physiological environment of the Complement proteins. This is an important point,

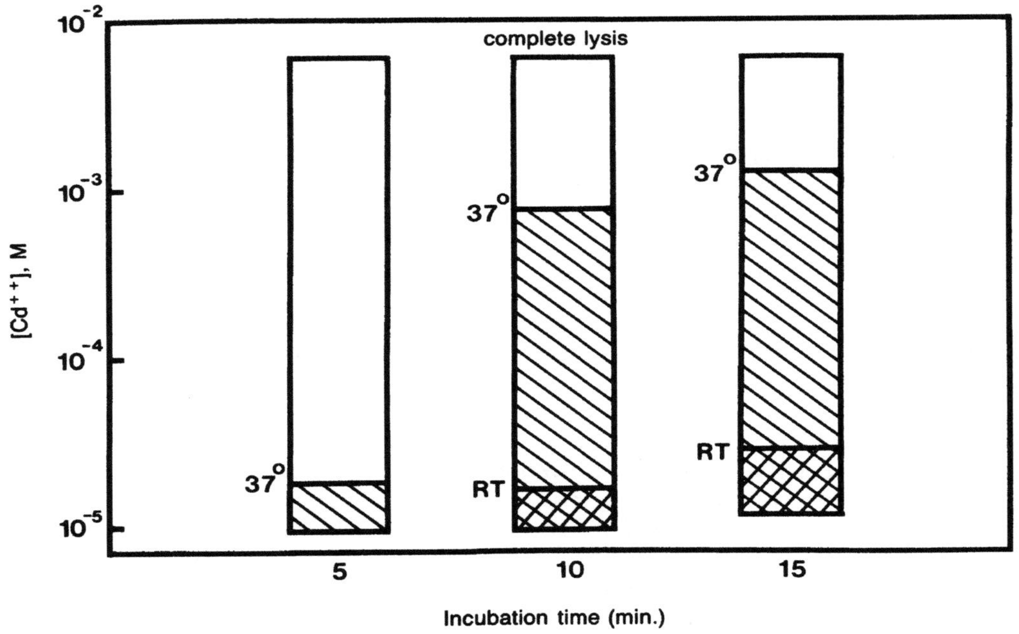

Figure 8. Measurement of Cadmium Ion Leakage From Phospholipid Liposomes (85 mole% phosphatidyl choline, 12 mole% cholesterol and 3 mole% cardiolipin) Using a Cadmium Ion Selective Electrode vs. SCE. One Hundred Microliters of the Liposome Suspension were Diluted to a Total Volume of 140 µl and the Cadmium Ion Concentration in Solution was Measured at Room Temperature and at 37 Degrees C as a Function of Time. Complete Lysis of the Liposomes was with Triton X-100.

especially if the kinetics of the Complement lysis process are to be followed. The enhanced leakage at the elevated temperature and the differences in the nature of the marker ions may account for the discrepancies observed between our work and that of Shiba.

Our potentiometric method, as discussed above, suffers from some disadvantages. The preparation of the sheep erythrocyte ghosts is a lengthy procedure. Moreover, the long term storage of these cells as a "reagent" is not presently feasible due to the leakage of marker ion. The lipophilic ion trimethylphenylammonium likely has an affinity for the cell membrane, enhancing its leakage rate from the cell. A stabilization effect was noted when the erythrocyte ghosts were sensitized with rabbit anti-sheep hemolysin. The ion leakage from these ghosts was noticeably less than the leakage from unsensitized ghosts. The layer of antibody coating the cell may present another diffusion barrier to the cation marker, slowing its leakage from the cell. Despite the leakage problem, it is possible to maintain the background level of marker ion at a low, constant value by occassional dialysis, however, the absolute amount of trapped marker decreases, changing the sensitivity of the system. Under the conditions described above, sensitized erythrocyte ghosts could be stored for up to two weeks while maintaining adequate sensitivity for our purposes. Lyophilization of antigen coated erythrocytes without damage to the cells has been demonstrated by Spona and Töpert (1976). The techniques employed, including hardening the erythrocytes with formaldehyde and the addition of glycine to protect the erythrocyte membrane during freeze drying may be employed here to obtain stable erythrocyte ghosts. Further investigation is required.

The ability of our method to measure low concentrations of immunoagents is an important consideration, particularly in the case of measurement of antigen concentrations. A major requirement of an immunoassay is a low limit of detection. The present configuration of our method is not optimized for low detection limits. The sigmoidal shape of the curve of Figure 4 does not permit resolution of Complement activities when approaching 100% lysis. Indeed, the shape of this curve from spectrophotometric Complement fixation assays is also sigmoidal and is described by the Von Krough Equation (Alper,1974). This equation, converted to its logarithmic form, is:

$$\log x = \log K + 1/n \log(y/1-y) \qquad (1)$$

and describes a line where x is the volume of guinea pig serum in milliliters at a stated dilution added to a sensitized cell suspension and y is the resulting percent lysis of the cells. Since $\log(y/1-y) = 0$ when $y = 0.5$, K is the volume of guinea pig serum producing 50% lysis of the cell suspension. 1/n is the slope of the line. The percent lysis of the cells, y, may be expressed in the case of the potentiometric Complement fixation studies as the ratio of the TMPA+ concentration present at the endpoint of the Complement lysis process ($TMPA+_{15}$) to the TMPA+ concentration upon complete lysis of an aliquot of cell ghosts ($TMPA+_{tot}$). Using the logarithmic relationship of the Nernst Equation, this may be written as:

$$y = TMPA+_{15} / TMPA+_{tot} = 10^{(E_{15} - E_{tot})/m} \qquad (2)$$

where E_{15} and E_{tot} are the electrode potentials at the endpoint of the Complement lysis process and upon complete lysis of the aliquot of cell ghosts, respectively, and m is the electrode response slope toward TMPA+ in mv/decade. The ΔE data of Figure 4 are converted to E_{15} values and a percent lysis, y, is calculated for each point on the potentiometric Complement titration curve using Equation 2 and experimentally determined values of −34.8 mv and 59.2 mv/decade for E_{tot} and m, respectively. A plot of percent lysis versus CH_{50} units (by the standard method) is shown in Figure 9 and demonstrates the sigmoidal shape of the potentiometric Complement titration curve when the electrode potential outputs (ΔE values) are converted to percent lysis values. The Von Krough Equation (Equation 1), when applied to this curve is written as:

$$\log x = \log K + 1/n \left[\frac{E_{15} - E_{tot}}{m} - \log(1 - 10^{(E_{15} - E_{tot})/m})\right] \qquad (3)$$

A plot of these data yields a line with a y-intercept of −2.53, corresponding to a K value of 0.0029. Therefore, 2.9 µl of undiluted guinea pig serum provides the quantity of Complement activity necessary to lyse 50% of the ghosts in the suspension used for our measurements. Application of the Von Krough Equation in the form of Equation 3 will provide the sensitivity required to resolve Complement activities to within ±0.1 units at points close to 100% cell lysis, allowing antibody determinations to be made when free Complement activities lie in this region of the Complement titration curve. This is represented by points at or below 0.25 µg of anti-BSA in the data of Figure 7 and Table 2.

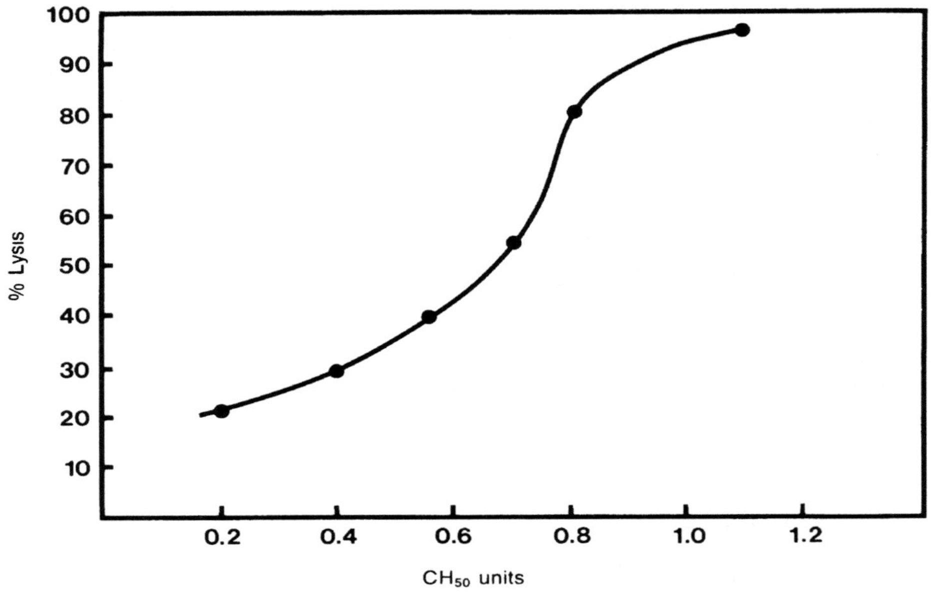

Figure 9. Potentiometric Complement Titration Curve Expressed as % Lysis of SRBC Ghosts vs. CH_{50} Units as Defined by the Standard Method.

SUMMARY

The application of a technology such as electrochemistry to immunodiagnosis must be viewed with respect to several parameters, including: specificity and sensitivity of the method, cost, and ease of use. The work described above has focused on these parameters to demonstrate that electrochemical sensors, and more specifically, ion selective electrodes, may be applied to Complement Fixation assays.

The high selectivities of many types of potentiometric membrane electrodes, when coupled with the inherent specificity of immunochemical reactions, may yield systems with potentially no interferents. The sensitivity of the electrodes is well known and it is possible to extend this sensitivity to immunochemical analysis through optimization of the "mediator" used to couple an electrode to the appropriate immune reaction. The optimum mediator should provide an amplification effect when coupling the two chemical systems (Czaban,1985). In the case of the work described in this report, marker loaded sheep erythrocyte ghosts serve as the mediator. An amplification effect occurs since each unit of the mediator, when turned over, releases multiple units of the marker into solution. Enzymes serve as another example of this concept, as described elsewhere in this volume, since each molecule of enzyme is capable of turning over multiple molecules of substrate.

Ease of use is an important factor in determining the practicality of a new procedure. New methods which require lengthy and complex sample or reagent pretreatment steps are likely to find little practical use. Likewise, the sensing system itself must not require extensive maintenance. Ion selective electrodes fit this description more closely than most electrochemical sensors owing to their ruggedness, ease of construction, and lack of need for electrode surface pretreatment. The ability to use these electrodes in a wide variety of sample matrices eliminates the "clean up" steps required with many other analytical techniques. An example of this is clearly shown in the above work in that the TMPA+ electrode may be used directly in suspensions of cell ghosts, avoiding the separation step required for optical methods.

The ultimate application of an electrochemical sensor to immunochemical analysis is the "immunoprobe", a sensor capable of performing an immunoassay directly in a sample without the addition of reagents. A probe for the potentiometric Complement Fixation test may be developed if it is possible to immobilize or regenerate the reagents at the electrode surface. A more short term application may be the incorporation of the electrochemical sensors and the associated reagents for performing immunoassays into clinical instrumentation. The reagent and sample delivery systems, electronics, and signal processing capabilities presently exist to make this achievable in the near future.

ACKNOWLEDGEMENT

Certain figures in the above test appeared courtesy of the American Chemical Society and Elsevier Science Publishers.

REFERENCES

Alper, C.A., 1974, "Complement" in Structure and Function of Plasma Proteins. Vol. 1, A.C. Allison, Ed., Plenum Press, New York, p. 195-222.

Czaban, J.D., 1985, "Electrochemical Sensors in Clinical Chemistry: Yesterday, Today, Tomorrow", Anal. Chem., 57, p. 345A-356A.

DeLoach, J. and Ihler, G., 1977, "A Dialysis Procedure for Loading Erythrocytes with Enzymes and Lipids", Biochim. Biophys. Acta, 496, p. 136-45.

Kabat, E.A. and Mayer, M.M., 1971, Experimental Immunochemistry, 2nd ed., Charles C. Thomas, Springfield, IL., p. 133-240.

Keleti, G. and Lederer, W.H., 1974, Handbook of Micromethods for the Biological Sciences, Van Nostrand Reinhold, New York, p. 147-48.

Kent, J.F. and Fife, E.H., 1963, "Precise Standardization of Reagents for Complement Fixation", Amer. J. Trop. Med. Hyg., 12, p. 103-16.

Kinsky, S.C., 1972, "Antibody-Complement Interaction with Lipid Model Membranes", Biochim. Biophys. Acta, 265, p. 1-23.

Mayer, M.M., 1978, personal communication.

Moody, G.J., Oke, R.B. and Thomas, J.D.R., 1970, "Calcium Sensitive Electrode Based on Liquid Ion Exchanger in a Poly(vinyl chloride) Matrix", Analyst, 95, p. 910.

Moody, G.J. and Thomas, J.D.R., 1971, Selective Ion Sensitive Electrodes, Merrow, Watford, England, p. 10-12.

Oesch, U., Dinten, O., Ammann, D. and Simon, W., 1985, "Lifetime of Neutral Carrier-Based Membranes in Aqueous Systems and Blood Serum" in Ion Measurements in Physiology and Medicine, M. Kessler, J. Hoper and D.K. Harrison, Eds., Springer-Verlag, Berlin, p. 42-47.

Rose, N.R., 1980, "Complement-Fixation Test" in Methods in Immunodiagnosis, N.R. Rose and P.E. Bigazzi, Eds., John Wiley and Sons, New York, p. 262.

Schwoch, G. and Passow, H., 1973, "Preparation and Properties of Human Erythrocyte Ghosts", Mol. Cell. Biochem., 2, p. 197-218.

Shiba, K., Umezawa, Y., Wantanabe, T., Ogawa, S. and Fujiwara, S., 1980, "Thin-Layer Potentiometric Analysis of Lipid Antigen-Antibody Reaction by Tetrapentylammonium (TPA+) Ion Loaded Liposomes and TPA+ Ion Selective Electrode", Anal. Chem., 52, p. 1610-13.

Spona, J. and Töpert, M., Oct. 19, 1976, U.S. Patent 3,987,159, "Stable Sensitized Erythrocytes and Preparation Means".
Stein, G.J. and Van Ngu, D.V., 1950, "Quantitative Complement Fixation Test: Titration of Luetic Sera by Unit of 50 Per Cent Hemolysis", J. Immunol., 65, p. 17-37.
U.S. Dept. of Health, Education, and Welfare, 1965, "Standardized Diagnostic Complement Fixation Method and Adaptation to Micro Test", Public Health Monograph No. 74, PHS Pub. No. 1228, U.S. Govt. Printing Office, Washington, D.C.

ANTIBODY-ANTIGEN PRECIPITIN REACTION MONITORING WITH AN ION SELECTIVE ELECTRODE

Peter W. Alexander

Department of Analytical Chemistry
University of New South Wales
P.O. Box 1, Kensington,
N.S.W., Australia 2033

INTRODUCTION

The development of ion selective electrode methodology for detection of immunochemicals is described in this chapter, where a solid state silver sulfide membrane electrode is utilized. The basis of this technique has been reported (Alexander and Rechnitz, 1974a, b and c), and was subsequently automated in a continuous flow system (Solsky and Rechnitz, 1978) with the same ion electrode as a potentiometric detector.

The detection of proteins with an ion selective electrode was first reported, however, by Alexander and Rechnitz (1974a), and was used as a method of detection with the silver membrane electrode in a continuous flow system for automated sampling (Alexander and Rechnitz, 1974b) and detection of immunochemical precipitates (Alexander and Rechnitz, 1974c). The technique is an heterogeneous immunoassay requiring separation of the precipitate formed in the classical "precipitin" method when an antigen reacts with its homologous antibody. The precipitate is then redissolved and total protein in the solution is determined in a continuous flow system with the silver electrode after hydrolysis of the protein in alkaline solution to generate thiol groups. The thiols are detected with the electrode after reaction with silver ions in the flow stream.

Since then there have been numerous valuable developments in the approach to immunoassay with electrochemical sensors, as outlined in this book, including both potentiometric assays with other types of ion selective electrodes and amperometric assays. The development of homogeneous immunoassay techniques is of particular interest when separation procedures can be eliminated, giving the possibility of fully automated electrochemical immunoassays with high sensitivity and specificity, as shown in studies by Gebauer and Rechnitz (1981) and Keating and Rechnitz (1985) with gas-sensing electrodes, and by Ngo *et al.* (1983, 1985) with an amperometric hydrogen peroxide electrode.

In the study described here with a silver sulfide electrode, the development of the detection method in a continuous flow automated sampling system is summarized from previously published reports on this technique (Alexander and Rechnitz, 1974a, b, and c), using the reaction between human serum albumin (HSA) as an antigen and its homologous antibody in antiserum as an example where a precipitate is formed.

The method used follows the classical precipitin procedure, as reviewed by Kabat *et al.* (1961) and originally introduced by Heidelberger *et al.* (1929, 1933, and 1935). The advantage of this approach is that there is no need for a label such as a radioactive label or an enzyme label, as commonly used in other immunochemical approaches. The method is therefore less expensive in terms of reagent costs and materials, requiring only an unlabelled antiserum which is commercially available, common chemical reagents, and is simpler to automate than the labelling techniques. In this report, the sensitivity of the method, interference effects, and theoretical aspects of the immuno-reaction are given.

EXPERIMENTAL

Instrumentation The continuous flow autoanalysis system for protein determinations is shown in Fig.1, as developed for total protein determinations (Alexander and Rechnitz, 1974b). A Technicon Sampler II and proportioning pump were used with the flow rates for each pump channel and the reagent concentrations shown. The detector flow-through assembly consisted of an Orion 94-16A silver sulfide electrode and an Orion double-junction reference electrode (Type 90-02-00), with a flow through cap fixed to the silver electrode, as previously described (Llenado and Rechnitz, 1973). The electrodes were connected to a Beckman Model 1055 pH recorder for monitoring of the electrode potential.

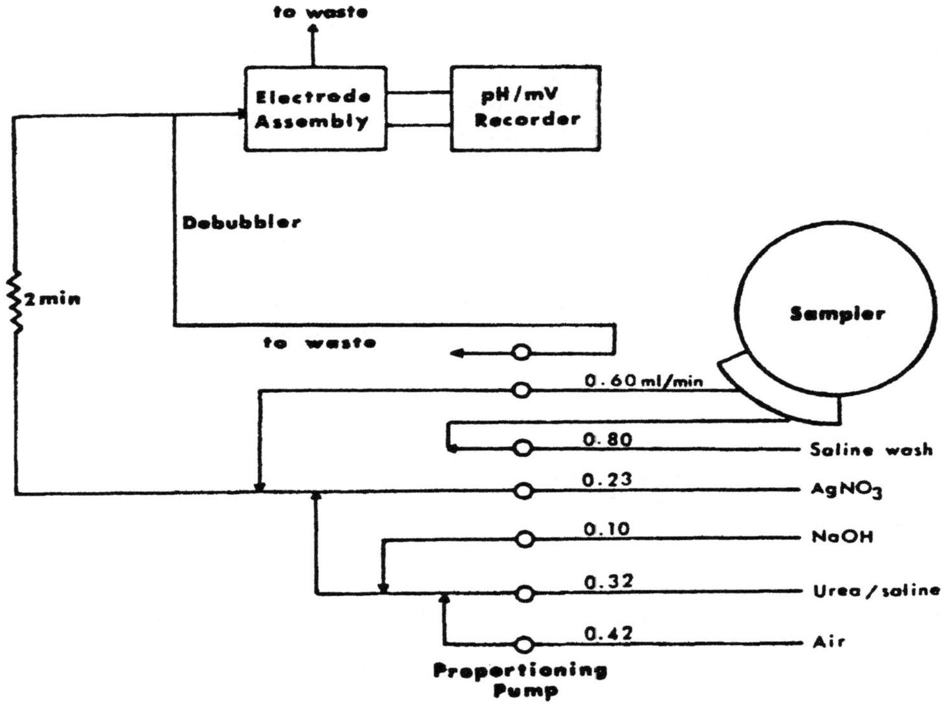

Figure 1. Schematic diagram of the continuous flow apparatus with the following reagent concentrations: silver nitrate ($1.0x10^{-4}$M), sodium hydroxide (4 M), urea (10 M),saline (0.9%) with 1:1 sample-to-wash ratio and 20 samples/hr. (Reprinted by permission of *Anal. Chem.*,46,860,(1974).Copyright 1974, American Chemical Society).

Reagents. Solutions of silver nitrate ($1.0x10^{-4}$M), sodium hydroxide (4 M), sodium chloride (0.9%), and urea (10 M) were prepared from AR grade reagents from Baker. Bovine serum albumin (BSA) was obtained from Sigma, human serum albumin (HSA) from Nutritional Biochemicals Corp., and human γ-globulin from Miles Research. Solutions of approx. 10 mg/ml of each protein were prepared in 0.9% saline solution and standardized by the biuret method (Kabat and Mayer, 1961) against a standard solution of HSA (100 mg/ml) obtained from Miles Laboratories.

The following antisera were obtained from Miles Laboratories: samples of the IgG fraction of goat antiserum to HSA (potencies 3.6, 2.2, and 0.5 mg antibody/ml); and whole goat antiserum to human albumin (potency 3.6 mg/ml). These samples were certified as monospecific to albumin. Other albumin antisera used for testing of supernatants were obtained from Nutritional Biochemicals and Hyland. Human blood serum used was Technicon reference serum (lot No. B2C101) containing 43 mg/ml albumin and 22 mg/ml total globulins. This was diluted by a factor of 100 in isotonic saline before reaction with the antisera.

Procedures. Aliquots of antigen solutions (2 - 100 µl), either HSA or serum, from stock solutions of approx. 0.4 or 4.0 mg/ml in isotonic saline, were added to a fixed volume of antiserum (0.1 ml) in 3-ml centrifuge tubes, keeping total volume constant at 0.20 ml by addition of saline. The antiserum used was either the whole antiserum to albumin for analysis of washed precipitates or the IgG fraction of antiserum for analysis of supernatants. The precipitates were allowed to incubate for 4 hr, and then centrifuged for 20 min, leaving the precipitates firmly packed in the base of the tubes. After washing three times with cold saline, the precipitates were redissolved in 1.0 ml of sodium hydroxide (0.1 M). The total protein content of this solution was then determined in the continuous flow system shown in Fig.1 by aspirating each sample solution into the flow system. The sampling rate was 20 per hr with a sample-to-wash ratio of 1:1 with 0.1M NaOH in saline as the wash.

For analysis of supernatants, the IgG fraction of the antiserum was used to determine the unreacted antibody in the supernatant after removal of the precipitate. Precipitation was carried out as described above. After precipitation, however, 1.0 ml of isotonic saline was added and mixed with the precipitate. After centrifuging off the precipitate, the supernatant was drawn off with disposable micropipets and transferred to the auto-analyser sample cups. The solutions were then analysed in the flow system as before at 20 samples per hour but with saline as the wash.

RESULTS

The determination of total protein precipitated by adding the HSA antigen to the antiserum solution was first done in the continuous flow system in Fig.1 by sampling supernatant solutions obtained after precipitation of the IgG fractionated antibody. The antisera of potencies 2.2 and 0.5 mg/ml were used for these studies, and it was found that the sample peaks recorded as shown in Fig.2 depended on the amount of antigen added to a fixed amount of the antiserum. The peak heights decreased as the amount of antigen (HSA) was increased due to precipitation of the antibody, resulting in a decrease of antibody content in the supernatant. However, as shown in Fig. 2, the peaks increased again when antigen was added in excess of the precipitable antibody content.

A plot of the peak potentials as a function of the HSA concentration in each solution in Fig.3(curve A) shows a minimum is observed at approx. 30 - 40 µg HSA/ml. The initial decrease in potential is due to the precipitation of the antibody, but is followed by an increase on addition of excess HSA antigen which is detected by the electrode. The curve B in Fig.3 was obtained for calibration of the HSA antigen alone showing an increase in potential from about the minimum for the precipitin curve. The antigen also is known (Heidelberger *et al.*, 1929,1933, and 1935) to have an effect on the solubility of the precipitate, and Fig.3 shows that the precipitin curve differs from thestandard HSA plot in the region of antigen excess due to formation of soluble antigen/antibody complexes.

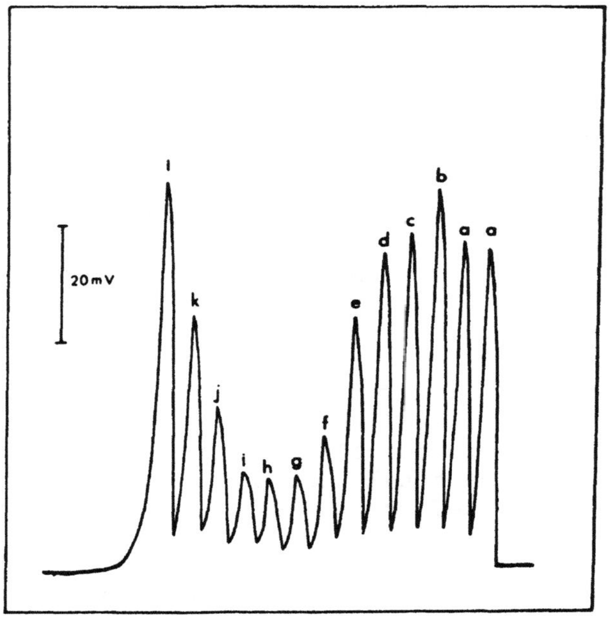

Figure 2. Recording of electrode potential for supernatant solutions after precipitin formation:(a)albumin 0.15 mg/ml; (b)antiserum blank; (c)2.4,(d)3.4,(e)6.8, (f)13.3, (g)25.8, (h)31.7, (i)47.5 (j)63.3,(k)79.2,(l)95.0 μg/ml HSA added to IgG fraction of anti-HSA, 2.2 mg/ml potency (reprinted by permission of *Anal.Chem.*,46,1253(1974).Copyright 1974, American Chemical Society).

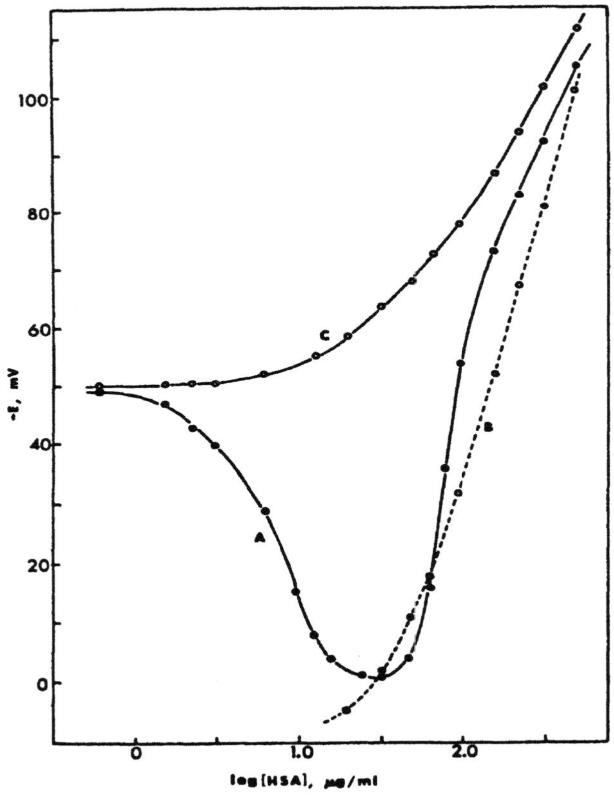

Figure 3. Peak potentials of supernatant sample solutions plotted as a function of total added HSA antigen concentration: (A) supernatant after precipitin formation; (B) antigen alone; (C) antigen plus γ-globulin (reprinted with permission from *Anal. Chem.*,46,1253,(1974).Copyright 1974, American Chemical Society).

Shape of Precipitin Curve

The data given in Fig.3 provide an alternative method of measuring the shape of the precipitin curve. Instead of measuring the antibody precipitated as in the conventional method (Kabat and Mayer, 1961), the quantity of unbound antibody and/or antigen was determined in the supernatant. Subtracting the blank γ-globulin electrode potential values (Fig.2, curve C) from the supernatant potentials (curve A) gave the true shape of the immuno-precipitin curve, as shown in Fig.4.

The results are shown for antisera with different potencies, viz. 2.2 and 0.5 mg/ml, and indicate the shape expected (Kabat and Mayer, 1961) with three zones. Zone A represents the region of antibody excess, zone B is the equivalence region, and zone C is the region of antigen excess. The potential changes for the more weakly potent antiserum were much smaller than for the high potency sample, with an antibody titre of approx. half the latter value. The silver selective electrode can therefore be used to detect quantitative changes during the titration of an antibody with an antigen.

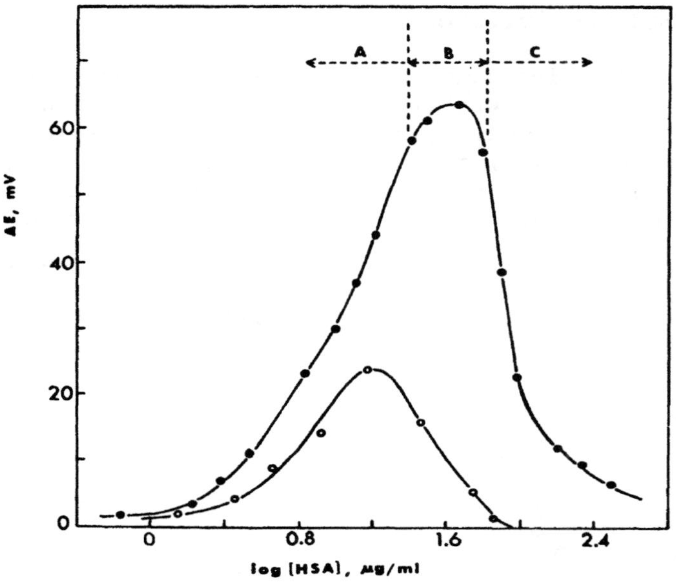

Figure 4. Potential difference between globulin blank solutions and supernatant solutions showing three precipitin zones: (A) antibody excess; (B) equivalence; (C) antigen excess - using anti-HSA serum with potencies (O) 0.5, (O) 2.2 mg/ml (reprinted by permission of *Anal. Chem.*, 46, 1253 (1974). Copyright 1974, American Chemical Society).

Interferences

The above method is valid, however, only in the absence of non-specific proteins which may also give a response at the electrode membrane. These proteins may lead to a source of high blank readings which in fact may completely obscure the small potential changes caused by the precipitation of the antibody. As an example, the effect of γ-globulin on the HSA peaks observed for supernatant samples is shown in Table 1. Peak height potentials for supernatant solutions were measured after precipitatiom of the HSA antibody in the presence of increasing quantities of the globulin. Table 1 shows that peak heights increased markedly at ratios of globulin to HSA > 5:1. This effect is due to the response of the electrode to the globulin remaining in the supernatant after precipitaion of HSA antigen-antibody complex.

Table 1. The Effect of Non-specific Protein on Peak Heights of Supernatants[a]

[HSA][b] µg/ml	[γ-G][b] µg/ml	[γ-G]:[HSA] Ratio	Peak Height (mV)
6.8	0	0	44
6.8	4.8	0.7	43
6.8	36.2	5.3	47
6.8	60.4	8.9	54
13.3	0	0	23
13.3	9.6	0.7	26
13.3	72.5	5.5	38
13.3	120.8	9.1	46

[a] Antiserum potency 2.2 mg/ml. [b] Total concentrations in final volume of 1.2 ml

Analysis of Washed Precipitates

The specific determination of an antigen in serum is shown in Fig.5, in which the analysis of the washed precipitates is necessary to avoid the interference from non-specific protein discussed above. To show the feasibility of this method, the HSA concentration in human serum control samples with certified HSA content was determined with whole goat anti-serum to HSA using 0.1 ml of antiserum (3.6 mg/ml potency).

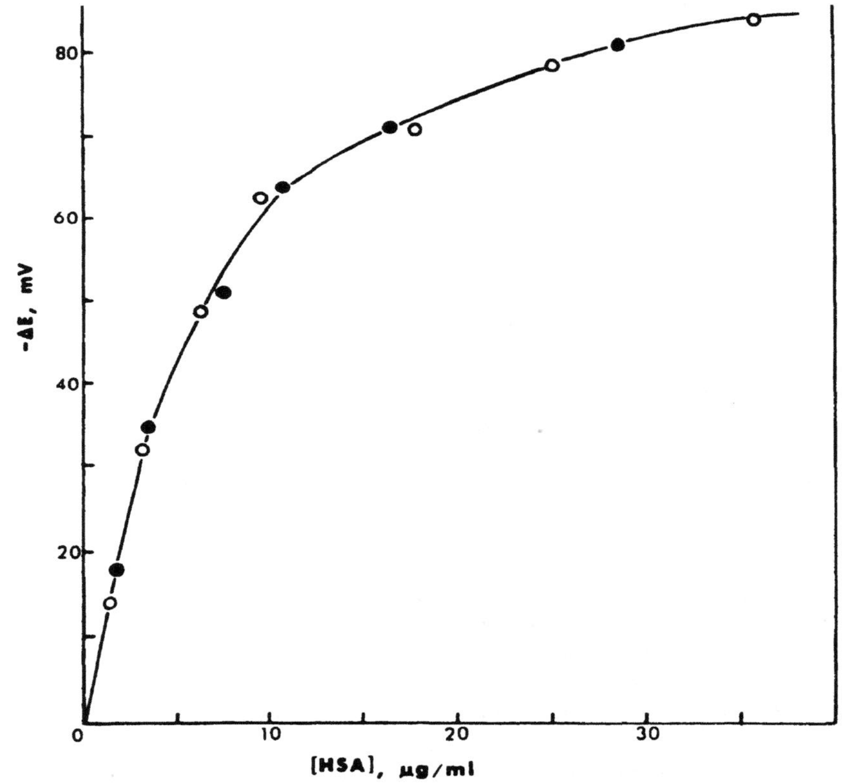

Figure 5. Calibration plot for determination of HSA antigen solutions by analysis of washed immuno-precipitates: (O)HSA in aqueous standards;(O)HSA in serum controls - determined with whole antisera, potency 3.6 mg/ml (reprinted by permission of *Anal.Chem.*, 46,1253,(1974).Copyright 1974, American Chemical Society).

For comparison, pure aqueous standard solutions of HSA were also determined with the same antiserum. The HSA was precipitated by reaction with the goat antiserum, the precipitates were centrifuged off and washed to remove unreacted protein, and then redissolved in alkali before determination in the continuous-flow system.

The data shown in Fig.5 give the peak heights observed for the serum and standard HSA samples. The peak heights were plotted against the known HSA content of the aqueous standards and the control serum samples. It is clear that the experimental data fall on a single curve showing good agreement between the two types of HSA samples analysed. This technique therefore by removing non-specific protein allows determination of low antigen concentrations in the range 0.5-30 μg/ml, and requires only small quantities of unlabelled antibody preparations (0.1 - 0.3 mg) for each determination.

DISCUSSION

The formation of precipitates in immunochemical reactions has been studied by many authors, as reviewed by Kabat and Mayer (1961) and Williiams and Chase (1971), and has been attributed to a simple chemical equilibrium for reaction of the antigen with the antibody:

$$x(AG) \quad + \quad y(AB) \quad \rightarrow \quad (AG)_x(AB)_y$$

where x moles of the antigen (AG) react with y moles of the antibody(AB). The response of the silver sulfide electrode depends on the hydrolysis of either the protein molecules in the immunoprecipitate, (AG)x(AB)y, or of the proteins remaining unprecipitated in the supernatant. In either case, the hydrolysis gives free thiol groups which change the silver ion activity in the reagent flow stream due to silver-thiol complex formation. The electrode therefore senses the remaining uncomplexed silver ion in the flow stream and gives a response on the strip chart recorder.

Fig. 6 shows the silver ion electrode response data obtained from the analysis of supernatant solutions where the weight ratio of antibody to antigen in the immuno-precipitates has been calculated as a function of the added weight of antigen.

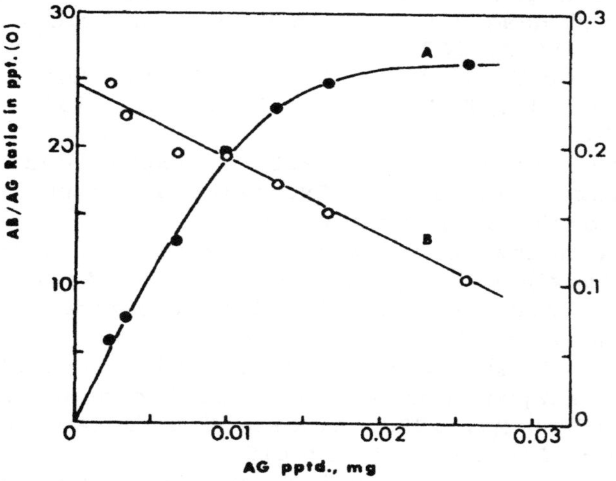

Figure 6. Effect of added antigen on the ratio of antibody to antigen in the precipitate formed by reaction of HSA with anti-HSA. (A) Antibody quantity precipitated; (B) Ratio of AB/AG in the precipitate (published with permission of *Anal. Chem.*, 46, 1253, (1974). Copyright 1974, American Chemical Society).

The plots in Fig.6 were calculated from mV data shown in Fig.3, and the AB/AG ratio in the precipitate is calculated on the assumption that the total antigen added prior to the equivalence zone is precipitated. It can be seen (plot B) that the data fall on a straight line, agreeing with the theory for precipitin formation developed by Heidelberger *et al.* (1935). The ratio of antigen to antibody in the precipitate formed has been shown (Heidelberger *et al.*, 1935) to vary according to the zone on the precipitin curve. When the antigen precipitated is plotted against the weight of antibody precipitated as shown in Fig.6 (curve A), then the curve reaches a maximum value as expected when the total antibody is precipitated.

The electrode data have therefore been shown to agree with concepts developed on precipitin formation, and allows the determination of antigens or antibodies in serum samples with a procedure considerably simplified in comparison to the more commonly used procedures such as radioimmunoassay (Williams and Chase, 1971), or enzyme-linked immunoassay (Wisdom, 1976). The technique described here depends on the physical separation of the immuno- precipitate and is therefore limited in sensitivity for detection of an antigen to approx. 0.5 μg/ml. Subsequent developments by Solsky and Rechnitz (1978) have shown that the silver electrode can be used in an automated flow system without the need to physically remove the precipitate from the sample prior to analysis. Labelling of the antibody with horseradish peroxidase has also been suggested (Alexander and Maitra, 1982) as a means for more sensitive detection of the precipitin formation by use of a fluoride selective electrode, as discussed in a later chapter of this book.

REFERENCES

Alexander, P.W. and Maitra, C., 1982, Enzyme-linked immunoassay of human immunoglobulin G with the fluoride ion selective electrode, Anal. Chem., 54: 68-71.

Alexander, P.W. and Rechnitz, G.A., 1974a, Serum protein monitoring and analysis with ion selective electrodes, Anal. Chem., 46: 250-254; 1974b, Automated protein determinations with ion selective membrane electrodes, Anal. Chem., 46: 860-864; 1974c, Ion-electrode based immunoassay and antibody-antigen precipitin reaction monitoring, Anal. Chem., 46: 1253-1257.

Gebauer, C.R. and Rechnitz, G.A., 1981, Immunoassay studies using adenosine deaminase enzyme with potentiometric rate measurement, Anal.Letters, 14:97-107.

Heidelberger, M. and Kendall, F.E., 1929, J.Exptl. Med., 50:809.

Heidelberger, M., Kendall, F.E. and Soo Hoo, C.M., 1933, J. Exptl. Med., 62:697.

Heidelberger, M. and Kendall, F.E., 1935, J. Exptl. Med., 62:697.

Kabat, E.A. and Mayer, M.M., 1961, "Experimental Immunochemistry", 2nd ed., C.C. Thomas, Springfield, Ill..

Keating, M.Y. and Rechnitz, G.A., 1985, Potentiometric enzyme immuno-assay for digoxin using polystyrene beads, Anal.Letters, 18:1-10.

Llenado, R.A. and Rechnitz, G.A., 1973, Ion electrode based autoanalysis system for enzymes, Anal. Chem., 45:826-830.

Ngo, T.T., Bovaird, J.H. and Lenhoff, H.M., 1985, Separation-free amperometric enzyme immunoassay, Appl.Biochem.Biotech,11:63-70.

Ngo, T.T. and Lenhoff, H.M., 1983, Amperometric assay for collagenase, Appl. Biochem. Biotech., 8:407-410.

Solsky, R.L. and Rechnitz, G.A., 1978, Automated immunoassay with a silver sulfide ion electrode, Anal. Chim. Acta, 99: 241-246.

Williams, C.A. and Chase, M.W., Ed., 1971, "Methods in Immunology and Immunochemistry", Vol.III, Academic Press, New York.

Wisdom, G.B., 1976, Enzyme immunoassay, Clin. Chem.,22:1243-1257.

ENZYME-LINKED IMMUNOASSAY WITH A FLUORIDE ION SELECTIVE ELECTRODE

Peter W. Alexander

Department of Analytical Chemistry
University of New South Wales
P.O. Box 1, Kensington
N.S.W., Australia 2033

INTRODUCTION

Enzyme-linked immunoassay (ELISA) techniques are of value as a means of sensitive detection of antigens or antibodies as an alternative to the use of a radioactive label in radioimmunoassay. Most of the ELISA methods have relied on spectroscopic techniques for detection of reaction products occurring when an enzyme labelled antibody reacts with an antigen, as reviewed by several authors (Wisdom, 1976; Schuurs *et al.*, 1977). The development of electrochemical techniques for the detection of antigens or antibodies by immunochemical reactions is discussed in the present book as offering certain unique capabilities which cannot be performed by spectroscopic methods, including possible applications as biosensors for *in vivo* monitoring, the development of sensors which are free from color and turbidity effects, and the development of simplified automated analysis systems.

In an earlier chapter, the measurement of the classical precipitin reaction was described using an indirect detection method with a silver ion selective electrode (Alexander and Rechnitz, 1974a,b,c; Solsky and Rechnitz, 1978). The use of an enzyme label, however, has been described (Alexander and Maitra, 1982) in order to obtain better sensitivity for detection of antigens after precipitation by reaction with a labelled antibody. This method is a heterogeneous immunoassay and depends on the precipitation of a labelled antibody, in this case anti-immunoglobulin G of human origin, using horseradish peroxidase (HRP) as the label. The enzyme label is detected by reaction with an appropriate substrate containing fluoride ion, such as p-fluoroaniline, and monitoring the fluoride ion released in the reaction.

The development of other enzyme-linked immunoassay procedures with ion selective electrodes has been reported by numerous authors, as reviewed in this book. Examples include the use of an iodide selective electrode by Boitieux et al.(1979, 1984), and a CO_2-electrode by Keating and Rechnitz (1985). The following sections give a summary of the detection method developed with the fluoride selective electrode, and include studies on the sensitivity, interferences to the enzyme method, and development of the flow analysis technique reported previously by Alexander and Maitra (1982).

EXPERIMENTAL

Potentiometric measurements were performed with an Orion mV meter, model 701A, connected to a Mace strip chart recorder, Model FBQ 100. The electrodes used were an Orion fluoride electrode, Model 94-09, provided with a flow-through cap of construction described (Alexander and Seegopaul, 1980), and an Orion double-junction reference electrode, Model 90-02. The electrodes were fitted into the flow system shown in Fig.1, with a constant temperature water bath at 36°C for control of the temperature in the flow system. The proportioning pump shown in Fig.1 was a Desaga pump, Model No. 100-364, with 6-channels. Flow rates in each channel are indicated in Fig.1.

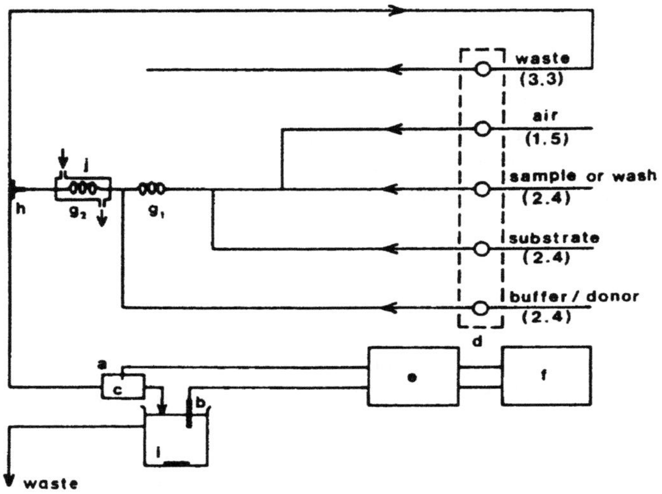

Figure 1. Schematic diagram of the flow system for HRP detection with flow rates (ml/min) marked in parentheses: (a) indicator electrode, (b) reference electrode, (c) flow-through cell, (d) proportioning pump, (e) mV meter, (f) recorder, (g) mixing coils, (h) debubbler, (i) buffer, (j) constant temperature water jacket (published with permission from *Anal. Chem.*, 54, 68 (1982). Copyright 1982, American Chemical Society)

Reagents The flow system for detection of the enzyme, HRP, was operated with the following reagents: hydrogen peroxide (4.45 x 10^{-3}M) as substrate, p-fluoroaniline (0.104 M) as hydrogen donor in 0.16 M sodium acetate - acetic acid buffer, pH 4.7, and horseradish peroxidase, Type II, with specific activity 100 units/mg (Lot No. 1259124, Boehringer Mannheim, Sydney).

The antiserum was HRP-labelled anti-human IgG conjugate (H and L chain) from a rabbit, Lot No. S589, Code No. 61-231, supplied by Miles-Yeda Ltd., Israel, with labelled enzyme activity of 520 units/ml. Unlabelled rabbit anti-human IgG was obtained from ORCM, Behringwerke, lot No. 03106, 1.7 mg/ml (equivalent to 19 I.U./ml). Human IgG standard serum samples were supplied by Behringwerke, Germany, batch No. a 375H, with potency of 193 mg/100ml (1 ampule with 81.47 mg of lyophilized serum contains 100 I.U. each of IgG, IgA, and IgM).

Procedures Standard HRP solutions were prepared by taking aliquots of 20 - 2000 µl of a stock HRP solution (27.6 units/ml) and diluting to 100 ml with water. The labelled antibody solution was standardised against these solutions. To test the enzyme activity of the labelled antiserum preparations, 100 µl was diluted to 10 ml, and aliquots of this solution from 20 - 1000 μl were then diluted to 10 ml. The diluted standards and antiserum were then pumped consecutively into the flow system shown in Fig.1 using a 30 sec sampling time and a 4-min water wash between samples. The concentration of fluoride generated by the enzymatic catalysis was monitored on the strip chart recorder, and peak heights for each standard were plotted as a function of HRP activity to construct a calibration curve. The peak heights for each diluted antiserum solution were measured and the concentration was then read off the curve.

Immunochemical reactions were carried out in the following way. Aliquots of the standard IgG serum sample (3 - 50 μl) containing 193 mg/100 ml were added to a fixed quantity (100 μl) of the undiluted HRP labelled antiserum of potency in the range 1 - 2 mg/ml in centrifuge tubes. The total volume was kept constant by adding distilled water to give 0.15 ml total volume. The precipitates were incubated for 1 h at 37°C and left overnight in a refrigerator at 4°C. The tubes were centrifuged for 20 min and the supernatant was drawn out and placed in a clean tube. The precipitates were washed twice with 1 ml of cold saline solution and once with 1 ml of cold water. After the final washing and decanting, the precipitates were dissolved in 1.0 ml of 0.16 M acetate buffer. The solutions were then quantitatively transferred to a 5.0 ml volumetric flask and diluted to volume with distilled water. The HRP activity of each solution was determined by aspirating the solutions into the flow system shown in Fig.1. The sampling rate was 12 per h, with a sample- to-wash ratio of 1:8, using a wash of 0.03 M acetate buffer, instead of water as used for the standard IgG solutions. Peak heights for each of the test solutions were recorded and plotted as a function of antigen concentration.

The supernatant was also analyzed by taking an aliquot of 30 μl and diluting to 25.0 ml total volume with distilled water. These samples were then fed into the continuous-flow system to determine enzyme activity remaining in the supernatant after immuno-precipitation.

For blank measurements, an unlabelled antiserum of potency 1.7 mg/ml was used for immuno-precipitation reactions as descibed above. A 20 μl aliquot of the standard IgG solution was added to 100 μl of the anti-serum, and the precipitate was treated as above, dissolved in 1 ml of 0.16 M acetate buffer, and diluted to 5.0 ml with water. This solution was then aspirated into the flow system to test for electrode response to the proteins in the absence of the enzyme label.

RESULTS

The classical precipitin technique was used for the studies discussed here, but with the antibody labelled with HRP. In the study with the silver sulfide electrode (Alexander and Rechnitz, 1974c), the antibody was not labelled and the total protein in the precipitated antibody-antigen complex was detected by hydrolysis of the protein in strong alkali. The silver electrode then responds to the thiol groups released after hydrolysis. The present technique with a fluoride electrode as sensor eliminates the need for the use of strong alkali by making use of an enzyme label. Horseradish peroxidase was employed as the label for the anti-IgG, and the enzyme was detected with a fluoride ion selective electrode, as shown in Fig.2.

The enzyme was detected after HRP-catalysed oxidation of p-fluoro-aniline with hydrogen peroxide according to the equation following:

$$2\ H_2N\text{-}C_6H_4\text{-}F\ +\ H_2O_2\ \rightarrow\ F\text{-}C_6H_4\text{-}N{=}C_6H_4{=}NH\ +\ H^+\ +\ F^-\ +\ 2H_2O$$

This reaction has been shown (Maitra, 1981) to have an optimum rate of F^- production at pH 4.7, and allows detection of HRP in the activity range from 0.005 - 0.80 units/ml. It can be seen from Fig. 2 that peak heights higher than 100 mV are obtained for the high enzyme activity in the flow analysis system with the fluoride electrode sensor. This is considerably higher than obtained in other ion selective electrode methods reported for enzyme-linked immunoassays (Boitieux *et al.*,1979,1984; Keating and Rechnitz, 1985), and may offer an advantage in terms of signal-to-noise ratio over other methods. However, the low pH required is not optimum for the immuno-precipitin technique, as discussed below.

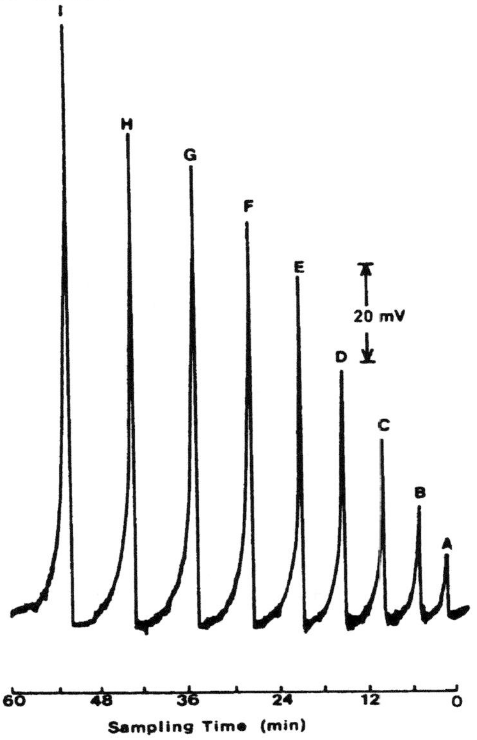

Figure 2. Sample peaks corresponding to various activities of HRP standard solutions sampled in the flow-system: (a) 5.5×10^{-3}, (b)1.4×10^{-2}, (c)2.8×10^{-2}, (d)5.5×10^{-2}, (e)1.1×10^{-1}, (f)1.7×10^{-1}, (g) 2.2×10^{-1}, (h) 2.8×10^{-1}, (i) 5.5×10^{-1} units/ml (reprinted with permission of *Anal. Chem.*,54,68(1982).Copyright 1982, American Chemical Society).

HRP Calibration Plot

The peak heights from Fig.2 were plotted as a function of HRP activity of standard solutions in order to obtain a calibration plot for the determination of the activity in the HRP-labelled antiserum solutions. Fig.3 shows the plot which indicates detectable enzyme activity from 0.005 to 0.80 units/ml. Peak heights ranged from approx. 10 - 120 mV with a linear range almost 2-orders of magnitude. A slope of approx. 30 mV was found for 1-decade change in HRP activity.

Precipitation of Labelled Antibody

The enzyme activity of the commercially supplied labelled-antiserum was tested first against the standard HRP peaks shown in Figs. 2 and 3. The antiserum was diluted by factors in the range 1: 1000 - 1:50,000 and then the enzyme activity was determined in the flow-system in Fig.1. The sample peaks are shown in Fig.4 indicating enzyme activites detectable from 0.002 - 0.12 units/ml. The sensitivity of this method therefore gives larger mV changes than the silver sulfide (Alexander and Rechnitz, 1974c) and iodide electrode (Boitieux *et al.*,1979,1984) methods for detection of electrode potential changes due to protein or enzyme reactions.

For the specific detection of the IgG protein in serum, the HRP-labelled antibody was precipitated by reaction with the IgG antigen. The precipitate was centrifuged, washed to remove unprecipitated antibody, and then redissolved in 1.0 ml of acetate buffer at pH 4.7.

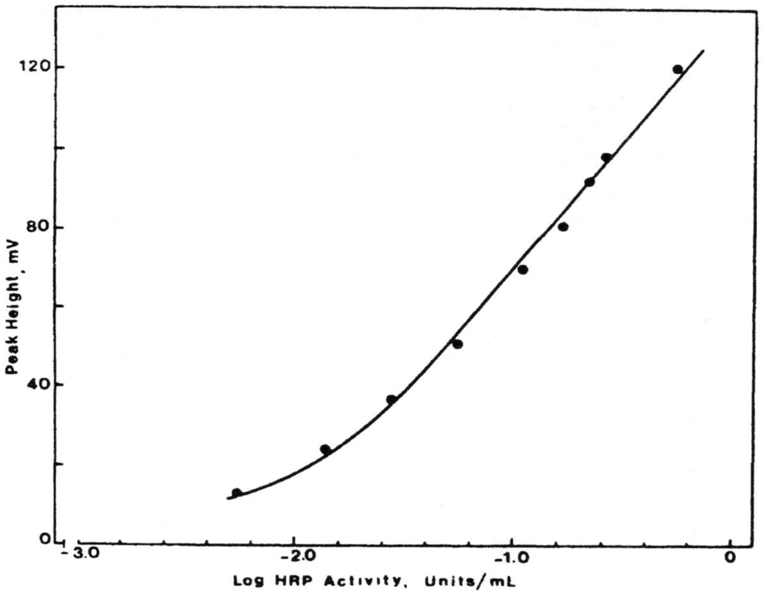

Figure 3. Calibration plot of peak height as a function of HRP activity in aqueous standard solutions shown in Fig.2 (published with permission of *Anal.Chem.*,54,68(1982).Copyright 1982, American Chemical Society)

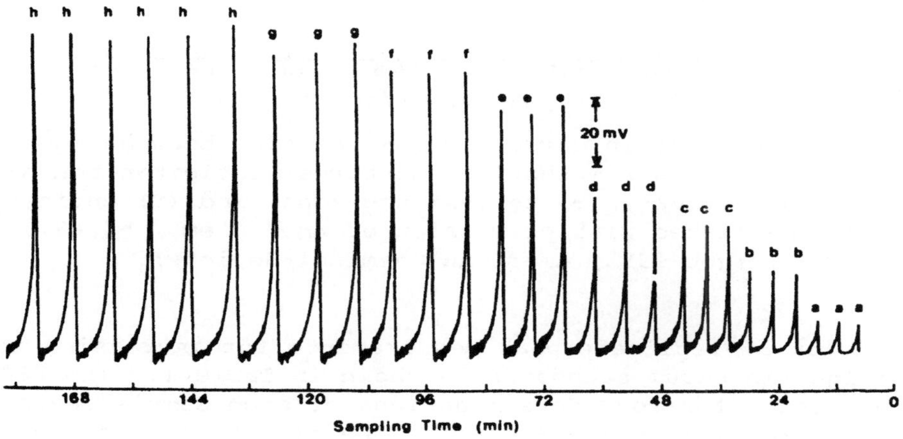

Figure 4.Sample peaks corresponding to various activities of HRP-conjugated anti-human IgG: (a) $2.0x10^{-3}$, (b) $7.5x10^{-3}$, (c) $1.5x10^{-2}$, (d) 2.1x10-2, (e) 0.05, (f) 0.08, (g) 0.10, (h) 0.12 units/ml. (published with permission of *Anal.Chem.*, 54,68(1982).Copyright 1982, American Chemical Society).

Due to the low pH necessary for optimum enzyme activity, the use of a homogeneous technique was not possible in the flow system used here because the precipitate redissolved at the pH in the flow system. The separation of the precipitate from the serum solution was therefore necessary to remove the unreacted enzyme-labelled antibody in order to prevent interference to the measurement of the precipitated enzyme-label. After separation and washing of the precipitate, the sample solution was then aspirated into the flow system to determine the concentration of enzyme in the precipitate solution.

The curve shown in Fig.5 is a result of plotting the enzyme peak height vs. IgG concentration added to the labelled-antiserum solution to precipitate the antibody label. The curve shows the typical classical precipitin shape (Alexander and Rechnitz, 1974c) with three precipitin zones labelled A, B, and C. The zone of antibody excess (A) shows increasing precipitin formation as IgG is added to excsess antibody, followed by the equivalence zone (B) and the zone of antigen excess (C) where the precipitin redissolves. Hence, the electrode method can detect changes in the antibody : antigen ratio over the entire curve and shows that the HRP enzyme label retains activity after immuno-precipitation and redissolution.

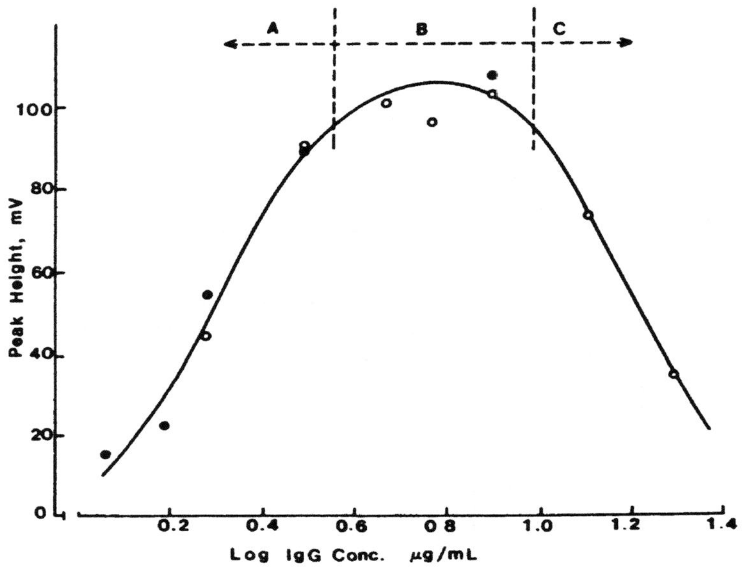

Figure 5. Precipitin curve for the reaction between IgG and the labelled antiserum showing the three precipitin zones: (A) antibody excess, (B) equivalence zone, and (C) antigen excess. (Published with permission of *Anal.Chem.*, 54,68, (1982).Copyright 1982, American Chemical Society).

The precision of replicate peak measurements for increasing concentrations of IgG in serum standards is shown in Table 1. The RSD values were 2.0% or better for four determinations of each sample containing IgG concentrations from 1.16 - 12.74 mg/ml. The results showed that precise analytical data are obtained for serum samples with certified IgG content.

Aqueous standards with pure IgG content are not required in this procedure because the serum components which may interfere are removed by washing of the precipitates. Direct analysis of the supernatant remaining after collection of the precipitate should therefore be possible, providing the HRP label is totally conjugated to the IgG antibody. Any unconjugated HRP would lead to high blanks.

Table 1. Peak Height Precision for Samples used for Measurement of the Precipitin Curve

IgG Conc. (μg/ml)	Peak Height* (mV)	Std.Dev. (mV)	RSD (%)
1.16	14.9	0.2	2.0
1.54	22.3	0.2	1.0
1.93	54.8	0.4	0.7
3.09	89.3	1.0	2.0
12.74	73.5	0.9	1.0

* Mean of four replicate determinations

Another source of high blanks may be caused by the response of proteins at the ion selective membrane. With the fluoride electrode, the blank reading was tested by precipitating unlabelled IgG antibody with the IgG antigen, redissolving the precipitate after washing, and then aspirating the solution into the flow system. A blank peak of less than 2 mV was recorded, indicating negligible interference from the protein on the fluoride electrode response under the test conditions used at pH 4.7. This effect would lead to blank readings when using the silver sulphide electrode method (Alexander and Rechnitz, 1974c) because hydrolysis of total protein is required in order for the electrode to respond by generation of thiol groups.

DISCUSSION

The fluoride electrode method described for enzyme-linked measurement of the immunoprecipitin curve has been shown to possess high sensitivity for detection of the HRP enzyme label. Dilutions of the commercially supplied antiserum have been detected down to 1:50,000 which is a dilution factor rivalling the sensitivity of radio-immunoassay techniques (Wisdom, 1977; Schuurs *et al.*, 1977). The advantages of the fluoride electrode method therefore are:

(1) High Sensitivity. The fluoride method gives high sensitivity for detection of HRP with peak heights > 100 mV. This is much more sensitive than other ion selective electrode methods, for example as proposed with the iodide electrode (Boitieux *et al.*,1979, 1984) where mV changes < 10 mV are reported.

(2) Rapid response. Detection of the IgG antigen down to 1.16 μg/ml has been demonstrated, with sampling rate of 12 per hour in the flow system, and with an antigen peak height detectable above the blank of approx. 0.5 μg/ml. This is about the same detection limit as the silver sulfide electrode method (Alexander and Rechnitz, 1974c) for the precipitin technique. However, both these methods rely on the physical separation of small quantities of the immuno-precipitate and this process is the limiting step in determining the sensitivity.

(3) Lack of interferences. The fluoride electrode is not responsive

to normal serum components or to protein molecules as is the silver sulfide electrode. The former electrode method is therefore less sensitive to potentially interfering organic and protein molecules normally present in serum, and may be applicable to supernatant analysis of unprecipitated enzyme label.

(4) Low Blanks. For the same reason as above, the fluoride electrode gives low blank readings when unlabelled antibody is collected and measured. Hence, non-specific protein molecules in serum samples will not interfere in the analysis.

The reason that the fluoride electrode could not be used for homogeneous immunoassay was the low pH 4.7 required for detection of HRP at optimum sensitivity. The low pH is due to the use of p-fluoro-aniline as substrate in the enzyme catalysed reaction. However, it may be possible in future studies to find a fluoride substituted substrate which will react optimally at higher pH and allow analysis of the supernatant for unprecipitated enzyme label to give a more sensitive and selective method than the present method without the need for physical separation of the immuno-precipitate.

REFERENCES

Alexander, P.W. and Maitra, C., 1982, Enzyme-linked immunoassay of human immunoglobulin G with the fluoride ion selective electrode, Anal.Chem., 54:68-71.

Alexander, P.W. and Rechnitz, G.A., 1974a, Serum protein monitoring and analysis with ion selective electrodes, Anal. Chem.,46:250-254; 1974b, Automated protein determinations with ion selective membrane electrodes, Anal. Chem., 46:860-864; 1974c, Ion electrode based immunoassay and antibody-antigen precipitin reaction monitoring, Anal. Chem., 46:1253-1257.

Alexander, P.W. and Seegopaul, P., 1980, Rapid flow analysis with ion selective electrodes, Anal. Chem., 52: 2403-2406.

Boitieux, J.L., Desmet, G. and Thomas, D., 1979, An antibody electrode, preliminary report on a new approach in enzyme immunoassay, Clin. Chem., 25:318-321.

Boitieux, J.L., Thomas, D. and Desmet, G., 1984, Un systeme potentio-metrique en phase heterogene pour le dosage enzymo-immunologique du 17 β-oestradiol, Clin. Biochem., 17:151-156.

Keating, M.Y. and Rechnitz, G.A., 1985, Potentiometric enzyme immunoassay for digoxin using polystyrene beads, Anal. Letters, 18:1-10.

Maitra, C., 1981, Continuous flow analysis with ion selective electrodes, Ph.D. Thesis, University of New South Wales.

Schuurs, A.H.W.M. and Van Weeman, B.K., 1977 Enzyme immunoassay, Clin. Chim. Acta, 81:1-40.

Solsky, R.L. and Rechnitz G.A., 1978, Automated Immunoassay with a silver sulfide ion selective electrode, Anal. Chim. Acta, 99:241-246.

Wisdom, G.B., 1976, Enzyme immunoassay, Clin. Chem., 22:1243-1257.

HETEROGENEOUS POTENTIOMETRIC ENZYME IMMUNOASSAY FOR ANTIGENS AND HAPTENS WITH IODIDE SELECTIVE ELECTRODE

Jean-Louis Boitieux*, Gerard Desmet** and Daniel Thomas*

* Laboratoire de Technologie Enzymatique, U.T.C.
60206 Compiegne, Cedex, France
** Laboratoire d'Hormonologie, Hospital-Sud
Chu d'Amiens, 80036 Amiens Cedex

INTRODUCTION

Numerous methods are currently proposed for accurate determination of antigens, haptens, and antibodies present in trace amounts in biological or industrial fluids. Among them, radioimmunoassay (RIA) and enzyme-immunoassay (EIA) are the most sensitive ones. Many of these immunoassays are being simplified to reduce costs and dependence on highly skilled personnel and to broaden their range of availability. Enzyme immunoassay techniques are extensively described and thoroughly discussed in many recent general reviews (Engvall, 1971) and(Pratt, 1978). Since the introduction of immunosorbents (Catt, 1967 ; Wide, 1967), many solid phase immunoassays have been described (Schuurs, 1977 ; Line, 1975) and are aimed at obtaining optimum results with enzyme-linked immunosorbent assays (ELISA), but for a long time little attention was given to protein supports. Recently, however, proteic membranes were introduced to original EIA techniques for determination of antigens and as hepatitis B surface antigen (HBsAg) and hapten.

Pure Immunoglobins (IgG) were immobilized on gelatin membranes using glutaraldehyde as the crosslinking. Optimal conditions for antibody immobilization were established using IgG labeled with horseradish peroxidase (HRP). Soluble HRP-labeled IgG were also used in a "sandwich" procedure, after immersion of the active membrane in a concentrated solution of highly purified antigen, to confirm the biological activity of the fixed antibodies. This estimation was carried out using a direct spectrophotometric method.

After preparing specific antibodies from antisera obtained in the rabbit according to a procedure recently described (Desmet, 1978), we have studied the immobilization of anti-HBsAg IgG on different proteic membranes and developed a new method using an iodide sensitive electrode which has been modified by fixation of the antibody membrane onto the crystal sensor. The EIA was realized according to the "sandwich" or "competitive" procedures by using respectively HRP-labeled antibodies or haptens.

The enzymatic activity of the membrane was detected by the electrode in the presence of substrate (H_2O_2) and potassium iodide (I^-) as proposed by Nagy et al.(1973) for the estimation of glucose in blood according to the following reaction : $H_2O_2 + 2\ I^- + 2\ H^+ \xrightarrow{HRP} 2\ H_2O + I_2$

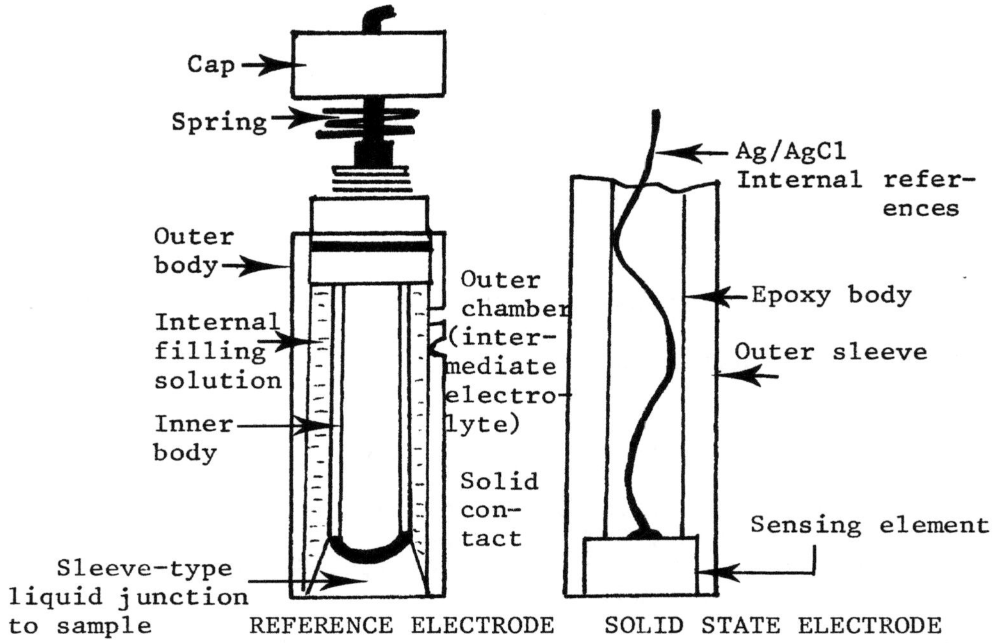

Fig. 1 : Schematic representation of Iodide sensitive and reference electrodes.

We also studied the effects of different variables such as incubation time, temperature, pH and other parameters, on the formation of the antigen-antibody complex and on the potentiometric kinetics before determining the sensitivity, reproducibility and reliability of the procedure for the quantitative estimation of HBsAg subtypes ad and ay and 17 β oestradiol. The quantity of immobilized antigens on the membrane is proportional to the membrane peroxidasic activity in the "sandwich" procedure, and is inversely proportional in the"competitive" method.

MATERIALS AND METHODS

1. Materials

Potentiometric measurements were made with a instrumental operational amplifier with a high impedence input. The amplified signal was fed into Orion Digital pH meter (Model 701) equipped with an iodide selective electrode (Orion 9453) and a reference electrode Orion 9001 (Figure 1).

Horse radish peroxidase (Type IV) and 17 β oestradiol were purchased from Sigma Chemical Co, St Louis, Mo. 63178. 1-fluoro-2-4 dinitrobenzene and 3,3'-dimethoxybenzidine (DMB) were supplied from Eastman Kodak Co, Rochester NY. 14650. Sodium m. periodate, ethylene glycol, sodium borohydride, and other reagents of analytical grade were obtained from Merck Darmstadt, FRG. Pig skin and ossein gelatins came from Rousselot Laboratory, 60400 Ribecourt, France. Oestradiol-17β and oestradiol-17 β -6-(o-carboxymethyl) oxime (oestradiol 6 CMO) were supplied by Steraloids. 3,3'-diaminobenzidine (DAB) was purchased from Eastman Kodak. HBsAg was prepared and purified from a pool of positive human sera as described by Boitieux and al (1978). Antisera against HBsAg raised in rabbits and IgG were prepared in our laboratory.

For the quantitative determination of HBsAg, we used the panel of the Reference Center for Virus Hepatitis, at the Institute of Hygiene of the University of Göttingen, and the reference was obtained by courtesy of Dr Gerlich. All other chemicals used were of analytical grade.

2. Methods

2.1. Preparation and purification of antibodies to HBsAg

Antisera were obtained after 2 cycles of injections of partly purified antigen emulsified in incomplete Freund's adjuvant, as described by Desmet (1977). During the first cycle, intraperitoneal injections of emulsion containing 5 mg antigen were made at 1 week intervals. At the end of the first series, after a rest of 7 days, the rabbits were submitted to a second cycle of injections. The blood was collected 7 days after the last injection by heart puncture.

Specific antisera were obtained by absorption of non-specific antibodies by proteins prepared from a pool of normal human sera. The antiserum was then submitted to a Sephadex G 200 preparation electrophoresis in a 0.02 M phosphate buffer pH 8.2. Cathodic fractions were collected and chromatographed on DEAE Cellulose with the same buffer. IgG were quantitatively eluted in the first peak.

2.2. Preparation of antibodies-HRP conjugates

Coupling was effected by the "Two step method" described by Avrameas (1969). The reaction between peroxidase and glutaraldehyde was realized during the first step and IgG was then chemically linked to the activated enzyme. Labeled IgG was purified by gel chromatography on Sephadex G 200. The labeling was checked by spectrophotometric measurement of peroxidase activity at 460 nm in presence of H_2O_2 and DMB. On the basis of the absorbance at 402 nm of the total peroxidase-IgG mixture and of the pooled material represented by the peak of peroxidase-labeled IgG, the amount of peroxidase bound to IgG was calculated ; it appears that two to three molecules of oxidized peroxidase can be attached to one IgG molecule.

2.3. Artificial proteic membrane

The study of the immobilization of a biologically active material has been vastly expanded in the past few years, notably in the field of enzyme research. These studies have resulted in numerous publications particularly concerning immobilized enzymes. We considered using proteic membrane used for enzyme immobilization by Thomas (1976) to fix antibodies and have principally studied 2 types of membranes.

The procedure for making these membranes is the same for each of the 2 types of collagen considered : alkaline-treated ossein collagen, (pi from 4.7 - 5.0) and acid hydrolysed pig skin collagen, (pi $\simeq$ 8.0). 10 g of collagen were dissolved in 100 ml distilled water for 24H at 37^{0}C. 1 ml from the solution is spread onto a 20 cm^2 surface polystyrene, a material chosen for its non-adherence. After air drying at room temperature, the membrane was peeled off and immersed in a freshly prepared solution of 5 % glutaraldehyde in 0.02 M phosphate buffer pH 6.8 for 5 min. The membrane was washed several times in distilled water, then saturated with specific antibodies for 30 minutes at 37°C and carefully washed again. To estimate the quantity of IgG fixed onto the membranes, we used HRP-labeled antibodies.

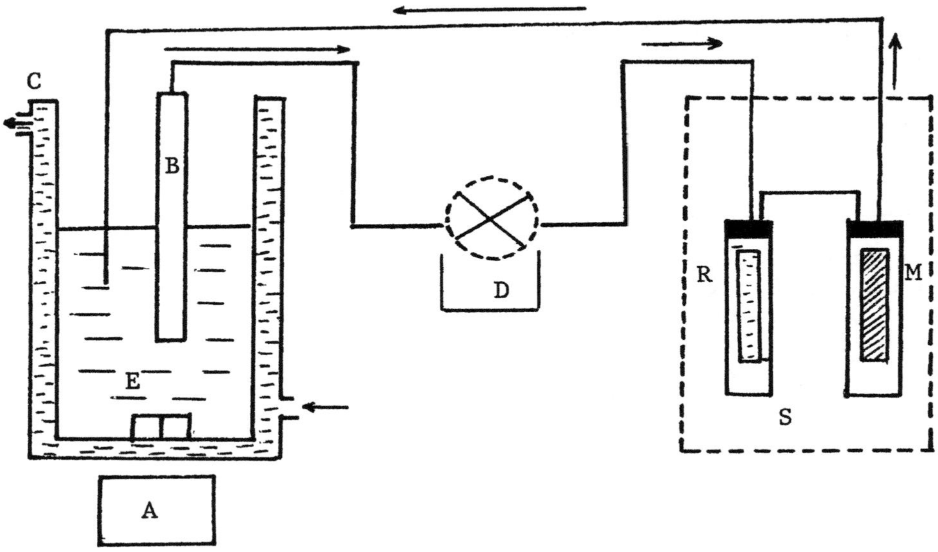

Fig. 2 : Schematic diagram of the system used for determination of enzymatic activity of membranes : A, stirrer ; B, trapping column of silica ; C, thermostated container ; D, peristaltic pump ; E, reaction medium ; M, membrane ; R, reference ; S, spectrophotometer

2.4. Determination of peroxidase activity by colorimetry (Figure 2)

The rate of decomposition of hydrogen peroxide by HRP in the presence of DMB was determined by measuring the rate of color development at 460 nm. 2,9 ml of 8.8 x 10^{-4} mol/l H_2O_2 containing 25 µl 1 % DMB were transfered into a 10 mm wide cuvette. After addition of 0.1 ml peroxidase solution to be determined, the absorption was read every 15 s for 2 min, the peroxidase activity being given by the rate of change of absorption.

When HRP-labeled antibodies were fixed on a membrane, the direct determination of the enzymatic activity was carried out in the presence of DAB (2.5 mM/l) and H_2O_2.

The membrane was placed in the center of the cuvette, perpendicular to the optical beam, then immersed into the reaction medium containing DAB in phosphate buffer 0.02 mMol/l), pH 6.8 . After base line stabilization reaction was started by addition of 20 µl 30 % H_2O_2. The slope of the curve obtained as a function of time is proportional to the quantity of HRP-labeled antibodies fixed onto the membrane (Figure 3).

2.5. Preparation of HRP-labeled oestradiol

Oestradiol 6 CMO-HRP conjugate was prepared as an enzyme-labeled antigen by the mixed anhydride method (Erlanger, 1957). The reaction mixture was dialyzed overnight against 0.01 mol/l phosphate buffer, pH 6.8, and then purified by gel filtration on Sephadex G 100 in the same buffer (Figure 4).

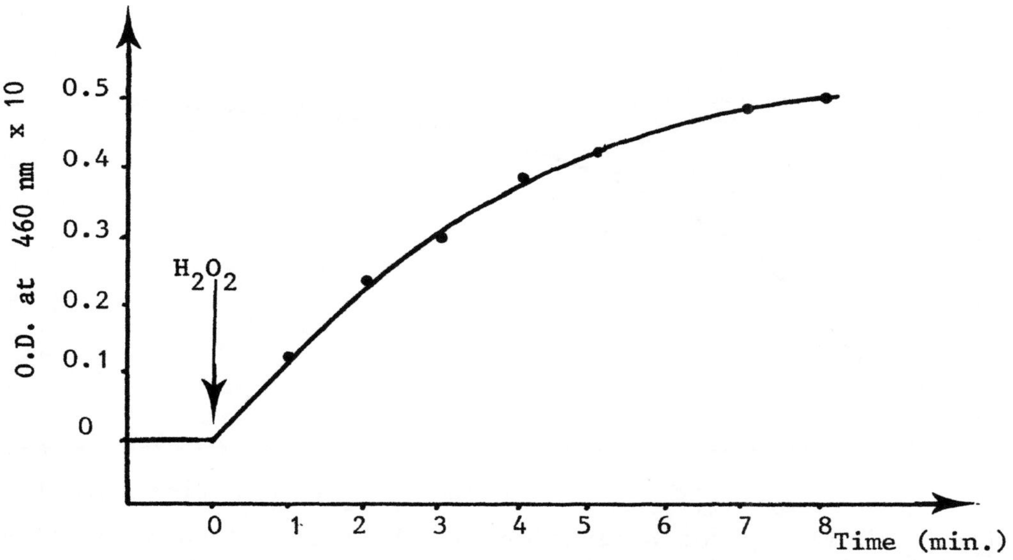

Fig.3 : Peroxidase activity of HRP-labeled antibodies in absorbance units as a function of time, in the presence of DAB and H_2O_2 in phosphate buffer, pH 6.8.

On the basis of the reading at 278 nm and 403 nm of eluates, the amount of HRP bound to haptens was calculated. It appears that 3 to 4 molecules of steroid can be attached to one HRP molecule. HRP-labeled oestradiol solutions may be stored at -20°C for several months.

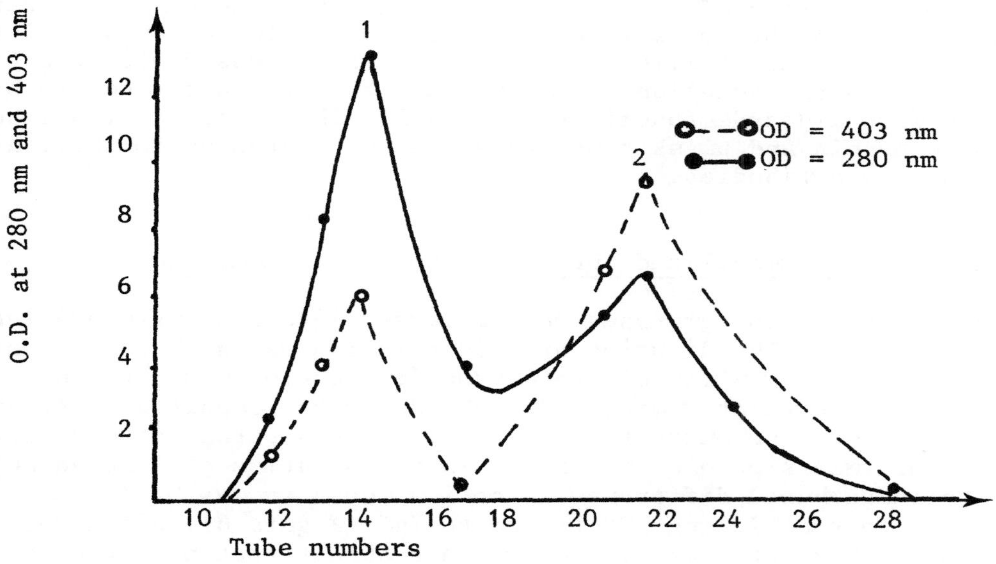

Fig.4 : Purification of HRP-labeled oestradiol by gel filtration on Sephadex G-100 columns (35 x 2.5 cm) in 0.01 mol/1 sodium phosphate buffer, pH 6.8. Peak 1 corresponds to enzyme-labeled steroid.

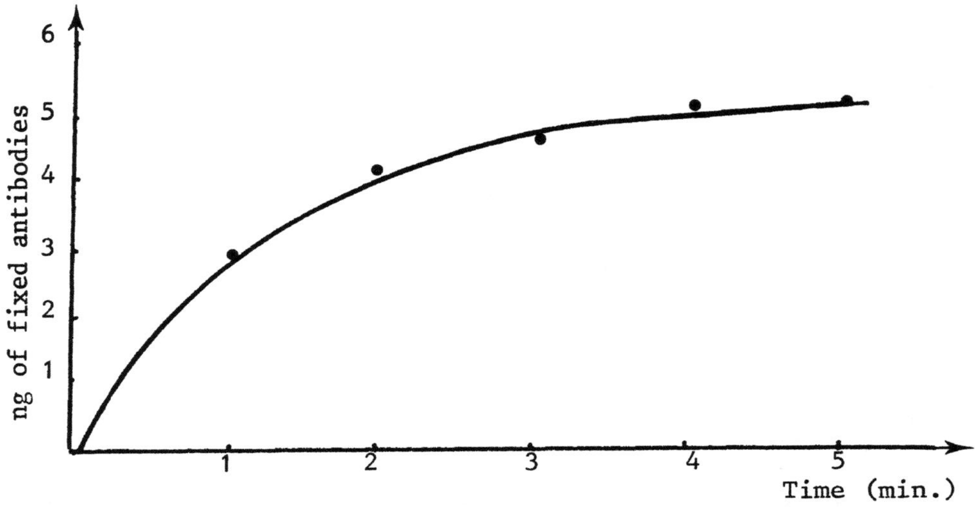

Fig. 5 : Effects of activation time by glutaraldehyde on binding of labeled antibodies to the artificial membranes.

RESULTS AND DISCUSSION

In a first step, we have studied the immobilization of antibodies on different proteic artificial supports.

The yield of fixation of antibodies was studied in relation to different parameters, such as collagen concentration, glutaraldehyde concentration and contact time with the bifunctional agent. It appears that the optimal concentration of collagen is 5 %, and that the maximal fixation occurs when the membrane is immersed 5 min in a solution of 5 % glutaraldehyde. The amount of immobilized antibodies increased with the time of contact with the solution to 4 min. Then the membrane seems to be saturated with aldehyde functions (Figure 5). The quantity of antibodies fixed onto ossein and pigskin membranes was estimated by immobilization of HRP labeled antibodies.

1. Estimation of immobilized antibodies on ossein membranes

Activated ossein membranes were immersed into a diluted HRP-labeled antibody solution. The dilution of this solution was such that absorption due to the colored product of the enzymatic reaction was between 0.1 and 0.5 absorbance units at 2 min, under the described conditions. After 30 min immersion, the enzymatic activity of the supernatant was determined. After immersion of 5 membranes in the solution of labeled antibodies the absorbance difference in optical density was 0.11, corresponding to the lost of 7.0 ng HRP linked to IgG (Figure 6). Since the HRP/IgG molecular ratio was 1 - 4, we were able to estimate the amount of immobilized antibodies corresponding approximately to 6.0 ng for each membrane (1 cm^2). Nevertheless, this was only an estimate because the amounts of labeled and free IgG which might have been immobilized on the membrane were not necessarily identical.

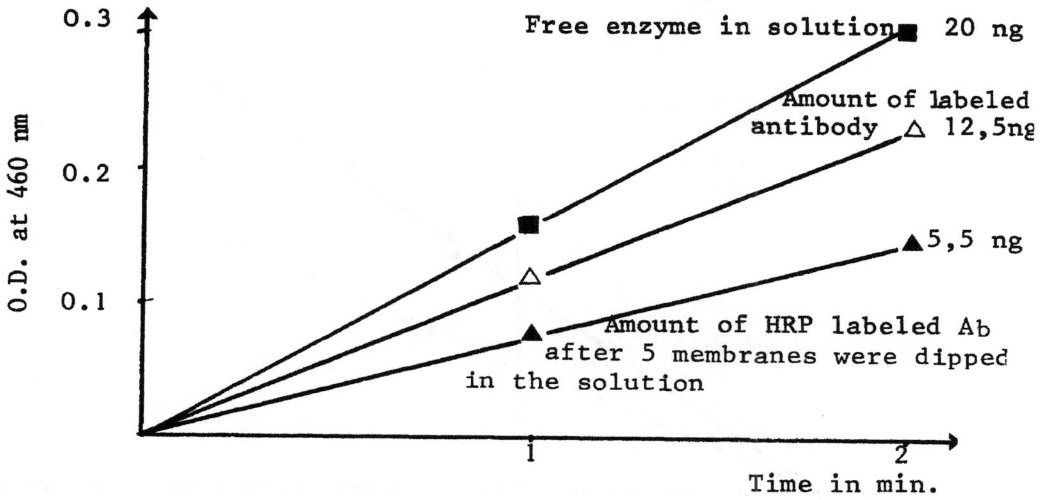

Fig. 6 : Determination of the amounts of HRP-labeled IgG immobilized on ossein membranes. 12.5 corresponds to the amount of antibodies labeled HRP in solution and 5.5 ng is the residual quantity of Ab coupled to HRP, after 5 membranes were dipped in the solution.

2. Estimation of immobilized antibodies on pig skin gelatin membranes

The quantity of labeled antibodies fixed onto these gelatin membranes in similar conditions was ≃ 26.5 ng/membrane (Figure 7). The variation of this amount was ± 0.4 ng after 15 assays. The spectrophotometric

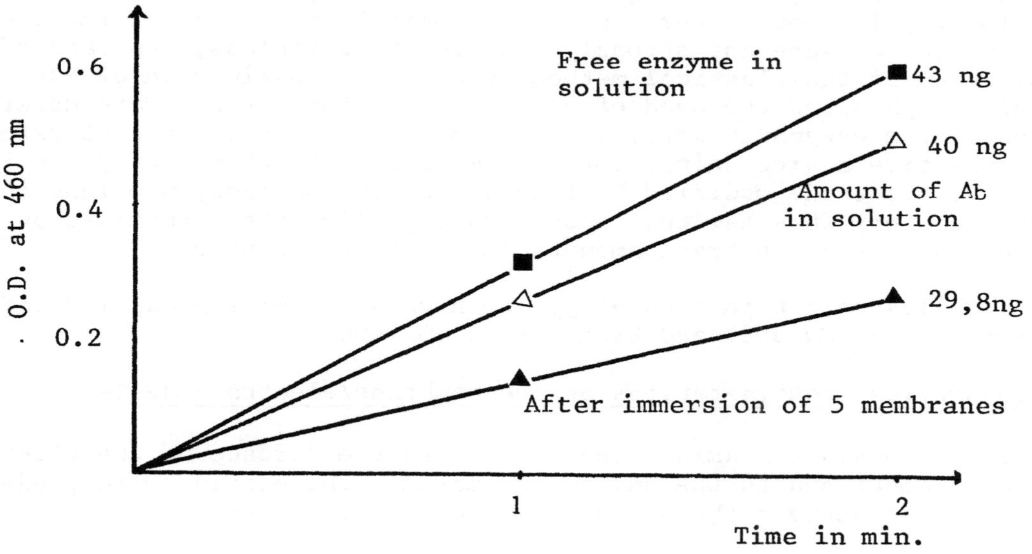

Fig. 7 : Determination of the amounts of HRP-labeled IgC immobilized on pig skin membranes. Amount of labeled IgC in solution was 40 ng and after 5 membranes dipped. The amount of IgC remaining was 29.8 ng.

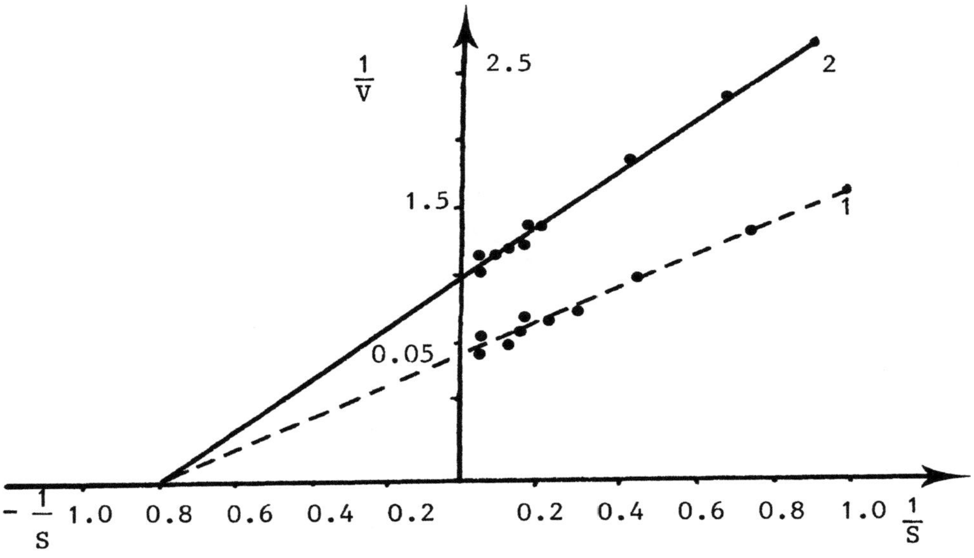

Fig. 8 : Determination of Michaelis constants (K_m and K_a) of peroxidase free in solution (1) and bound to IgG (2), according to the Lineweaver Burk representation. V_m = ΔOD/min and S = concentration of H_2O_2 (substrate) in the range 0.2 to $1.0.10^{-4}$ mol/l.

study of the enzymatic kinetics showed that HRP fixation to IgG and, indirectly, to membranes, did not significantly modify the K_m of the enzyme (Figure 8), which allowed a correct evaluation of the quantity of immobilized IgG.

After immobilization of anti-HBsAg antibodies, pig skin collagen membranes were immersed in concentrated solution of purified antigen for 30 min, then washed several times, and allowed to react for 30 min with labeled antibodies, according to the "sandwich" procedure. After several experiments, it appears that the immunological properties of the immobilized antibodies were not strongly altered. Nevertheless, the lack of sensitivity of the classical method used for the estimation of HRP activity emphasised the need of another technique for accurate determination of the enzymatic activity of membranes. Recently we elaborated a more sensitive device using, for determination of HRP activity, an iodide selective electrode modified by fixation of the antibody membrane on the crystal sensor. This new tool should easily allow the determination of substances present in trace amounts in biological fluids.

To illustrate this technology, we have used "biological models": oestradiol 17β (MW 300) and HBsAg (MW 2 400 000).

3. Theoretical considerations on the membrane/electrode design

If we consider x axis perpendicular to the surface of the electrode sensing element and to the antibody membrane, the origin corresponds to the interface between the sensor and the proteic membrane.

$S_1 = H_2O_2$

$S_2 = I^-$

e_1 is the thickness of the proteic membrane

e_2 is the thickness of the undisturbed limit layer

In applying a realist hypothesis for which the thickness of the active layer is such that it exists a unity of concentrations at this level, diffusion coefficient being assumed identical in the fields I and II (Figure 9), we can write :

Field I :
Evolution equations are :

$$\frac{\delta S_1}{\delta t} = D_1 \frac{\delta^2 S_1}{\delta x^2} \quad \text{and} \quad \frac{\delta^2 S_2}{\delta t} = D_2$$

$$\frac{\delta^2 S_2}{\delta x^2} \quad \text{with} \quad \frac{S_1}{x} = 0 \quad \text{and} \quad \frac{S_2}{x} = 0 \text{ if } x = 0$$

D_1 is the diffusion coefficient of H_2O_2 and D_2 the coefficient of I^-.

This corresponds to the expression of a molecular flow equal to zero through the sensitive surface and

$$(D_1 \frac{\delta S_1}{\delta x_1})_{xI = e_1} = (D1 \frac{\delta S_1}{\delta x_1})_{xII = e_1} + J_M f\ (S_1 e_1 S_2 e_1)$$

J_M is the maximal biochemical flow (number of transformed molecules per surface area and unit time in the course of the enzymatic reaction) and $f(S_1 e_1 S_2 e_1)$ is a function ranging from 0 to 1 which relies on enzyme activity and concentrations of both substrates.

Field II :
The equations are similar to those of Field I but the conditions at the limits if $x = e_1$ and e_2 is given by Dirichlet. S_1 and S_2 remain constant and equal to $S_1 e_2$ and $S_2 e_2$.

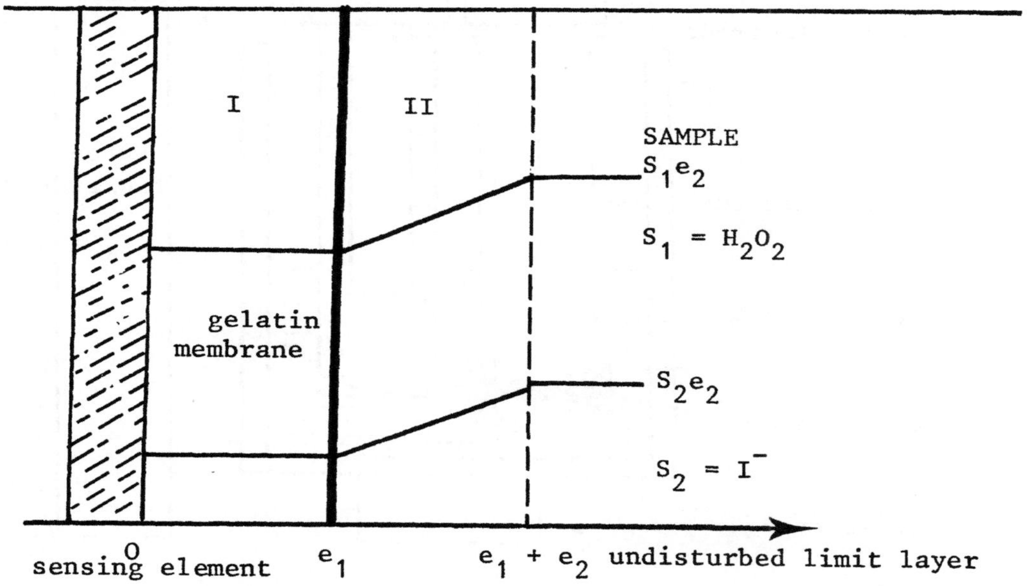

Fig. 9 : Schematic representation of concentration in space at the level of different layers of the proteic membrane
$S_1 e_2$: stationary state

In the stationary state, the equations are :

$$S_{1.0} = S_1e_1 \text{ and } S_{2.0} = S_2e_2$$

$$D_1 = \frac{S_1e_2 - S_1e_1}{e_2} = J_M f(S_1e_1, S_1e_2)$$

$$D_2 = \frac{S_2e_2 - S_2e_1}{e_2} = 2\ J_M f(S_1e_1, S_1e_2)$$

Knowing reaction term, diffusion coefficients and electrode response, we can determine J_M and, subsequently the amount of enzyme molecules immobilized on the proteic membrane.

4. Determination of HBsAg

4.1. Assay procedure

Discs measuring 10 mm in diameter and about 0.05 mm in thickness were cut out of the activated proteic membrane and saturated with a solution of anti-HBsAg antibodies (1 mg/1) for 60 min. They were then washed several times in distilled water, immersed for 30 min in the diluted solution of HBsAg, immobilized IgG was in excess with respect to the antigen, and thoroughly washed again.

The membrane was next incubated for 2hr.in 1 ml of solution containing 100 μg of peroxidase-labeled IgG, according to the "sandwich" method.

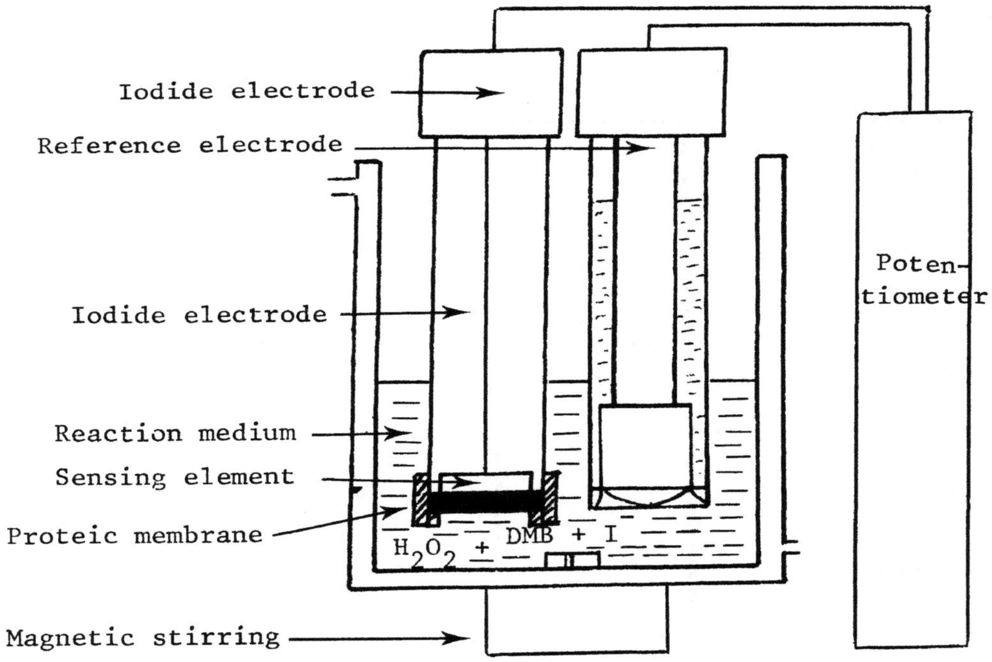

Fig. 10 : Schematic representation of the antibody electrode and the initial system for the potentiometric measurements of antigens in biological fluids.

After incubation, the membrane was washed with 0.1 mol/l sodium phosphate buffer, pH 5.0, and fixed onto the active surface of the iodide sensitive electrode, being kept in position by a rubber ring (Figure 10).

The enzymatic membrane activity was evaluated in the presence of hydrogen peroxide and iodides, under optimal conditions as described below, the electrode potential being a function of the antigen concentration in the biological fluid. The variation of the electrode potential after 1 min will be proportional to the enzymatic activity, and subsequently to HBsAg concentration, expressed in micrograms per liter. We determined the limit of saturation of the antibody membrane by the antigen by using concentrated solutions of HBsAg submitted to dilution series, which gives the maximal enzymatic activity of the membrane after realization of the "sandwich" procedure.

We first established the optimal conditions for detection of the enzymatic activity, studying the main factors affecting the sensitivity of the assay.

4.2. Effect of incubation time

Figure 11 shows the role of incubation time during the first incubation of the antibody membrane with the surface antigen. The duration varied from 5 to 60 min while the second incubation with HRP-labeled antibodies was maintained constant for 6 hr. This experiment was performed with a 4 µg/l solution of HBsAg (subtype ad). We observed an increase in binding of the antigen to the active membrane for 60 min.

In the second experiment, the first incubation time was constant (60 min), and the second varied from 1 to 9 hr.(Figure 12). In these conditions we selected incubation periods of 30 min for the first incubation and 2 hr. for the second.

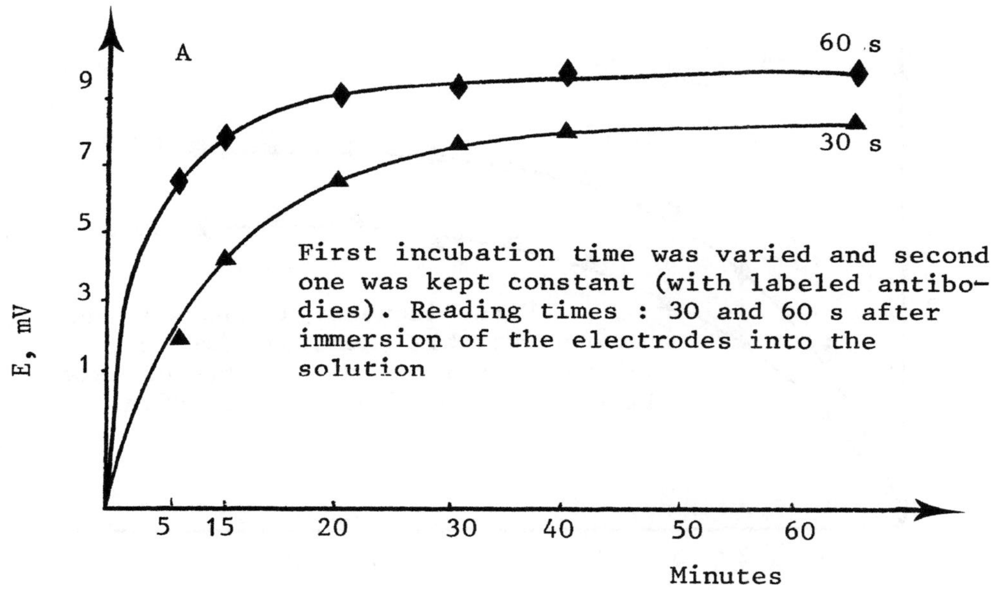

Fig. 11 : Kinetics of the binding antigens by antibodies immobilized on gelatin membrane to HBsAg in diluted solution.

4.3. Temperature dependence

We studied the variations in potential differences between reference and sensitive electrode with respect to temperature, varying from 10 to 40°C (Figure 13) and observed a maximal activity at 27°C.

This temperature dependence corresponds to the temperature dependent rate of the peroxidase catalyzed reaction limited by enzyme denaturation.

4.4. pH dependence

The pH dependence of the reaction rate was studied at 27°C between pH 4.5 and pH 7.0 for free peroxidase in aqueous solution (final concentration, 4 μg/l) in the presence of potassium iodide, hydrogen peroxide and3 3'dimethoxybenzidine , in final concentrations corresponding, respectively, to 0.1 mmol/l, 4.4 mmol/l and 50 mg/l. The optimum pH for peroxidase in solution is pH 6.5 but for the antibody-bound enzyme the reaction rate vs, pH shows a pH optimum at 5.0. A possible explanation for this pH shift may be the presence of ionizable groups borne by proteins, H^+ ions being necessary to dimethoxybenzidine activity and increasing the peroxidase reaction rate (Figure 14).The DMB was used as hydrogen donor.

4.5. Effect of hydrogen donor

We have used for this experiment three different hydrogen donors: ortho-phenylene-diamine (DAB), 3.3' dimethoxybenzidine (DMB) and ascorbic acid. The presence of DMB was used for increasing the quantity of hydrogen (H^+) which increases the rate of the reaction.

Figure 15 shows the maximal potential values are obtained with DMB as the hydrogen donor.

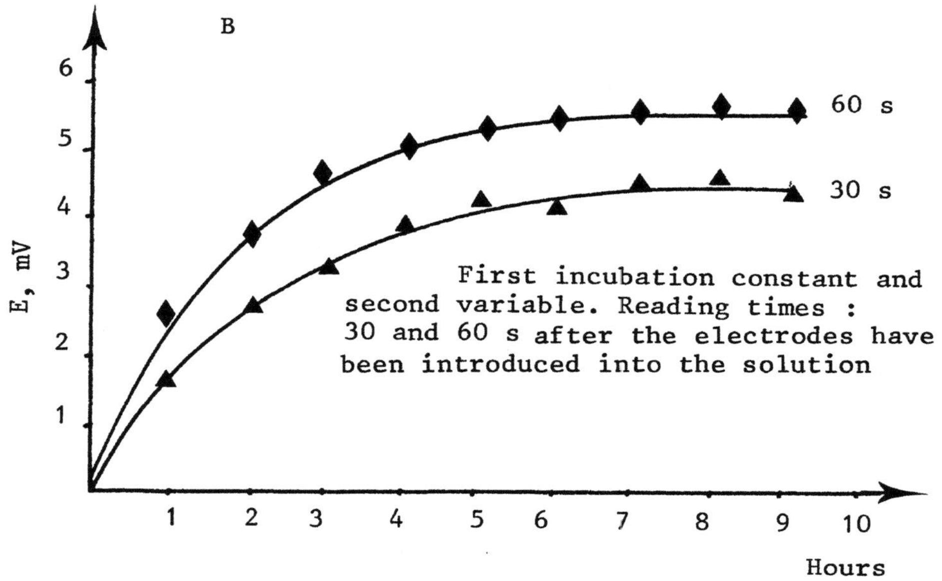

Fig. 12 : Kinetics of the binding of labeled antibodies to antigens complexed with antibodies immobilized on gelatin membrane.

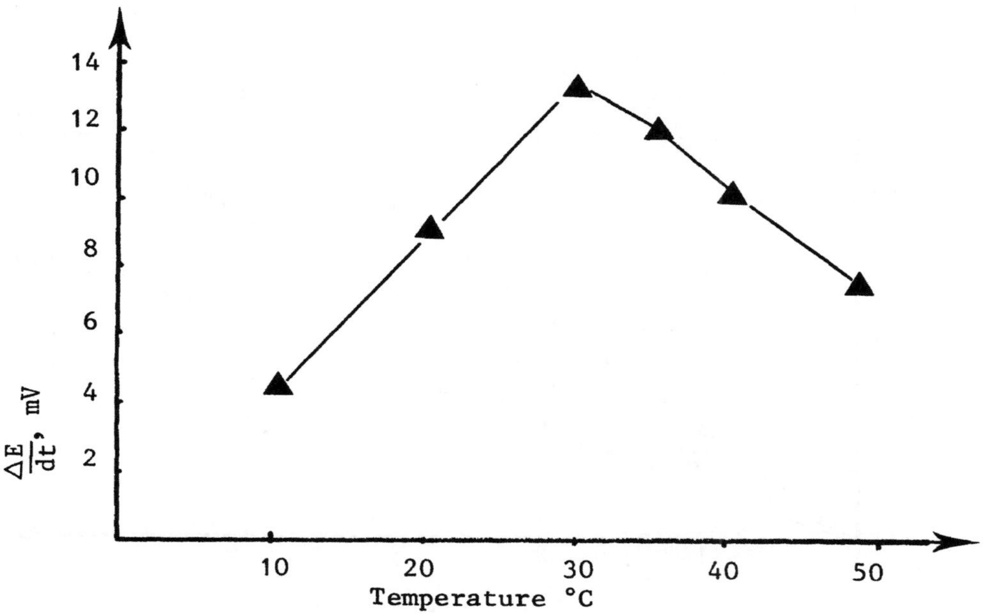

<u>Fig. 13</u> : Influence of temperature on the reaction rate of HRP-labeled antibody b ound to the membrane. Final concentrations : horseradish peroxidase 4 µg/l, H_2O_2 4.4 mol/l, KI : 0.1 mol/l, pH 5.0, reaction time : 60 s.

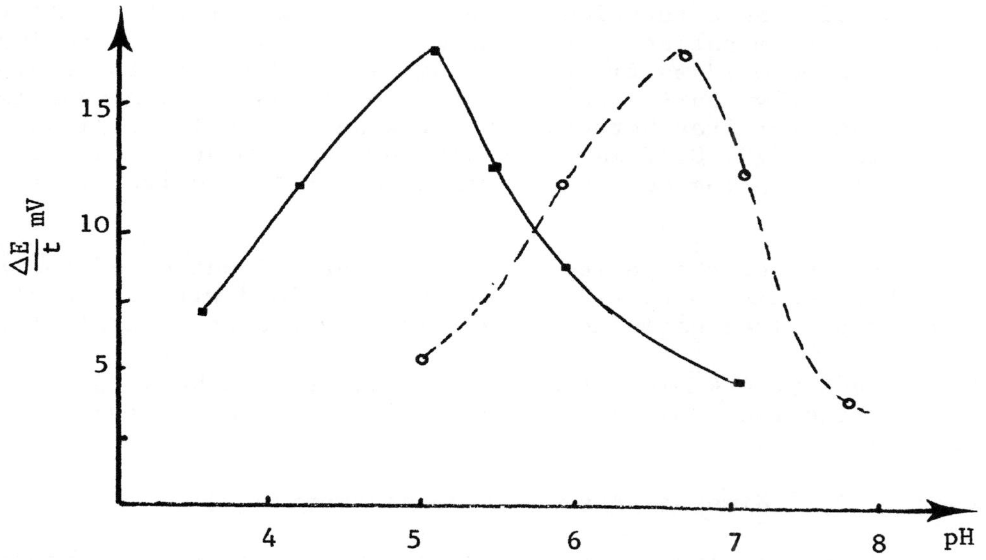

<u>Fig. 14</u> : Effect of pH dependence on the DMB-HRP reaction studied with the iodide electrode for the free-HRP O...O and HRP antibodies conjugates complexed on the membrane ■———◆. Both reactions are measured by potentiometric system.

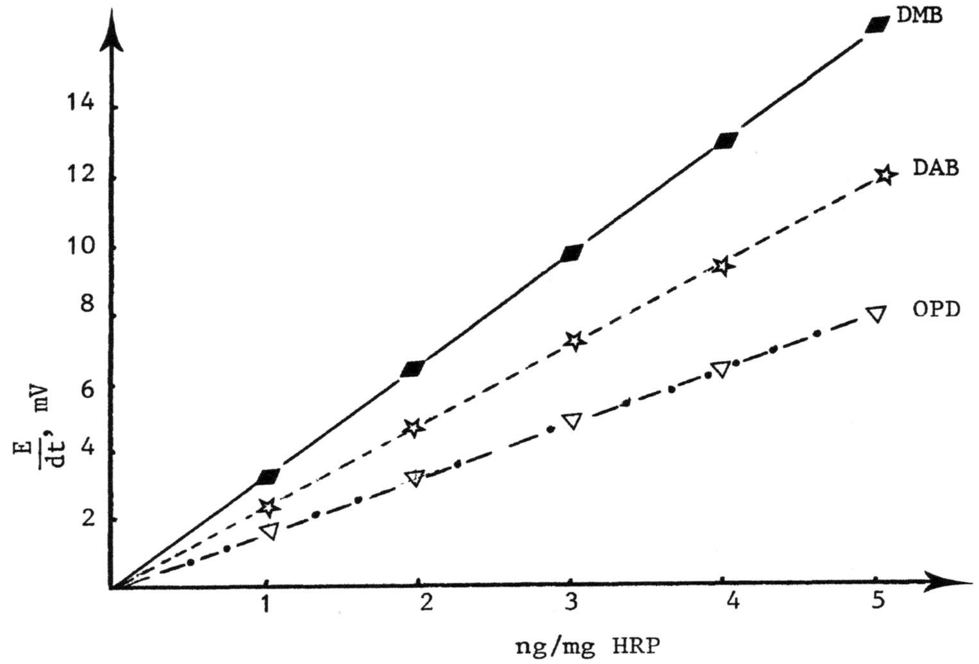

Fig. 15 : Effect of different hydrogen donors (OPD, DAB, DMB) on the reaction rate of free HRP.

4.6. Effect of H_2O_2 concentration

The activities of free peroxidase and horseradish peroxidase bound to IgG were studied as a function of substrate concentrations. After potentiometric determination of the enzymatic kinetics, we established that horseradish peroxidase fixation to IgG and indirectly to membranes did not significantly modify the K_m of the peroxidase. The Michaelis constant obtained for free horseradish peroxidase is 0.1 mmol/l and, for peroxidase labeled IgG, 0.12 mmol/l under our experimental conditions showing that the coupling does not strongly alter the active sites of the enzyme.

The response of the antibody electrode as a function of the reaction rate shows that maximal potential values are obtained with a reaction time of 60 s, which provides satisfactory sensitivity and reproductibility.

Consequently, 4.4 mmol of H_2O_2 per liter should be used as substrate solution, and reaction times of 60 s and 120 s as reading times for electrode potentials.

4.7. Effect of 3,3' dimethoxybenzidine concentration

Peroxidase activity was observed as a function of DMB concentrations at 30 and 60 s.

The maximum reaction rate was obtained with a 50 mg/l concentration of DMB under our experimental conditions (Figure 16).

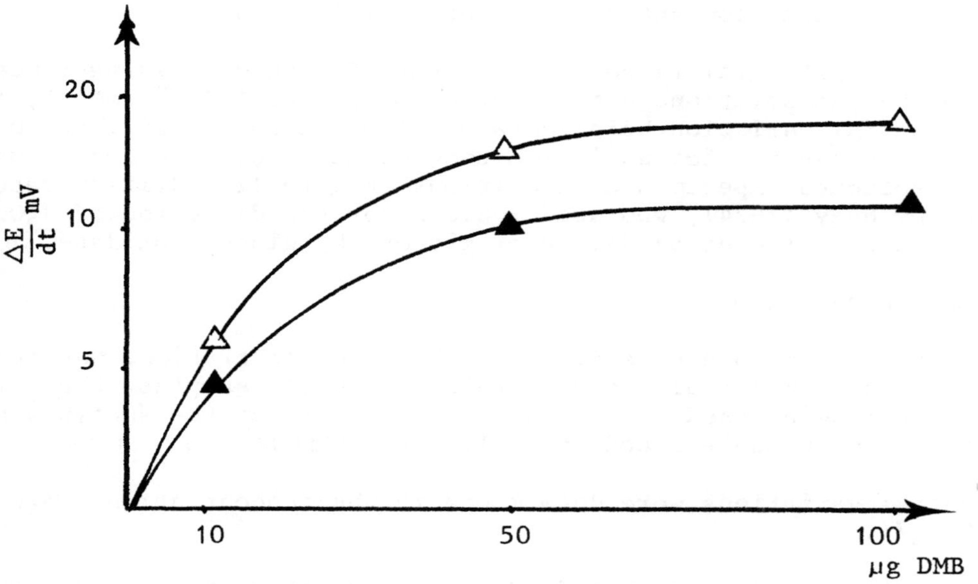

Fig. 16 : Effect of DMB concentration on the kinetic iodide oxydation kinetics at differnt time are studied. ▲ : 30 s, Δ : 60 s. The final concentration of H_2O_2 is 4.4 10^{-3} mol/l, I^-, 1.10^{-4} mol/l and for HRP 4,0 ng/ml.

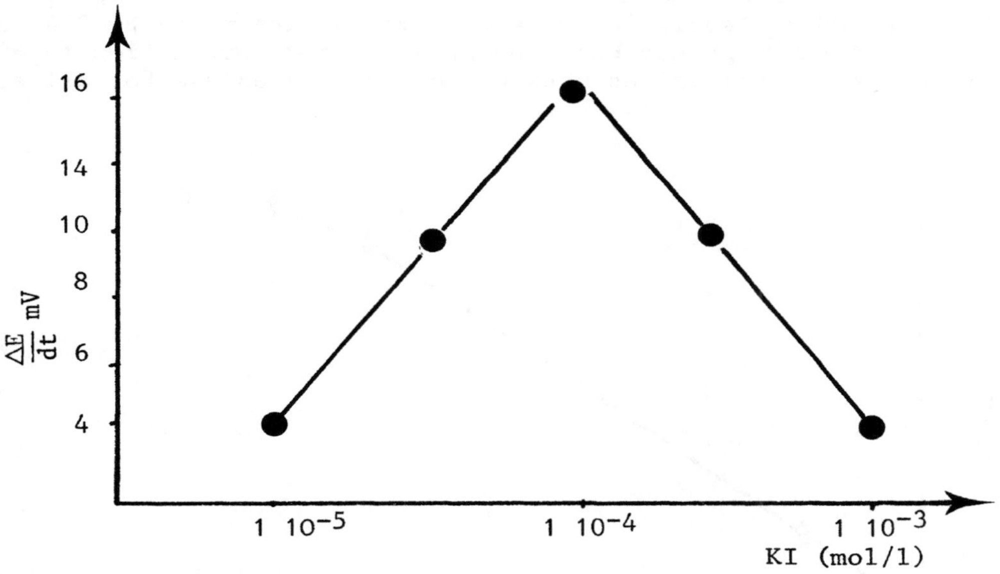

Fig. 17 : Effect of the iodide concentration on the variation of potential electrode, pH 5.0, temperature 27°C, time of response : 1 min.

4.8. Effect of iodide concentration on potential variations

Increasing potential response was obtained with decreasing concentrations of iodide solutions between 0.10 mmol/l and 1.10^{-3} mmol/l, the maximal potential variation being obtained 1 min after stabilization of the electrode potential for an iodide concentration of 0.10 mmol/l under the above mentioned experimental conditions (Figure 17). Similar results were found by Nagy (1974), who used iodides for the detection of H_2O_2 produced in the course of oxidation of glucose by glucose oxidase.

4.9. HBsAg concentration

The electrode response vs different HBsAg concentrations was read on a calibration curve realized with reference HBsAg samples, the erman reference panel being used as a standard (100 µg/l for the subtypes ad and ay) as shown in Table I and Table II, by BOITIEUX et al (1979).

Potential variations were determined in the concentration range 0.5 to 50 µg/l.

Calibration curve constructed at room temperature for the HBsAg electrode is linear in this range which is good enough to cover both the normal and anormal serum HBsAg range (Figures 18 and 19).

4.10. Analytical criteria

The limit of sensitivity of the method is near 0.5 µg/l. The working curve obtained for assays is perfectly linear between 0.5 and 50 µg/l. The calibration points were determined with 10 separate sets of assays of the same sample. The reproducibility is satisfactory, the coefficient of variation varying from 15 to 20 % in the linear part of the curve.

An unknown sample of HBsAg positive serum (subtype ad) was analyzed by the present procedure and by radio immunoassay with good agreement with each other. The HB_sA_g levels in 10 serum samples determined by RIA and this potentiometric technology has been compared. A good correlation (σ =0.970) was observed between the values obtained by the two assays for all samples.

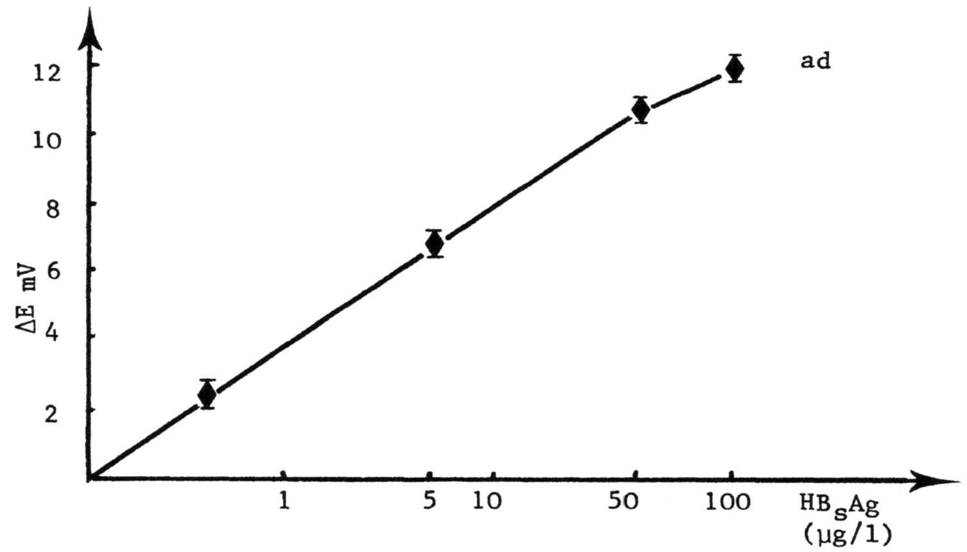

Fig. 18 : Standard curves showing the reproducibility of 10 assays, done in two separate sets, for the determination of HBsAg subtype ad.

Table I : Calibration values for HBsAg subtype ad

HBsAg concentration subtype ad (ng/ml)	0.1	0.5	5	50
E(mV) on the expanded scale at 1 min	0.4	1.2	4.9	8.8

Table II : Values obtained with different concentrations of HBsAg subtype ay.

HBsAg concentration subtype ay (ng/ml)	0.1	0.5	5	50
E (mV) on the expanded scale at 1 min	0.3	1.1	4.6	8.1

Thirty five different measurements of HBsAg subtype ad, performed with the same serum by the described method, gave a mean value of 85.0 ± 14.2 µg/l and, by radioimmunoassay, 100 µg/l. For similar concentrations the relative error will not exceed 16 %. These results are encouraging because the limit of sensitivity found by Wolters et al. (1976) with an enzyme immunoassay technique was at least 10.5 µg/l for subtype ad and at least 2.4 µg/l for subtype ay, whereas for Gerlich et al. (1976) using a R I A procedure, the limits were 3 to 4 µg/l for ad (Table III) and 5 to 10 µg/l for ay (Table IV).

Specificity has not yet been studied because of the difficulty of obtaining pure antigen from the standard samples. Nevertheless, the use of immunodiffusion methods revealed the good specificity of the anti-HBsAg antibodies.

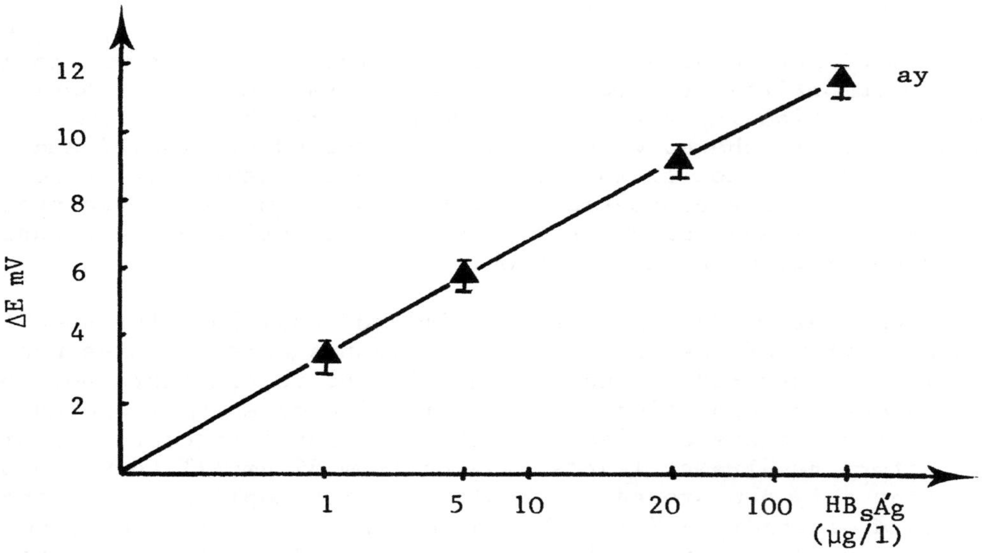

Fig. 19 : Standard curves showing the reproducibility of 10 assays, done in two separate sets, for the determination of HBsAg subtype ay.

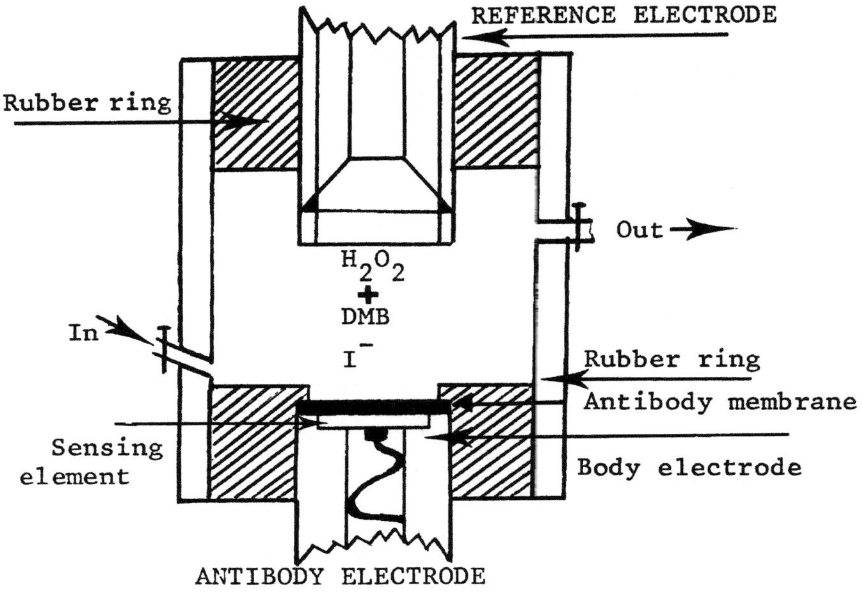

Fig. 20 : Schematic representation of the new device used for potentiometric enzyme immunoassay.

From these preliminary experiments with the antibody electrode device, we believe that this new enzyme immunoassay technique compares well with radio immunoassay procedures. To confirm these results we have tried another molecule of low molecular weight, estradiol-17 β .

5. Determination of estradiol-17β

This approach consists to use the technology previously described for the determination of small molecules. An hapten as estradiol-17β is taken as second "biological model" for elaborating semi-automatic system.

This potentiometric determination of oestradiol-17β is an application of the competitive enzyme-linked immunoassay technique. Anti-oestradiol-17 β antibodies are immobilized on a pig skin gelatin membrane, which is incubated with peroxidase labeled oestradiol and oestradiol. After fixation of the membrane onto the sensor of an iodide sensitive electrode, the enzymatic activity was evaluated in the presence of hydrogen peroxide and iodide, the electrode potential being a function of hapten concentration in the solution.

In this chapter we determine the optimized conditions for specificity and sensitivity of the antibody-coated membrane in the presence of peroxidase labeled oestradiol and oestradiol, and to eliminate possible interferences due to adsorption or ionic fixation of enzyme labeled steroids. The first tests carried out with oestradiol standard solutions gave satisfactory results at levels ranging from 57 pmol/l to 9.2 nmol/l, suggesting that this new procedure should find its application in the determination of oestradiol-17 β in biological fluids. Till now, membranes could not be re-used because of fixation of colored complexes limiting the diffusion of measured ions.

Table III : Reproducibility and precision for HBsAg subtype ad

Concentration of HBsAg subtype ad	0.10	0.50	5.0	50.0
Assay replicate	10	10	10	10
Values in mV (expansed scale)	0.44	1.25	4.87	8.7
Standard deviation	0.13	0.32	0.43	0.46
Coefficient of variation (%)	2	9	16	19

Table IV : Reproducibility and precision for HBsAg subtype ay

Concentration of HBsAg subtype ay	0.10	0.50	5.0	50
Assay replicate	10	10	10	10
Values (mV) (expansed scale)	0.33	1.01	4.51	8.09
Standard deviation	0.21	0.29	0.36	0.52
Coefficient of variation (%)	4	8	13	26

5.1 Potentiometric determination of HRP activity

For oestradiol assays as well as for determination of the enzymatic activity of the antibody membrane saturated or half saturated with HRP-labeled hapten, the systems were incubated at room temperature for 4 H. The membrane was then thoroughly washed with distilled water and 0.1 mol/l phosphate buffer, pH 5.0, before determination of HRP activity.

For preliminary tests, a series of dilutions in 0.01 mol/l phosphate buffer, pH 6.8 was prepared from a solution of 3.7 µmol/l oestradiol in absolute ethanol, giving standard solutions between 57 pmol/l and 9.2 nmol/l.

HRP activity was determined with the help of an iodide-sensitive electrode in a reactive medium made of hydrogen peroxide and iodide, in the presence of (DMB) as electron donor, starting the oxidation of iodides. In a first series of experiments, we used, for determining free and bound HRP activities, a solution of potassium iodide, hydrogen peroxide and DMB (methanolic solution) in 0.1 mol/l sodium phosphate buffer, pH 5.0, in final concentrations corresponding respectively to

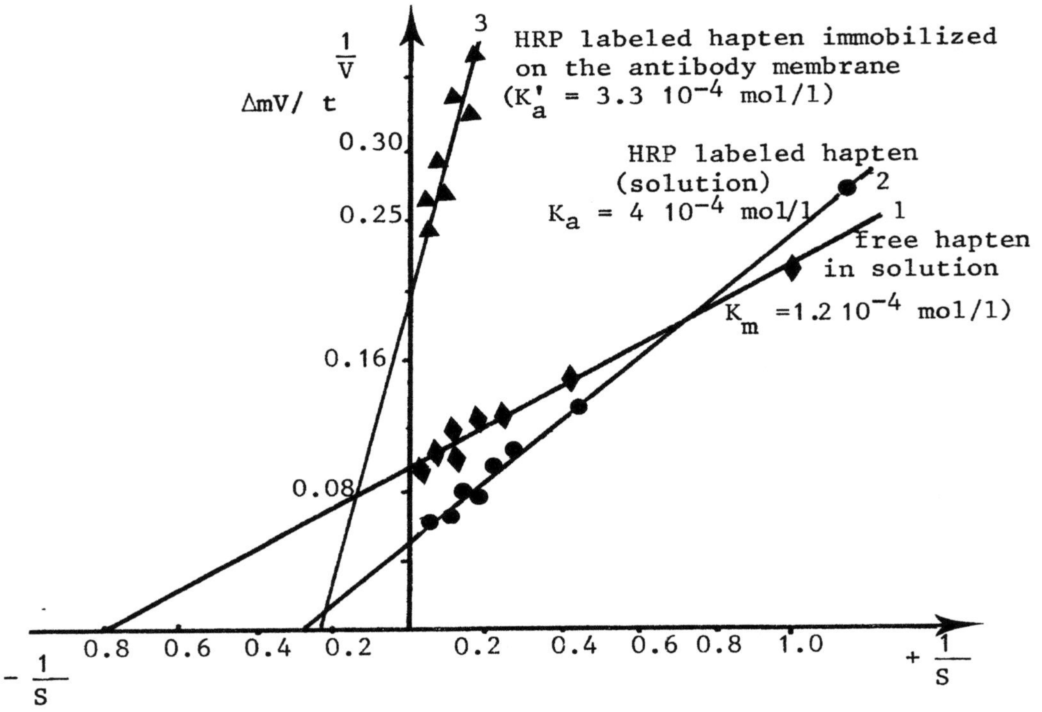

Fig. 21 : Potentiometric determination of K_m of free HRP (1), of K_a of HRP coupled to oestradiol (2) and of K_a of HRP bound to the antibody membrane (3), at 2 min, 20° X. The substrate (S) used for this determination was H_2O_2 in the concentration range 0.2 to $1.0.10^{-4}$ mol/l.

0.1 mmol/l, 4.4 mmol/l and 0.2 mmol/l.

To determine the enzymatic activity on membranes after incubation in the presence of HRP-labeled haptens, we made a new device using a 5 ml flow cell enclosing the reference electrode at the upper side and, at the lower side, the membrane-coated selective electrode, which may be easily removed (Fig. 20).

The enzymatic activity is evaluated in the presence of hydrogen peroxide and iodides, under optimal conditions as previously described. A total volume of 5 ml reactive medium is sufficient for the potentiometric measurement. After a stabilization time of 1 min, to allow ionic diffusion through the membrane, we determined the variations of the electrode potential every 15 s for 2 min, after addition of 50 µl of 41 mmol/l DMB methanolic solution. The potential variations as a function of time ($\Delta E/dt$) are proportional to iodide oxidation and, consequently, to the enzymatic activity.

Various delays to the stabilization of the electrode potential were observed, owing to the diffusion rate of ions through the antibody membrane.

The authors believe that the lack of stabilization of the electrode potential may be due to the decomposition of H_2O_2 as soon as the electrodes are immersed in the reactive medium, before addition of DMB. For these reasons, we used in a second series of trials a 0.1 mmol/l iodide solution in 0.1 mol/l sodium phosphate buffer, pH 5.0, for stabilization of the

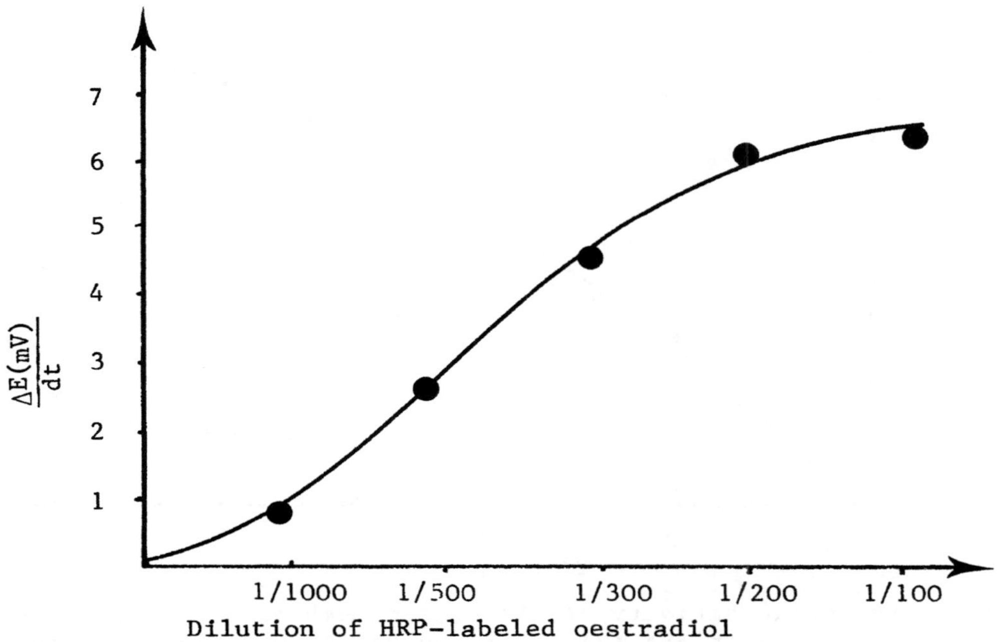

Fig. 22 : Fixation of HRP-labeled oestradiol onto the antibody membrane. Determination of the dilutions of labeled oestradiol solution (30 nmol/1) allowing membrane saturation and half-saturation (reading at 2 min.).

electrode potential adding in a second step the same volume of another solution containing iodides, H_2O_2 and DMB in order to obtain final concentrations as previously described.

b) Fixation of HRP labeled oestradiol onto the antibody coated membrane

We showed that the coupling of HRP to oestradiol and indirectly to the antibody membrane did not significantly modify the K_m of the enzyme (Figure 21). Free enzyme K_m was evaluated at 1.2×10^{-4}mol/1 H_2O_2, K_a of HRP-labeled oestradiol corresponding to 3.8×10^{-4}mol/1 and K_a of HRP bound to the membrane to 3.5×10^{-4} mol/1. Consequently, it seemed possible to use the experimental conditions previously established for free HRP potentiometric assay and to determine the enzyme activity of the HRP-labeled oestradiol solution. It corresponded to the activity of a solution containing 300nmol/1 of free enzyme.

In preliminary tests fixation of HRP-labeled oestradiol onto the antibody membrane and estimation of enzymatic activity were carried out at the same pH (pH 6.8 and pH 5.0). Owing to the lack of reproducibility of the different assays the authors used another system involving incubation in 0.01 mol/1 sodium phosphate buffer, pH 6.8 washing with distilled water and 0.01 mol/1 sodium phosphate buffer, pH 5.0 and determination of HRP activity in the same buffer.

Serial dilutions of HRP-labeled oestradiol solution were tested for the immunochemical reaction. In our system, 0.5 ml of labeled oestradiol solution diluted 1 : 200 in 0.01 mol/1 sodium phosphate buffer, pH 6.8, was sufficient for saturation of the antibody-coated disks. We then determined the final HRP-labeled oestradiol concentrations required to bind approximately 50 % of antibody immobilized on the protein membrane, according to conditions usually observed for radioimmunoassay techniques (RIA).

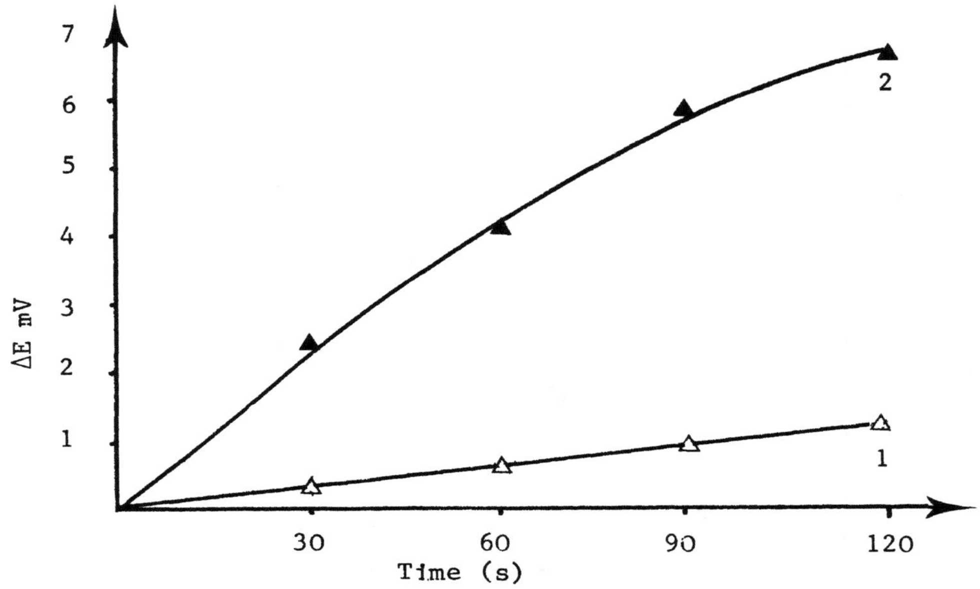

Fig. 23 : Enzymatic activity of BSA-coated membrane (1) and antibody membrane saturated with BSA (2) after incubation with HRP-labeled oestradiol. Determination of potentials in 0.01 mol/l sodium phosphate buffer, pH 6.8.

0.5 ml of labeled oestradiol solution diluted 1 : 300 was needed for half saturation of the antibody-coated disks (Fig. 22).

c) Determination of optimal conditions for fixation of HRP-labeled hapten onto the antibody membrane

Colorimetric determination of HRP activity after reaction of enzyme-labeled haptens against immobilized anti-oestradiol-17β antibodies suggested that interferences may occur in the course of the immunological reaction. It was shown that labeled haptens were capable of reacting with polystyrene in the absence of antibody. Numerous tests using the same HRP-labeled steroid solution under similar conditions gave unsatisfactory results with respect to reproducibility and sensitivity. Saturation of microtitration plates with 10 g/l BSA solution for 1 h after fixation of antibody onto polystyrene wells did not give satisfactory results even by varying pH and ionic strength of buffers.

The results obtained with microtitration plates involving antibody-coated polystyrene wells suggested the possibility of interferences in the course of the immunological reaction when using an antibody-coated protein membrane.

For that reason we investigated a BSA-coated membrane in order to evaluate possible non-specific reactions of HRP-labeled hapten with the antibody membrane. Figure 23 shows the results obtained with the BSA-coated membrane incubated overnight at 4° C in the presence of a 1 : 200 diluted solution of HRP-labeled steroid in 0.01 mol/l sodium phosphate buffer, pH 6.8, potentials being determined with the same buffer after washing in distilled water. However, no enzymatic activity could be observed when BSA-coated membranes incubated with labeled oestradiol were thoroughly washed in distilled water, and then in 0.1 mol/l sodium phosphate buffer, pH 5.0.

On the other hand, we observed that variations of electrode potential for reading times of 1 min and 2 min were significantly higher when antibody membranes were not saturated with BSA, potentiometric determination being carried out at pH 5.0 in 0.1 mol/l sodium phosphate buffer.

d) Preliminary results of oestradiol-17β competitive enzyme immunoassay

Standard curves were obtained (Fig. 24) by plotting bound enzyme activity represented by the variations of electrode potential (ΔE) against different working dilutions of oestradiol-17β in the presence of the labeled tracer solution required to bind approximately 50 % of the membrane antibody sites. It appears that oestradiol concentrations may be satisfactorily determined in the range 50 pmol/l to 10 nmol/l. The calibration points were determined with 10 separate sets of assays of the same sample and standard deviations (S.D.) were calculated. The reproducibility was satisfactory, the coefficient of variation ranging from 11 to 18 % (Table V).

The optimal conditions for determination of the enzymatic activity were studied in a recent paper but the specificity of the antibody-coated protein membrane against enzyme-labeled haptens had not been established. The main objectives of the present work were to confirm the best conditions for fixation of the HRP-labeled steroid onto the antibody sites of the antibody membranes, avoiding interferences such as absorption and ionic fixation of enzyme-labeled haptens onto the artificial membrane.

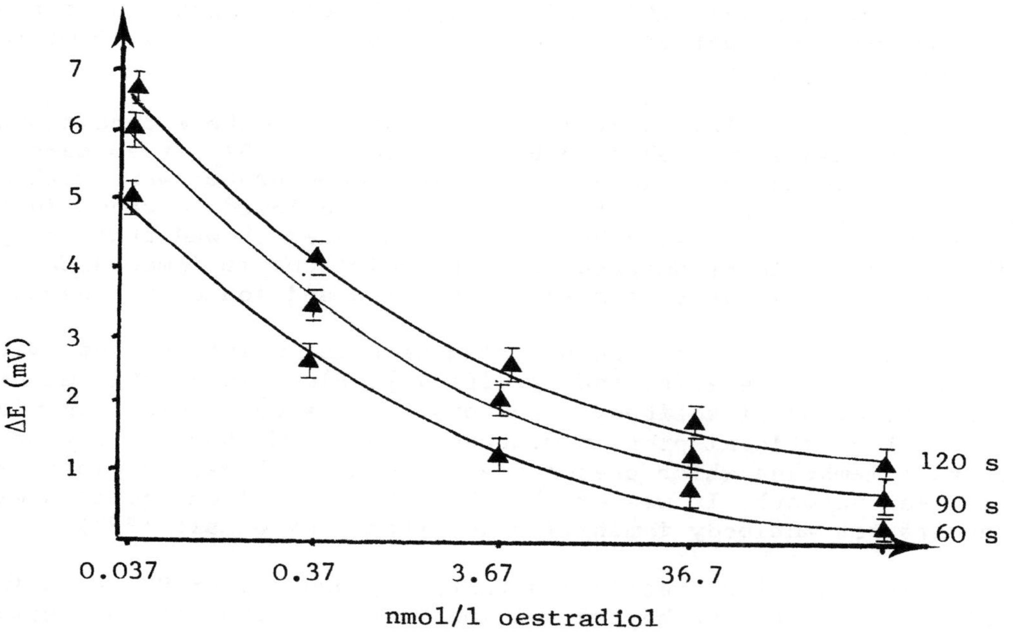

Fig. 24 : Standard curves obtained with oestradiol-17 β dilutions. Readings at pH 5.0 in 0.1 mol/l sodium phosphate buffer. The vertical bars represent ± 2 S.D. of each value.

Table V : Calibration curves for oestradiol 17β in solution

Concentration of oestradiol 17 β (pg/ml)	15,5	31	156	625
Assay number	10	10	10	10
Values on expanded scale (mV)	6.0	5.2	3.9	2.4
Standard deviation	1	0.8	0.6	0.3
Coefficient of variation %	17	15	14	12

The lack of reproducibility of enzyme-immunoassays carried out on microtitration plates suggested the possibility of absorption of HRP-labeled oestradiol onto polystyrene and also of ionic exchange processes when wells were saturated with BSA (Boitieux et al., 1984).

It seemed that similar interferences might occur when using antibody-coated membranes. It was first suggested that amino groups of HRP coupled to oestradiol could bind to active groups remaining on the artificial protein membrane after glutaraldehyde treatment and fixation of antibody. However, saturation of these active groups with BSA dis not afford satisfactory results when incubation with enzyme-labeled hapten and determination of the enzymatic activity were carried out at pH 6.8 in 0.01 mol/1 sodium phosphate buffer.

This could be explained by the fact that under these conditions HRP coupled to the hapten was able to bind to membrane BSA. It appeared that ionic fixation could be avoided when incubated membranes were washed in 0.1 mol/1 sodium phosphate buffer, pH 5.0, and allowed to react in the same buffer. Results obtained with BSA-coated membranes showed that the protocol described above could be adopted for oestradiol enzyme immunoassay, eliminating interferences due to adsorption effects and ionic exchanges.

The nature of the membrane has also been taken into account with respect to its binding capacity and specificity. Pig skin gelatin membrane was chosen for fixation of antibody. Previous work based on electron microscopy studies of artificial protein membranes showed the homogeneity of the pig skin gelatin membrane which presents a uniform thickness, a glossy surface and a regular network. It was established that 120 Blooms gelatin gave the best results for antibody immobilization (Boitieux et al. 1978).

Determination of Michaelis constants (K_m and K_a) of HRP free in solution, of HRP bound to the hapten and indirectly to the antibody membrane showed that the coupling and the fixation onto the membrane did not strongly alter the active sites of the enzyme. They also brought the optimal substrate concentrations of the reaction medium.

For estimation of oestradiol-17β in solution, we elaborated a new device for determination of HRP activity, as described above, using a new approach to semi-automation of the enzyme immunoassay. The preliminary work shows that the limit of sensitivity is satisfactory (50 pmol/l) and we believe that, despite the difficulty of a complete automation, this new procedure which is rapid, practical and sensitive, should be well-suited to routine clinical laboratories.

From these experiments on the antibody electrode device we believe that this new enzyme immunoassay technique compares well with radio immunoassay procedures. The advantages over the latter are evident : it is less time consuming, more practicable, more sensitive ; it requires only small amounts of antibodies and no purified antigen for the sandwich method (HB_sAg) for example. This potentiometric system eliminates most non-specific interfering factors, which is not the case with spectrophotometric determinations.

This type of enzyme immunoassay should be extended to the assay of antibodies, antigens or haptens, after linkage to horseradish peroxidase, and might be used on a large scale in medical laboratories.

Difficulties met for automation of the new 2 electrodes device led us, despite of satisfactory results already obtained, to consider another type of electrochemical sensor.

This new sensor used a pO_2 electrode in an electrochemical detection system according to Boitieux et al., (1984).

The use of the same membrane for every enzyme immunoassay owing to the reversible fixation of immunocomplexes through the ligand is being carried out in our laboratory and preliminary results were already presented (Boitieux et al., 1985).

SUMMARY

We report here a study of the immobilization of specific antibodies on different proteic membrane and experiments on a new, sensitive, and reliable procedure for enzyme immunoassay of various antigens in biological fluids. The method, developed from two biological models "hepatitis B surface antigen/antibodies" and "oestradiol 17β antibodies" are less time consuming than most immunochemical techniques and eliminates many inconveniences arising from use of isotopes. We use a solid phase for the "sandwich" and "competition" procedures.

The antibodies being immobilized on gelatine membranes, we determined antigen concentration with the help of an iodide-sensitive electrode modified by fixing the active membrane onto the cristal sensor. We have studied the effects of different variables which give the maximal enzymatic activity of the membrane after using different immunological procedures.

Analytical criteria of this method have been etablished and compared with the solid phase radio immunoassay for the hepatitis B surface antigen in dilution series.

One tenth microgram of antigen per liter can be reproducibly detected with our method. For the determination of hapten as oestradiol-17 β, anti-oestradiol 17 β antibodies are immobilized on a pig skin gelatin membrane, which is incubated with peroxidase-labeled steroid and oestradiol. After fixation of the membrane onto the sensor of an iodide sensitive electrode, the enzymatic activity was evaluated in the presence of hydrogen peroxide and iodides, the electrode potential being a function of hapten concentration in the solution. The preliminary studies consisted of determining the optimized conditions for specificity and sensitivity of the antibody coated membrane in the presence of peroxidase labeled oestradiol and oestradiol, and eliminating possible interferences due to adsorption or ionic fixation of enzyme labeled steroid. The tests carried out with oestradiol standard solutions gave satisfactory results at levels ranging from 57 pmol/l to 9.2 nmol/l, suggesting that this new procedure should find application in the determination of oestradiol-17β in biological fluids. Use of the antibody electrode can easily be extended to assay of other antigens and haptens, viruses, antibodies, that usually are determined by radio immunoassay or other technics.

Notes

REFERENCES

Avrameas, S., Ternynck, T., 1971, Peroxidase labeled antibody and fat conjugates with enhanced intra cellular penetration, Immunochemistry 8 : 1175-1179.

Boitieux, J.L., Desmet, G. and Thomas, D. 1978, Détermination potentiométrique de l'antigène de surface de l'hépatite B dans les liquides biologiques, Clin. Chim. Acta 88 : 329-338.

Boitieux, J.L., Desmet, G. and Thomas, D. 1979, An "Antibody electrode". Preliminary report on a new approach in Enzyme Immunoassay, Clin. Chem, 25 : 318-321.

Boitieux, J.L., Lemay, C., Desmet, G. and Thomas, D. 1981, Use of solid phase biochemistry for potentiometric enzyme immunoassay of oestradiol 17 β, Clin. Chim. Acta, 113 : 175-182.

Boitieux, J.L., Aubry, N., Thomas, D. 1984, A computerized enzyme immuno sensor : application for determination of antigens. Clin. Chem. Acta, 136 : 19-28.

Boitieux, J.L., Desmet, G. and Thomas, D. 1985, Immobilized inhibitors as ligand for the development of a reversible bienzymatic system in immuno enzymatic technics, Ann. Ac. Sci. N.Y. In Press.

Catt, K.J. and Trejear, G.W. 1967, Radio immunosorbent assay for proteins chemical coupling of antibodies coated tubes, Science, 158 : 1570-1578.

Desmet, G. and Boitieux, J.L. 1977, Préparation et purification de l'antigène de surface du virus de l'hépatite B, Clin. Chim. Acta, 74 : 59-72.

Engvall, E., Perlmann, P. 1971, Enzyme linked immunosorbent assay (ELISA) Quantitative assay of immunoglobulin G, Immunochemistry, 8 : 871-880.

Erlanger, B.F., Borek, F., Beiser, S.M. and Lieberman, S.M. 1957, Steroid protein conjugates. I. Preparation and characterization of conjugates of bovine serum albumin with testosterone and cortisone, J. Biol. Chem. 228 : 713-727.

Gerlich, W., Stamm, B. and Thomssen, R. 1976, Quantitative determination in the detection of hepatitis B surface antigen : results of collaborative study involving 74 laboratories, J. Biol. Stand. 4 : 189-197.

Line, W.F. and Becher, M.J. 1975, Solid phase immunoassay, in :"Immobilized enzymes, antigens, antibodies and peptides", H.H. Weetall, ed., Marcel Dekker Inc. New York NY pp. 497-566.

Nagy, G., Von Storp, L.H. and Guilbault, G.C. 1973, Enzyme electrode for glucose based on an iodide membrane sensor, Anal. Chim. Acta 66 : 443-450.

Nakane, P.K. and Kawaoi, A. 1974, Peroxidase labeled antibody, a new method of conjugation, J. Histochem. Cytochem. 22 : 1084-1098.

Pratt, J.J. 1978, Steroid immunoassay in Clinical Chemistry, Clin. Chem. 24 (11) : 1869-1890.

Schuurs, A.H.W.M. and Van Weemen, B.K., 1977, Enzyme immunoassay, Clin. Chim. Acta, 81 : 1-12.

Thomas, D., 1976, Artificial enzyme membranes transport, memory and oscillatory phenomena, in : "Analysis and control of immobilized enzyme systems", D. Thomas and J.P. Kernevez, ed., North-Holland, American Elsevier, New York NY pp. 115-147.

Wolters, G., Kuijpers, Kaccadi, J. and Schuurs, A. 1976, Solid phase enzyme immunoassay for detection of hepatitis B surface antigen, J. Clin. Path 29 : 873-892.

Wide, L., Axen, B. and Porath, J. 1967, Radio immunosorbent assay for proteins chemical coupling of antibodies to insoluble Dextran, Immunochemistry 4 : 381-388.

ENZYME IMMUNOASSAY USING THE AMMONIA GAS-SENSING ELECTRODE

Carl R. Gebauer

Technicon Instruments Corporation
511 Benedict Avenue
Tarrytown New York 10591

INTRODUCTION

Potentiometric membrane electrodes have been applied to immunological measurements both as direct immunosensors and as detectors of enzyme label activity in enzyme immunoassays (EIA). In the latter case, no universal potentiometric sensor is available for EIA use but rather a distinct electrode type is matched with the class of enzyme label it is best suited to detect. The ammonia gas-sensing electrode is one example and has the capability of detecting enzymes that catalyze ammonia producing reactions. Meyerhoff and Rechnitz (1979) first described the application of this gas-sensing electrode to EIA measurements through the use of a urease label for bovine serum albumin (BSA), a model protein analyte, and for cyclic AMP (adenosine monophosphate). In this chapter the general application of the ammonia gas-sensing electrode to immunoassay methods employing deaminating enzyme labels is examined for both a model hapten, dinitrophenyl (DNP), and one of clinical interest, cortisol.

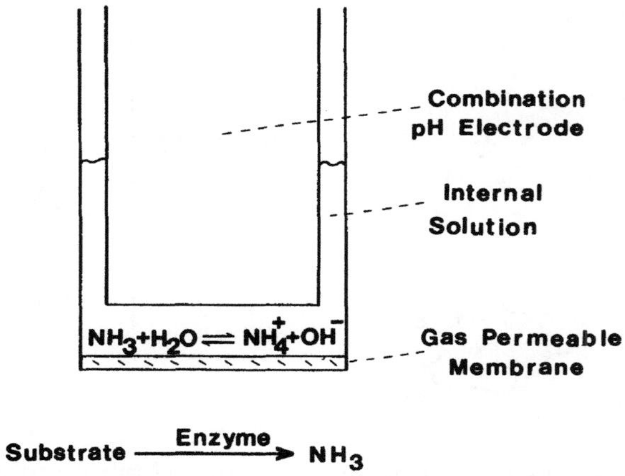

Figure 1. The ammonia gas-sensing electrode.

TABLE I

Examples of Enzymes Detectable by the Ammonia Gas-Sensing Electrode

ENZYME	EC Number	REFERENCE
Adenosine deaminase	3.5.4.1	Hjemdahl-Monsen et al., 1977
Amine oxidase	1.4.3.4	Meyerson et al., 1978
D-Amino acid oxidase	1.4.3.3	Guilbault and Hrabankova, 1971
L-Amino acid oxidase	1.4.3.2	Guilbault and Hrabankova, 1970
AMP deaminase	3.5.4.6	Papastathopoulos and Rechnitz, 1975
Asparaginase	3.5.1.1	Guilbault et al., 1969
Creatinine deiminase	3.5.4.21	Meyerhoff and Rechnitz, 1976
Glutamate dehydrogenase	1.4.1.2	Davies and Mosbach, 1974
Glutaminase	3.5.1.2	Huang, 1974
Guanase	3.5.4.3	Nikolelis et al., 1981
Histidine ammonia-lyase	4.3.1.3	Ngo, 1975
Methionine lyase	4.4.1.11	Fung et al., 1979
Nitrite reductase	1.6.6.4	Hussein and Guilbault, 1975
Phenylalanine ammonia-lyase	4.3.1.5	Hsiung et al., 1977
Urease	3.5.1.5	Katz, 1964

A simple schematic representation of an ammonia gas-sensing electrode is shown in Figure 1. Ammonia produced in an enzyme catalyzed reaction diffuses through a gas permeable membrane, usually non-wetting Teflon, and changes the pH of a thin layer of NH_4Cl solution (Ross et al., 1973). This pH change is measured as a change in potential by the internal element, a combination pH electrode. Alternatively, the ammonia diffusing through the membrane can be detected in the form of ammonium ion by a polymer based internal sensing element (Meyerhoff, 1980). In both cases the gas

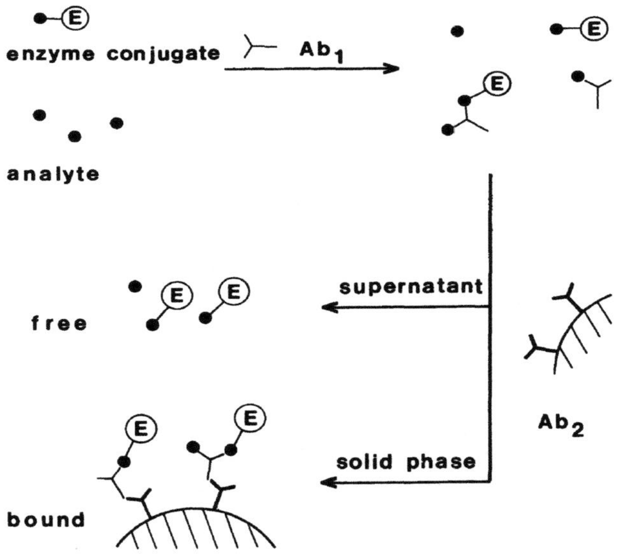

Figure 2. Heterogeneous enzyme immunoassay scheme. Second antibody is immobilized on agarose beads.

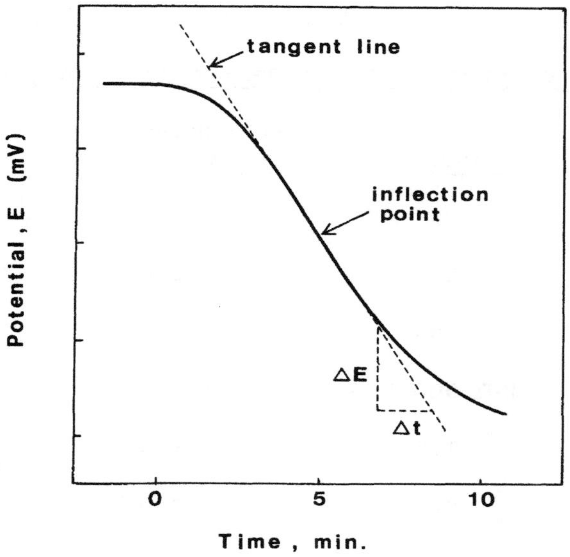

Figure 3. A typical potentiometric rate curve obtained with an ammonia gas-sensor, enzyme (or substrate) added at t = 0.

permeable membrane excludes ionic or non-volatile species that could potentially interfere with the measurement. This ionic exclusion characteristic is one of the electrode's major advantages over other potentiometric sensors.

Deaminating Enzyme Labels for EIA

There are numerous enzymes that catalyze reactions where one of the products is ammonia or ammonium ion. Some of these are listed in Table I along with their Enzyme Commission number and early references linking them with potentiometric measurement. In addition to these enzymes, whose activities can be determined directly with the ammonia electrode, there are many other enzymes that can be coupled with a second, ammonia producing, enzyme for potentiometric measurement (Gebauer et al., 1979). For immunoassay measurements, one of these enzyme labels would be covalently linked to a hapten analog similar or equivalent in structure to the analyte of interest. This enzyme conjugate competes with analyte in the sample for the available binding sites on antibody raised against that hapten. At this point the antibody bound enzyme label can be separated from the free label. In a heterogeneous EIA of the type shown in Figure 2 the activity of the free label increases with higher analyte concentration while the activity of the bound enzyme label decreases. Examples in this chapter use a second antibody, raised in another mammalian species against the first antibody and attached to insoluble agarose beads, to separate bound from free enzyme label. Measurement of activity on the solid phase poses no problem with a potentiometric detector as turbidity of the sample causes no interference. Homogeneous, separation free, enzyme immunoassays have not yet been demonstrated with deaminating enzyme labels. Those labels that have been studied show no inhibition of enzymatic activity due to binding of antibody.

TABLE II

Characteristics of Three Deaminating Enzyme Labels

Characteristics	Urease	Adenosine Deaminase	Asparaginase
Activity			
U/mg	800	230	300
U/nmole	380	7.8	41
Molecular Weight	480,000	34,000	136,000
pH Optimum	7.5	7.0 - 8.5	8.5
with electrode	8.0	9.0	10.0
Accessible $-NH_2$	30	4	52

Potentiometric Rate Measurement

Measurements of the enzyme-label activity are made at high substrate concentrations where the rate is zero order with respect to sustrate and pseudo first order with respect to enzyme concentration. While the rate of production of ammonia in the enzyme catalyzed reaction is constant at any enzyme concentration, the response of the ammonia gas-sensing electrode is governed by a logarithmic relationship:

$$E = E_c + \frac{RT}{nF} \ln [NH_3]$$

The rate of potential change as ammonia is enzymatically generated is:

$$\frac{dE}{dt} = \frac{1}{[NH_3]} \frac{RT}{nF} \frac{d [NH_3]}{dt}$$

Under initial rate conditions (no product, NH_3) the second term of this equation changes much more rapidly than the $1/[NH_3]$ term. This gives a rate of potential change proportional to the rate of ammonia production and hence proportional to the enzyme concentration (Camman, 1977). A typical strip chart recording of such a measurement is shown in Figure 3. This response can be generated by adding enzyme to a substrate solution in which the electrode has been equilibrated or by adding a concentrated substrate solution to the enzyme containing solution. Rates are easily measured graphically by drawing a line tangent to the inflection point of this curve.

Properties of the Deaminating Enzyme Labels

In making a choice among the available deaminating enzymes for use as labels in EIA, several characteristics had to be considered. First among these is commercial availability. While many deaminating enzymes are known, not all are commercially available at a reasonable quantity, quality and cost. Stability, solubility, high specific activity, and retention of activity after coupling to hapten are other important enzyme label characteristics. Based on these criteria, three enzymes were chosen to be examined as deaminating enzyme labels for immunoassay with potentiometric detection.

Urease, adenosine deaminase, and asparaginase are all available in high purity and reasonable cost from commercial suppliers. From their high specific activities it was anticipated that they could be detected at very low concentrations. This detection at low levels is perhaps the most important characteristic as it will determine the sensitivity and the detection limits of the immunoassay itself. For highest sensitivity, high activity on a per molecule level is the best indicator since detection of the enzyme label bound by antibody is also on a molecular basis. These and some other pertinent characteristics of the three deaminating labels are listed in Table II.

Assay conditions for the measurement of the activity of urease, adenosine deaminase and asparaginase are a compromise between the optimum conditions for the ammonia gas-sensing electrode and conditions that give the highest enzymatic activity. Since the pH optimum of the ammonia gas sensor is dependent on the ammonia-ammonium equilibrium and is above pH 10, this will tend to increase the overall enzyme/sensor pH optimum from that of the enzyme alone. Such combined pH optima are depicted in Figure 4 for the three selected enzyme labels at 25°C: asparaginase, pH 10.0 (borate buffer); urease, pH 8.0 (Tris-HCl buffer with 5 mM EDTA); and adenosine deaminase, pH 9.0 (borate buffer). Calibration curves for these enzymes obtained at their respective pH optima are shown in Figure 5. Linearity was observed with rates as high as 20 mV/min for this particular ammonia sensor configuration. At some point the rate of ammonia production will exceed the response of the electrode itself and result in a curve with less sensitivity at higher enzymatic activity. Detection limits of 30, 35, and 160 fmole (defined as the amount of enzyme giving a 1 mV/min rate) were obtained for asparaginase, urease, and adenosine deaminase respectively. In terms of sensitivity, the same enzymes gave slopes of 0.022, 0.029, and 0.005 mV min^{-1} $fmol^{-1}$. It is suggested from these data alone that asparaginase and urease, with the two best sensitivity and detection limit values, should be superior to adenosine deaminase as enzyme labels for potentiometric immunoassay. The other qualities described above as well as the available hapten coupling sites on each enzyme will, however, influence the utility of these labels for enzyme immunoassay.

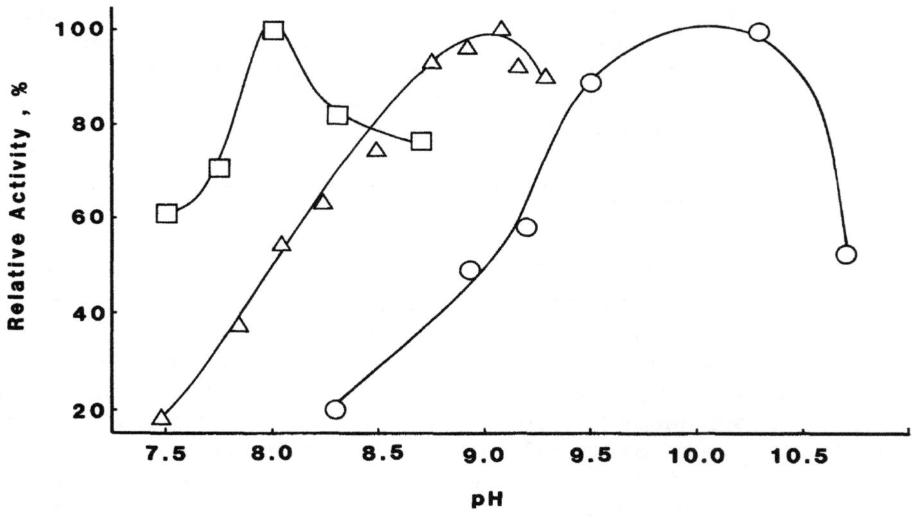

Figure 4. Combined enzyme and gas-sensor pH optima for urease (squares), adenosine deaminase (triangles) and asparaginase (circles) at 25°C.

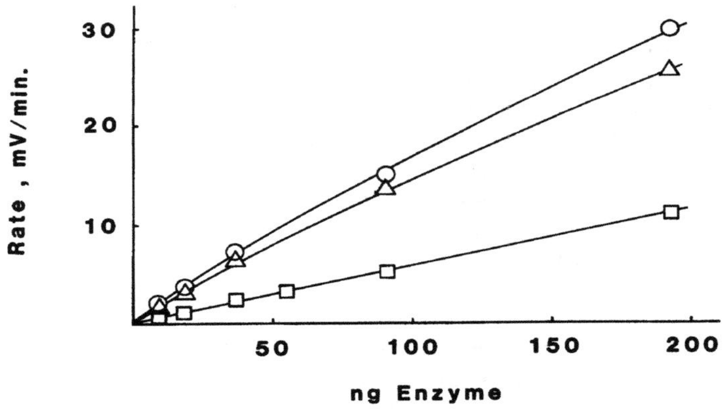

Figure 5. Calibration curves for urease (squares), adenosine deaminase (triangles) and asparaginase (circles).

MATERIALS AND METHODS

Equipment and Reagents

Orion 95-10 ammonia gas sensing electrodes were used in all experiments under thermostated conditions provided by a Haake FS circulating bath. Potentiometric rate data were obtained with a Corning 12 Research pH meter and recorded on a Heath Schlumberger Model SR204 strip chart recorder. Antibody incubations were performed at room temperature with gentle agitation provided by a Bronwill vortex mixer.

Urease type VII (EC 3.5.1.5), adenosine deaminase (EC 3.5.4.4), all steroids, asparagine, and adenosine were obtained from Sigma Chemical Company. Asparaginase (EC 3.5.1.1) was a product of Merck, Sharpe & Dohme. Anti-DNP/BSA antisera (rabbit) anti-cortisol-21-thyroglobin antisera (rabbit), protein A, and agarose bound IgG (goat) fraction of anti-rabbit IgG, were purchased from Miles Laboratories. Aminocaproic acid, 2,4-dinitrofluorobenzene, and N-ethyl-5-phenylisoxazolium-3'-sulfonate (Woodward's reagent K) were obtained from Aldrich Chemical. All other chemicals used were of reagent grade or better quality.

Preparation of the Adenosine Deaminase-DNP Conjugates

Adenosine deaminase, 0.2 ml (1mg protein, 240 units) was diluted with 7.8 ml of pH 7.5 Tris-HCl (I=0.1M) buffer and reacted with 0.5 ml of 2,4-dinitrofluorobenzene with vigorous stirring at 4°C for one hour. The aqueous portion was extensively dialyzed vs. pH 7.5 Tris buffer (I=0.1M) containing 0.02% sodium azide. A 1:1 ratio of the colored DNP groups to adenosine deaminase molecules was indicated by spectrophotometric measurements. Adenosine deaminase-DNP conjugates prepared in this manner retained greater than 90% of their original enzymatic activity. Another indication that the kinetic properties of adenosine deaminase were not altered by this procedure is the unchanged Michaelis constant, Km. The experimentally determined Km value of 6×10^{-5}M for both native and DNP coupled adenosine deaminase agrees with the literature value (Zielke and Suelter, 1971).

Preparation of the Asparaginase-DNP Conjugate

Asparaginase, 0.85 mg protein, was dissolved in 2.9 ml of phosphate buffer, pH 7.5, and reacted with 40 ul of dinitrofluorobenzene at 4°C. The product was then dialyzed vs. Tris-HCl buffer, pH 7.5, to remove any unreacted DNP groups. The concentration of protein and of DNP groups in this preparation were too low to allow a reasonable estimation of the number of DNP groups per molecule of asparaginase. The conjugate showed retention of approximately 75% of its original activity with 48% of this remaining activity capable of being bound by antibody. The kinetic properties of this enzyme were unaltered as indicated by the retention of activity upon coupling to hapten and unchanged Michaelis constant, K_m. Approximate K_m values of 6×10^{-5} M were obtained for both the conjugate and native asparaginase, slightly higher than the 1.25×10^{-5} M literature value (Wriston and Yellin, 1973) obtained under different buffer conditions. Asparaginase-DNP conjugates prepared with DNP-aminocaproic acid (DNP-ACA) and either Woodward's reagent K (Bodlaender et al., 1969) or 1-ethyl-3-(3-dimethyl-amino propyl) carbodiimide-HCl (EDAC) (Schuurs and Van Weeman, 1977) as the coupling agent exhibited high non-specific adsorption to the solid phase and were therefore unacceptable (Gebauer, 1982).

Preparation of the Enzyme-Cortisol Conjugates

Coupling procedures for cortisol protein conjugates were reported by Erlanger (1973). Conjugates in this work were generally prepared in 2.5 ml volumes containing the same protein concentration (mg/ml) for each enzyme. In order to obtain conjugates with different cortisol-to-enzyme ratios, cortisol and coupling reagent concentrations as well as reaction times were varied. The sodium salt of cortisol-21-hemisuccinate was used for the carbodiimide-coupling reaction while the free acid form was used in the mixed-anhydride procedures in accord with their respective solubility properties in aqueous and organic solvents. Conjugate designations were by enzyme, coupling procedure, and order prepared (e.g., asparaginase CDI(1): CDI, carbodiimide; MA, mixed anhydride). Immediately after preparation, each conjugate was dialyzed five times vs. 4 liters of phosphate buffered saline (PBS) pH 7.4, each time overnight at 4°C. This reduces the concentration of any unreacted cortisol to a level that will cause no interference with the assays.

Asparaginase CDI(1), adenosine deaminase CDI(1), and urease CDI(1). For each enzyme, cortisol-21-hemisuccinate, 4.9 mg in 250 ul pH 7.4 PBS buffer was reacted with 2.3 mg 1-ethyl-3-(3-dimethylaminopropyl)-carbodiimide-HCl (EDAC), also in 250 ul PBS buffer, pH 7.4, for 30 min at room temperature. This solution was then added to 2 ml of PBS buffer, pH 7.4 (no azide), containing 0.73 mg of enzyme protein, and allowed to react for 2 h at room temperature followed by 4 h at 4°C.

Asparaginase MA(1). Cortisol-21-hemisuccinate, 5 mg, was dissolved in 600 ul dioxane and 5 ul tributylamine. Isobutylchloroformate, 5 ul was added and the mixture was allowed to react at room temperature for 20 min before the organic solvent was removed under reduced pressure. Asparaginase, 2.5 mg (0.73 mg protein) in 2.5 ml carbonate buffer, pH 9.5, was then added and reacted overnight at 4°C.

Asparaginase CDI(2), adenosine deaminase CDI(2), and urease CDI(2). For these conjugates the initial step in the carbodiimide reaction was carried out at pH 5 instead of the pH 7.4 used for the CDI(1) conjugates. A greater activation of the cortisol-21-hemisuccinate is expected since the carbodiimide reaction is acid catalyzed. Cortisol-21-hemisuccinate, 19.6 mg, and EDAC, 9.2 mg, were dissolved in 2 ml deionized water, adjusted to pH 5. and allowed to react at room temperature for 15 min. One-half milliliter of this solution was then added to each enzyme solution (2 ml, PBS buffer, pH 7.4, with 0.73 mg protein) of asparaginase, adenosine deaminase, or urease. The pH was adjusted to 5 with dilute HCl, and reacted for 3 h at 4°C.

Asparaginase MA(2), adenosine deaminase MA(1), and urease MA(1). The mixed anhydride of cortisol-21-hemisuccinate was prepared using the same amounts and same protocol as in asparaginase MA(1). After evaporation of solvent, 5 ml of carbonate buffer, pH 9.5, was added to dissolve the residue. This solution was added, 0.5 ml aliquots, to 2 ml of each enzyme (0.73 mg protein) in carbonate buffer, pH 9.5, and allowed to react at 4°C overnight with stirring.

Asparaginase MA(3), adenosine deaminase MA(2), and urease MA(2). A higher concentration of cortisol-21-hemisuccinate was used in an attempt to increase the ratio of cortisol to enzyme. For each enzyme, 15 mg of cortisol-21-hemisuccinate was dissolved in a solution of 600 ul dioxane, 5 ul tributylamine, and 200 ul tetrahydrofuran (used to prevent the solution from freezing at 4°C). After cooling to 4°C, 5 ul of isobutylchloroformate was added and allowed to react for 1 h. The organic solvents were then removed by evaporation under reduced pressure. Enzyme solution at 4°C, (2.5 ml deionized water containing 0.73 mg of enzyme protein adjusted to pH 9.4 with dilute NaOH) was added to the mixed anhydride and allowed to react at 4°C with stirring over a period of 4 h. In the case of urease, 80 ul of 6 M urea was also added since the presence of substrate was reported to help retain activity during the mixed anhydride coupling of this enzyme (Meyerhoff and Rechnitz, 1979).

Determination of the Cortisol-to-Enzyme Ratio

A nondestructive spectrophotometric method employing two simultaneous equations was used to estimate the concentration of both the cortisol groups (CA) and protein (E).

$$A_{248} = a_{CA(248)}\,[CA] + a_{E(248)}[E].$$

$$A_{280} = a_{CA(280)}\,[CA] + a_{E(280)}[E].$$

Absorptivities, a, of each of the three highly purified enzymes (E) and cortisol-21-acetate (CA) were obtained at both 280 and 248 nm. Cortisol-21-acetate absorbs at 248 nm and was used due to its similarity in structure to cortisol and its bridging group. The cortisol to enzyme (C/E) ratios were obtained from the calculated concentrations.

Determination of Anti-DNP Antibody

Conjugates of adenosine deaminase-DNP or asparginase-DNP were used to label the available binding sites of anti-DNP antibody. In this procedure, 50 ul of a 1:10 dilution of enzyme-DNP (in Tris-HCl buffer, pH 7.5) was incubated with 150 ul of antibody solution for 1 h at room temperature. This was followed by the addition of a 1:1 dilution of anti-rabbit IgG antibody attached to agarose beads to allow a quick separation of antibody-bound label from label free in solution. A 2 h room temperature incubation allowed this immobilized second antibody to bind the anti-DNP antibody and remove it from solution along with the enzyme label to which it was bound. After washing twice with Tris buffer, pH 7.5, the beads were resuspended either in 2 ml of borate buffer, pH 10.0, and the rate obtained by adding 20 ul of 0.1 M asparagine, or in 2 ml of pH 9.0 Tris-HCl buffer with the reaction started by adding 20 ul of 0.2 M adenosine.

Determination of DNP Hapten

The hapten analog 6-(N-dinitrophenyl)amino caproic acid (DNP-ACA) was used. Enzyme-DNP and DNP-ACA compete for the binding sites on the anti-DNP antibody molecule resulting in an inverse relationship between activity bound and DNP groups present. To obtain the hapten curve in Figure 8, 1 ml samples of DNP-ACA were incubated with 50 ul of 1:10 dilution of the appropriate enzyme conjugate and 50 ul of 1:100 anti-DNP antibody (0.8 ug) for 1 h at room temperature. This was followed by a 2 h incubation with 50 ul of 1:1 dilution of agarose bound second antibody to separate the bound phase from the free. Enzymatic activity bound to the solid phase was then measured potentiometrically.

Cortisol Enzyme Immunoassay Conditions

The amount of primary antibody (Ab_1) that will bind 50% of 5-10 pg of tritiated cortisol (50 ul of stock solution) in a standard radioimmunoassay was used in most measurements. More primary antibody gives higher rates but would prove uneconomical for routine measurements. An amount of each enzyme-cortisol conjugate was used that resulted in 80% of its activity being bound by the 50 ul of Ab_1. This was determined by obtaining conjugate dilution curves for each conjugate preparation using 50 ul of Ab_1. The optimum conjugate concentration for each preparation was the amount at the break point in the curve times 1.25.

In these procedures, cortisol (or cortisol acetate) standards were prepared in PBS buffer pH 7.4 by serial dilution from 1 mg/ml (in ethanol) stock solutions and immediately added (100 ul) to tubes containing 50 ul Ab_1 solution in 500 ul 0.5% BSA, PBS buffer, pH 7.4. This mixture was allowed to incubate at room temperature 1/2 h. An appropriate dilution of the enzyme-cortisol conjugate was added and was followed by a 1 h, room temperature incubation. This sequential saturation order of reagent addition should provide the greatest sensitivity (Zettner and Duly, 1974).

Separation of bound from free enzyme-cortisol conjugate was accomplished with 50 ul of a 1:1 dilution of Ab_2 on agarose beads. This amount will bind 50 ng of rabbit IgG and is more than sufficient to remove all of the primary antibody from solution. An incubation time of 1 h at room temperature is adequate for the binding reaction to go to completion. After washing twice with 1 ml volumes of PBS buffer, pH 7.4, with centrifugation at 600xg the pellet of agarose beads was resuspended in 2 ml of pH 10.0 borate buffer, (I = 0.1 M), and transferred to the thermostated measurement cell (30°C). When the ammonia electrode immersed in this stirred suspension (400 rpm stir rate) reached a steady potential, 20 ul of 0.1 M asparagine (in pH 10.0 borate buffer) was added to start the enzymatic reaction. Rates, obtained as previously described, were converted into percentage values and plotted vs. cortisol concentration.

RESULTS

Model Hapten System

The effect of pH on the measurement of anti-DNP antibody is readily seen in Figure 6 where the adenosine deaminase-DNP label was used. The antibody curves were obtained at the pH optimum for the enzyme/gas-sensor system, pH 9.0, and at a lower pH, 8.25. The sub-optimal pH curve has reduced sensitivity with a maximum activity only 75% that of the pH 9.0 curve. A higher detection limit at pH 8.25 is the result. A similar antibody titration curve is obtained with the asparaginase-DNP conjugate (Figure 7). In this plot, 100% activity corresponds to a rate of about 20

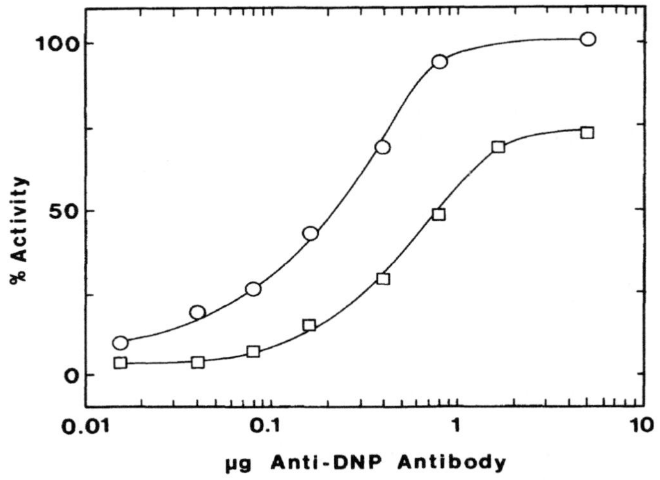

Figure 6. Antibody titration curve of adenosine deaminase-DNP conjugate, pH 9.0 (circles), pH 8.25 (squares).

mV/min. while the 100% activity value with adenosine deaminase conjugate was about 12 mV/min.

The higher sensitivity of the asparaginase conjugate over the adenosine deaminase conjugate is expressed as higher rates over a narrower range for both the antibody titration curves and for measurements of the water soluble hapten analog, DNP-ACA. Hapten curves for both enzyme conjugates are superimposed and shown in Figure 8. While the asparaginase conjugate has higher sensitivity due to its higher specific activity, the conjugate suffers from higher detection limits (adenosine deaminase, 10^{-10}M DNP

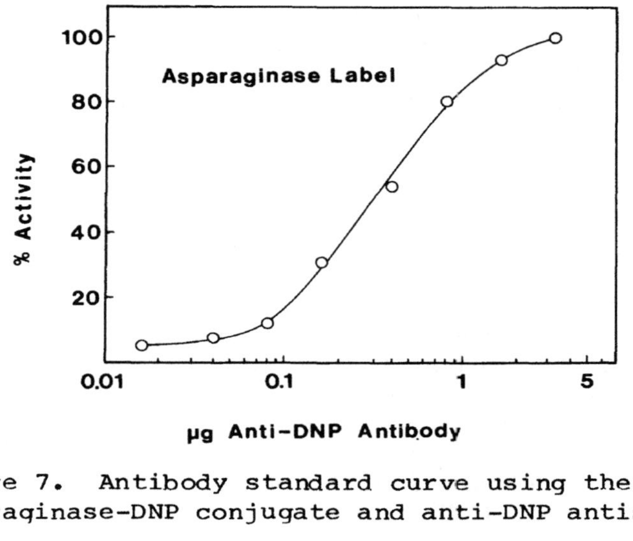

Figure 7. Antibody standard curve using the asparaginase-DNP conjugate and anti-DNP antisera.

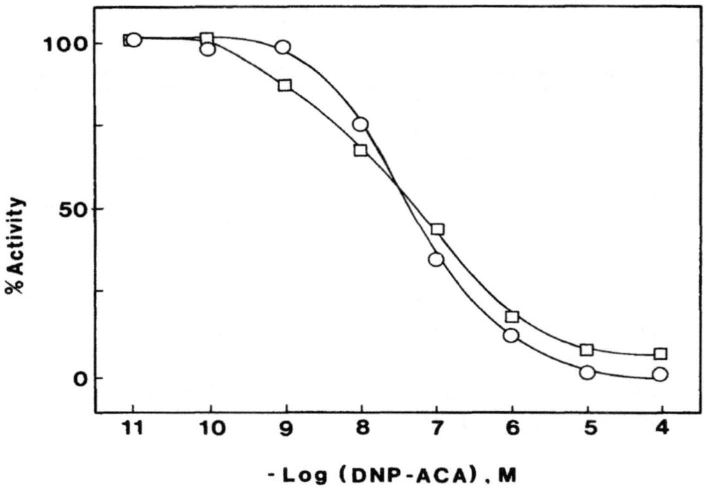

Figure 8. Hapten standard curve for dinitrophenyl aminocaproic acid (DNP-ACA) using the asparaginase-DNP conjugate (circles) and the adenosine deaminase-DNP conjugate (squares).

hapten: asparaginase, 10^{-9}M DNP hapten) for both antibody and hapten. This effect is likely due to the greater number of hapten groups that may be coupled to the enzyme; 52 available lysine residues out of 92 total on asparaginase (Matsushima et al., 1980) vs. 22 total lysine residues on adenosine deaminase (Phelan et al., 1970). A slightly higher free hapten concentration would then be required to displace an equivalent amount of asparaginase label than adenosine deaminase label.

In comparison to the urease label described by Meyerhoff and Rechnitz (1979) for cyclic AMP mesurements, both asparaginase and adenosine deaminase have better detection limits for DNP hapten than the urease-cyclic AMP system had for cyclic AMP (10^{-8}M cyclic AMP). Adenosine deaminase was superior to asparaginase in terms of both DNP hapten (10^{-10}M vs. 10^{-9}M) and anti-DNP antibody (50 ng vs. 80 ng) detection limits.

Comparisons of different assay systems however, are not always easy to assess. For example, urease conjugates prepared with cyclic GMP rather than cyclic AMP allow measurements of the latter to be made as low as 10^{-9}M (Meyerhoff and Rechnitz, 1979). For an effective comparison of these three enzyme labels, all three must be used to measure the same species with the same hapten to enzyme coupling procedures. It is toward this end that urease, asparaginase and adenosine deaminase were examined as labels for the EIA of cortisol.

Stability and Utility of Enzyme-Cortisol Conjugates

In this particular system a full comparison of successfully coupled hapten-enzyme conjugates was not possible owing to the loss of activity of the urease conjugates prepared by carbodiimide or mixed anhydride coupling. Table III summarizes the conjugates prepared as described in the Methods section. Unfortunately, all urease-cortisol conjugates showed an unaccept-

TABLE III

Characteristics of Enzyme-Cortisol Conjugates

Enzyme Conjugate	Percent Activity Retained	C/E Ratio	Activity Bound[a] rate, mV/min.	Percent Activity Immunoreactive
Asparaginase				
CDI(1)	72	0.52	9.1	47
CDI(2)	41	5.5	8.1	79
MA(1)	22	6.5	5.2	65
MA(2)	97	5.3	0.55	b
MA(3)	59	1.9	4.1	21
Adenosine deaminase				
CDI(1)	67	0.39	0.93	b
CDI(2)	31	2.5	1.4	57
MA(1)	76	0.53	0.73	b
MA(2)	55	2.2	1.2	b
Urease				
CDI(1)	15	6.4	0	b
CDI(2)	0	12	0	b
MA(1)	6.3	0	0	b
MA(2)	7.2	15	0	b

a: activity bound by 50 ul Ab_1, excess Ab_2
b: not determined

able loss in enzymatic activity upon coupling to cortisol. Urease is known to be susceptible to denaturation by organic solvent, which may account for the poor performance of the mixed anhydride conjugates; however a similar activity loss was observed with coupling procedures using a water soluble carbodiimide. Special care to remove all organic solvent in these reactions did not prevent the observed loss of activity. The tendency to denaturation by organic solvent may be agravated by the hydrophobic nature of the hapten since similar coupling procedures with cyclic AMP resulted in active conjugates that were stable and analytically useful.

Adenosine deaminase and asparaginase, which retained similar and acceptable activities upon coupling to cortisol, are known to be relatively hardy enzymes from their use with DNP hapten. Adenosine deaminase is a monomeric enzyme that does not readily lose activity under the mild conditions employed here. Asparaginase, which is subject to loss of activity when dissociated into its four subunits by drastic pH or denaturing reagents, exhibited good stability and activity retention on coupling with cortisol. Apparently this substitution on the lysine residues does not significantly affect the tertiary structure of the enzyme or interfere with the active sites.

Once prepared the conjugates were examined for stability over a three month period of refrigerated storage. Asparaginase conjugates showed negligible loss in activity while the adenosine deaminase and urease conjugates retained an average of 52% and 40% of their activity, respectively.

Effects of Cortisol Incorporation

Cortisol-to-enzyme ratios (C/E) are the average number of cortisols per molecule of enzyme. This value along with the maximum percentage activity bound gives a better indication of the distribution of cortisol residues over the enzyme molecules. For example, two theoretical conjugate preparations both have a cortisol-to-enzyme ratio of 1.0 but have maximum bindable activities of 50 and 100%. It can then be said that each enzyme molecule of the 100% maximum bindable activity preparation has one cortisol attached while in the other preparation it is likely that half of the enzyme molecules have two cortisol attached and the other half none.

Best "maximum percentage activity bound" characteristics were obtained with conjugates having the higher cortisol-to-enzyme ratios, asparaginase CDI(2) and MA(1), It is more likely with these conjugates, having ratios of 5.5 and 6.5, respectively, that each enzyme molecule will have at least one cortisol attached, than with those conjugates having ratios of 0.52 or 1.9. Those conjugates prepared under more rigorous conditions (e.g., all the CDI(2) preparations, asparaginase MA(1), MA(3), adenosine deaminase MA(2), and urease MA(2) preparations) showed a greater loss in activity after coupling but had higher activity bound to antibody. The preparations with a high degree of activity retained, e.g., the asparaginase MA(2), adenosine deaminase MA(1), and urease MA(1) conjugates, were prepared under

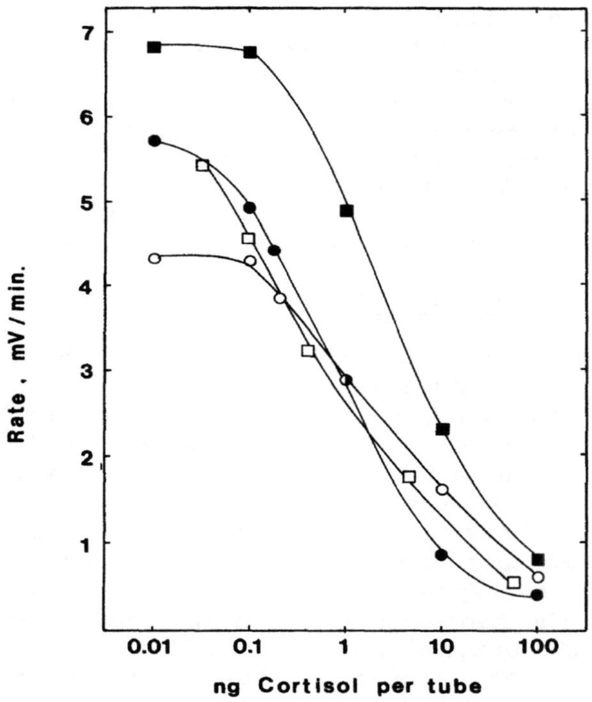

Figure 9. Cortisol standard curves using the four asparaginase conjugates listed in Table III. Closed circles, CDI(1); closed squares, CDI(2); open circles, MA(1); open squares, MA(3).

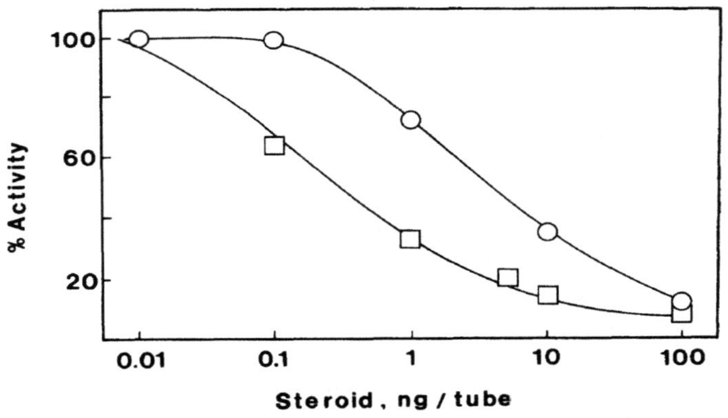

Figure 10. Steriod standard curves showing the effect of higher antibody affinity for cortisol-21-acetate (squares) than for cortisol (circles).

milder conditions and show very little activity bound by antibody, indicative of a low level of cortisol incorporation.

Cortisol Standard Curves

Of the five conjugates listed in Table III only the asparaginase conjugates gave useful cortisol standard curves (Figure 9). Adenosine deaminase CDI(2) had a maximum rate too low to provide even minimal sensitivity. Examination of Figure 9 reveals that the conjugates with the higher cortisol-to-enzyme ratios, asparaginase CDI(2) and MA(1), have higher detection limits than the conjugates with lower cortisol to enzyme ratios, e.g., asparaginase CDI(1) and MA(3). Ideally, enzyme-hapten conjugates for use in heterogeneous EIA systems should have a 1:1 hapten-to-enzyme ratio so as to maximize the activity bound by a fixed amount of antibody. Asparaginase CDI(1) with its superior detection limit and sensitivity of its cortisol standard curve appeared to be the best conjugate preparation of those studied here for use in the immunoassay of cortisol.

Site of Conjugation and Bridging Group Effects

The immunoassay system employed for cortisol involved coupling the cortisol to the enzyme in the same manner as it was coupled to thyroglobin. Thyroglobin-cortisol conjugate was the antigenic molecule used to stimulate antibody production. This homologous immunoassay employs cortisol linked at its 21-carbon position via a succinate bridge to a lysine residue on the protein. The nature of the hapten linkage to enzyme label can have a profound influence on the response. In the DNP system which uses an antibody raised to DNP-BSA this can be seen in the relative response toward DNP-ACA (essentially the hapten plus its bridging group) and dinitroaniline (similar to the hapten group alone). The anti-DNP antibody has a greater affinity to the enzyme-label and the hapten with the bridging group than the hapten alone. The amount of dinitroaniline required to displace the same amount of enzyme label is thus about three orders of magnitude higher than with DNP-ACA.

The effect of the bridging group contribution to antibody affinity is again seen in Figure 10 obtained with asparaginase CDI(2) conjugate. The higher affinity of antibody for cortisol-21-acetate than for cortisol is apparent from the offset of the cortisol acetate curve toward lower concentration.

DISCUSSION

Examination of the ammonia gas-sensing electrode as a detector for immunoassay measurements has hinged on the utility of deaminating enzyme labels for EIA purposes. All three of the labels examined are easily detected by the ammonia sensor at low levels through potentiometric rate measurements. A particular advantage of potentiometric detection over standard spectrophotometric means is the ability to follow initial rate reactions in turbid media. Such conditions are encountered with activity measurements on the solid phase used to separate the bound from free fractions in heterogeneous EIA. The gas permeable membrane employed in the sensor provides a barrier to possible interferences such as ions and other non-volatile species. Certain enzymes (AMP deaminase), conjugate preparations, or adsorbant media (Protein A) have been observed to adsorb on the membrane and cause difficulty in repetitive measurements. Steps can be taken to attenuate this phenomena (0.5% BSA is helpful), however care must be taken that the membrane not lose its hydrophobic nature through wetting by surfactants.

Asparaginase, adenosine deaminase (Gebauer, 1982) and urease (Meyerhoff and Rechnitz, 1979) have been shown to be effective EIA labels for potentio-metric detection of model haptens, and in the urease example, cyclic AMP. In a direct comparison for the immunoassay of the steriod hormone cortisol employing similar hapten coupling procedures for each enzyme label only the asparaginase conjugates proved useful. Adenosine deaminase conjugates had insufficient activity to allow even minimal sensitivity, while urease suffered from an unacceptable loss in activity upon coupling to cortisol.

The utility of asparaginase as an immunoassay label was demonstrated with cortisol assay curves that showed a 50% displacement of enzyme label by between 1 and 5 ng/tube of cortisol. Allowing for a 1:10 sample dilution these standard curves would be useful for determination of cortisol since normal human plasma has a cortisol concentration of 5-20 mg/100 ml.

Precision of potentiometric immunoassay measurements is limited by the precision of the enzyme activity measurement. Enzyme assays by potentiometric rate methods generally yield values with relative standard deviations (RSD) of 2-3%. These values will be among the best that can be expected from measurements in the more complex immunolabel assay systems. Studies with the asparaginase label for antibody determinations gave relative standard deviations ranging from 2 to 12% over the useful portion of the standard curve, with the lower RSD at values near the curve's inflection point and the higher RSD at the high antibody (high rates) portion of the curve. Similarly, RSD values of 5-7% were observed for hapten concentration from 10^{-11} to 10^{-7} M and up to 15% outside that range. In the regions of the standard curves that would be used for analysis, the reproducibility of rate measurements is acceptable and comparable to other EIA systems.

Rates of potential change that allow the completion of a measurement in four to six minutes were the sensitivity goal of this work. If a longer time period were available for measurement, either in a rate format or in a fixed time measurement on an automated system, better sensitivity and detection limits could be obtained from these same deaminating enzyme labels.

ACKNOWLEDGEMENT

The assistance of P. Tisza in the preparation of this chapter is greatly appreciated.

REFERENCES

Bodlaender, P., Feinstein, G., and Shaw E. The Use of Isoxazolium Salts for Carboxyl Group Modification of Proteins. Trypsin. Biochem. 8, 4941-4948 (1969).

Camman, K. Bio-Sensors Based on Ion-Selective Electrodes. Fres. Z. Anal. Chem. 287, 1-9 (1977).

Davies, P., and Mosbach, K. The Application of Immobilized NAD^+ in an Enzyme Electrode and in Model Enzyme Reactors. Biochim. Biophys. Acta 370, 329-338 (1974).

Erlanger, B. F. Principles and Methods for the Preparation of Drug Protein Conjugates for Immunological Studies. Pharm. Rev. 25, 271-280 (1973).

Fung, K. W., Kuan, S. S., Sung, H. Y., and Guilbault, G. G. Methionine Selective Enzyme Electrode, Anal. Chem. 51, 2319-2324 (1979).

Gebauer, C. R., Meyerhoff, M. E., and Rechnitz, G. A. Enzyme Electrode-Based Kinetic Assays of Enzyme Activities. Anal. Biochem. 95, 479-482 (1979).

Gebauer, C. R., and Rechnitz, G. A. Immunoassay Studies Using Adenosine Deaminase Enzyme with Potentiometric Rate Measurement. Anal. Lett. 14, 97-109 (1981).

Gebauer, C. R., and Rechnitz, G. A. Deaminating Enzyme Labels for Potentiometric Enzyme Immunoassay. Anal. Biochem. 124, 338-348 (1982).

Gebauer, C. R. Enzyme and Immunoassay Measurements with Ion-Selective Electrodes. Ph.D. Thesis, Univ. of Delaware (1982) University Microfilm, Ann Arbor.

Guilbault, G. G., Smith, R. K., and Montalvo, J. G. Use of Ion Selective Electrodes in Enzymic Analysis - Cation Electrodes for Deaminase Enzyme Systems. Anal. Chem. 41, 600-605 (1969).

Guilbault, G. G., and Hrabankova, E. An Electrode for Determination of Amino Acids. Anal. Chem. 42, 1779-1783 (1970).

Guilbault, G. G., and Hrabankova, E. New Enzyme Electrode Probes for D-Amino Acids and Asparagive. Anal. Chim. Acta 56, 285-290 (1971).

Hjemdahl-Monsen, C., Papastathopoulos, D. S., and Rechnitz, G. A. Automated Adenosine Deaminase Enzyme Determination with an Ammonia-Sensing Membrane Electrode. Anal. Chim. Acta 88, 253-259 (1977).

Hsiung, C. P., Kaun, S. S., and Guilbault, G. G. A Specific Enzyme Electrode for L-Phenylalanine. Anal. Chim. Acta 90, 45-49 (1977).

Huang, Y-Z. A New Method for the Assay of Glutaminase Activity: Direct Measurement of Product Formation by an Ammonia Electrode. Anal. Biochem. 61, 464-470 (1974).

Hussein, W. R., and Guilbault, G. G. Nitrate and Ammonia Ion-Selective Electrodes as Sensors. Part II. Assay of Nitrate Ion and Nitrite Reductases in Stationary Solutions and Under Flow-Stream Condition. Anal. Chim. Acta 76, 183-192 (1975).

Katz, S. A. Direct Potentiometric Determination of Urease Activity. Anal. Chem 36, 2500-2501 (1964).

Matsushima, A., Nishimura, H., Ashihara, Y., Yokota, Y., and Inada, Y. Modifications of E. coli Asparagainase with 2,4-bis(O-methoxypolyethylene glycol)-6-chloro-s-triazine (Activated PEG_2); Disappearance of Binding Ability Towards Anti-Serum and Retention of Enzymatic Activity. Chem. Lett. 1980, 773-776 (1980).

Meyerhoff, M. E., and Rechnitz, G. A. An Activated Enzyme Electrode for Creatinine. Anal. Chim. Acta 85, 277-285 (1976).

Meyerhoff, M. E., and Rechnitz, G. A. Electrode-Based Enzyme Immunoassays Using Urease Conjugates. Anal. Biochem. 95, 483-493 (1979).

Meyerhoff, M. E. Polymer Membrane Electrode-Based Potentiometric Ammonia Gas Sensor. Anal. Chem. 52, 1532-1534 (1980).

Meyerson, L. R., McMurtrey, K. D., and Davis, V. E. A Rapid and Sensitive Potentiometric Assay for Monoamine Oxidase Using an Ammonia-Selective Electrode. Anal. Biochem. 86, 287-297 (1978).

Ngo, T. T. Ion-Selective Electrode-Based Enzymatic Determination of L-Histidine. Int. J. Biochem. 6, 371-373 (1975).

Nikolelis, D. P., Papastathopoulos, D. S., and Hadjiiannau, T. P. Construction of a Guanine Enzyme Electrode and Determination of Guanase in Human Blood Serum with an Ammonia Gas Sensor. Anal. Chim. Acta 126, 43-50 (1981).

Papastathopoulos, D. S., and Rechnitz, G. A. Highly Selective Electrode for 5'-Adenosine Monophosphate. Anal. Chem. 48, 862-864 (1976).

Phelan, J., McEvoy, F., Rooney, S., and Brady, T. G. Structural Studies on Adenosine Deaminase from Calf Intestinal Mucosa. Biochim. Biophys. Acta 200, 370-377 (1970).

Ross, J. W., Riseman, J. H. and Krueger, J. A. Potentiometric Gas Sensing Electrodes. Pure Appl. Chem. 36, 473-487 (1973).

Schuurs, A. H. W. M., and Van Weemen, B. K. Enzyme Immunoassay. Clin. Chim. Acta 81, 1-40 (1977).

Wriston, J. C., and Yellin, T. O. L-Asparaginase: A Review. Adv. Enzymol. Relat. Areas Mol. Biol. 39, 185-248 (1973).

Zettner, A., and Duly, P. E. Principles of Competitive Binding Assays (Saturation Analyses) II. Sequential Saturation. Clin. Chem. 20, 5-14 (1974).

Zielke, C. L., and Suelter, C. H. IV. 5'-Adenylic Acid Aminohydrolase, in The Enzymes, Vol. V, Boyer, P.D., ed. Academic Press, NY pp 64-73 (1971).

HOMOGENEOUS AND HETEROGENEOUS ENZYME IMMUNOASSAYS MONITORED WITH CARBON DIOXIDE SENSING MEMBRANE ELECTRODES

C. A. Broyles and G. A. Rechnitz

Department of Chemistry
University of Delaware
Newark, Delaware 19716

Immunochemical potentiometric and amperometric electrodes are in the infant stages of development, yet the literature is already replete with diversified efforts to design increasingly sensitive and specific immunoelectrodes (Eggers et al., 1982; Guilbault, 1983; Boitieux et al., 1984; Keating and Rechnitz, 1984). The carbon dioxide gas-sensing probe, however, has played only a limited role in this research. Despite all the schemes designed for immunosensors, there is a conspicuous absence in the use of the carbon dioxide probe. Why this might be so is suggested by the constraints within which use of the carbon dioxide probe is practical as well as the limiting characteristics of the immunochemical reaction being studied. Additional restrictions result from the presence of the enzyme when developing an enzyme immunoassay [EIA]. Despite this, the enzyme is a favorite "transducer", or label, which produces a signal related to the immunochemical reaction and can be recognized by the electrochemical detector. This integration of enzyme chemistry, immunochemistry and probe characteristics is a critical factor in determining, during method development, the feasibility of selecting a carbon dioxide probe as the analytical detector. These key ingredients will be discussed in this overview together with possibilities for the future of the carbon dioxide sensor in immunochemical applications. Improvements in the probe and broader applications will highlight this future.

Within the broad spectrum of immunochemical substances, qualitative and quantitative determinations of antigens and antibodies have predominated in the application of electrochemical devices. The challenge in quantifying antigens and antibodies arises because these substances do not usually lend themselves to measurement by simple and ordinary chemical means. Their inherent lack of a specific signal, whether separate or in combination with one another, usually leads to indirect multi-step methods. Certainly, a gas-sensing potentiometric probe, in and of itself, would be of little value. But, recalling the successful history of biocatalytic electrodes (Freiser, 1978, 1980) and taking an example from enzyme immunoassay (Blake and Gould, 1984; Monroe, 1984), electrochemical sensors have entered the burgeoning field of immunochemical analysis (Sittampalam and Wilson, 1984).

Promising applications for ion selective electrodes (under which gas-sensing probes are traditionally classified) have been expounded on for years. A seemingly endless list of possibilities offers hope for simple, inexpensive and accurate analytical tools. Indeed, these goals have been achieved in certain circumstances. Epitomized by the fluoride ion selective electrode, cation and anion selective electrodes have been established in industrial and clinical applications (Jackson et al., 1981; Analytical Chemistry Reviews). One is hard-pressed to match that record with gas-sensing probes, particularly that for carbon dioxide. Nevertheless, these gas sensors together with various electrodes remain the focus of efforts to provide reliable and straightforward analytical instruments in many areas of research, including immunochemistry. Thus, potentiometric enzyme immunoassays have been reported using the ammonia gas-sensing probe (Brontman and Meyerhoff, 1984; Gebauer and Rechnitz, 1981) as well as the carbon dioxide probe (Keating and Rechnitz, 1985; Fonong and Rechnitz, 1984). The potpourri also includes use of iodide (Boitieux et al., 1979), fluoride (Alexander and Maltra, 1982) and pH (Mascini et al., 1982) electrodes. An assortment of other immunoelectrochemical methods have been developed (Sittampalam and Wilson, 1984; Janata, 1975; Wehmeyer et al., 1983) as have different labels (Cais, 1983; Doyle et al., 1982).

CARBON DIOXIDE PROBE

One of the few commercially available gas-sensing probes, the carbon dioxide sensor is relatively simple in design and easy to operate. It consists of a pH electrode enclosed in a sleeve of electrolyte solution with the pH sensitive glass flush against a carbon dioxide gas permeable membrane at the base of the sleeve. The electrolyte (usually bicarbonate and sodium chloride) forms a very thin film between the glass surface and the outer membrane (e.g. Teflon, silicon rubber). A general diagram of a carbon dioxide probe is shown in Figure 1. When the probe tip comes in contact with solution, carbon dioxide gas present in that solution will diffuse through the membrane, altering the pH of the electrolyte layer. This pH change is sensed by the inner electrode causing a potential change. The resulting Nernstian response of a typical carbon dioxide probe extends from 10^{-2}M to slightly less than 10^{-4}M. Thorough discussions of the theory and operation of gas-sensing probes have been presented (Covington, 1979; Midgley, 1975). The high selectivity of the membrane has made the carbon dioxide probe, as well as other gas sensors, a very attractive analytical tool.

A second type of carbon dioxide probe is of the "air gap electrode" family. Fashioned after the conventional membrane gas sensors, the primary difference lies in the replacement of the membrane with an air space between the sample and the electrode. A thin film of electrolyte is applied directly to the pH membrane of the glass electrode. Carbon dioxide in the sample, which is contained in a sealed cell open to the probe tip, diffuses through the air gap to the electrolyte and the pH electrode (Covington, 1979).

After almost thirty years, the carbon dioxide probe's primary justification for existence remains that for which it was originally intended --the determination of the partial pressure of carbon dioxide in blood (Stow et al., 1957; Severinghaus and Bradley, 1958; Wimberley et al., 1983; Kost et al., 1983). Several commercial instruments tailored for clinical use measure pCO_2 potentiometrically with probes based on the design just described. Similar inroads into other fields have not kept pace despite the fact that numerous analytical applications have been suggested throughout the past decade, especially since the advent of biocatalytic sensors (Freiser, 1978, 1980; Analytical Chemistry Reviews).

A biocatalytic electrode, in its simplest form, is an electrochemical probe used in conjunction with a biocatalyst. The probe acts as the

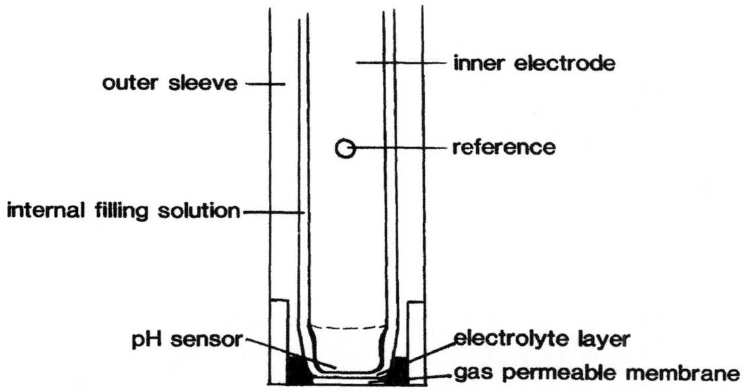

Fig. 1. Carbon dioxide gas-sensing probe. Typical of commercially available sensors with a gas permeable membrane.

detector while the biocatalyst, immobilized on the probe, mediates some reaction. The physical construction of the typical carbon dioxide probe is well-suited for conversion into a biocatalytic device. In such a configuration, a biocatalyst (e.g. enzyme, bacterial cells) is immobilized in such a way as to form a layer in contact with the gas permeable membrane. The biocatalytic material participates in a reaction which consumes or produces carbon dioxide which is measured by the electrochemical sensor. A logical extension of such progress would be an immunochemical sensor based on the carbon dioxide probe as the detector. That this has been realized to a limited degree may be less a reflection of the immunochemistry than of the carbon dioxide probe itself as indicated when one reviews carbon dioxide probe immunoelectrochemical methods.

ENZYME IMMUNOASSAYS WITH CARBON DIOXIDE PROBES

The carbon dioxide probe has been used in both a homogeneous and a heterogeneous immunochemical assay. The latter method (Keating and Rechnitz, 1985) measures digoxin via a typical competitive assay. Free digoxin (the analyte) and digoxin-BSA coated on polystyrene beads compete for anti-digoxin in solution. The antibody is conjugated to horseradish peroxidase(HRP). The more digoxin in the sample, the less anti-digoxin-HRP available to bind to the digoxin on the beads. The beads are separated from the reaction solution by centrifugation and transferred to the assay mixture. Carbon dioxide produced by the HRP-pyrogallol system is then monitored by the carbon dioxide probe. The lower the enzyme activity on the beads, the greater the antigen concentration in the original sample. The standard curve for the digoxin EIA shown in Figure 2 indicates a sensitivity in the nanogram range.

The second carbon dioxide immunochemical method reported is a homogeneous one for the determination of human immunoglobulin-G (IgG) (Fonong and Rechnitz, 1984). The method involves a reaction between human IgG and anti-human IgG which is conjugated to the enzyme chloroperoxidase. Enzyme activity is inhibited in the presence of the antigen. When increasing concentrations of IgG in solution bind with the anti-IgG the enzyme activity was shown to be proportionately decreased. The enzymatic reaction (with reactants β-ketoadipic acid, bromide and peroxide) yields carbon dioxide as one of the products. The resulting calibration curve

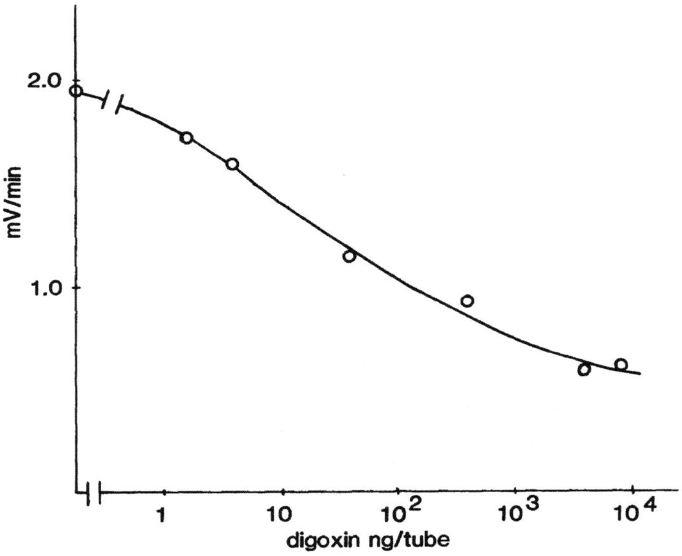

Fig. 2. Digoxin enzyme immunoassay standard curve. Potentiometric heterogeneous assay using antigen-coated polystyrene beads.

is shown in Figure 3. Similar to the heterogeneous method, an inverse relationship exists between enzyme activity and antigen concentration.

In both of the potentiometric immunochemical methods just described, the carbon dioxide probe is participating only in the final phase of the total system, but coupling the probe in this manner is not as straightforward as one might think. For example, in the heterogeneous method the

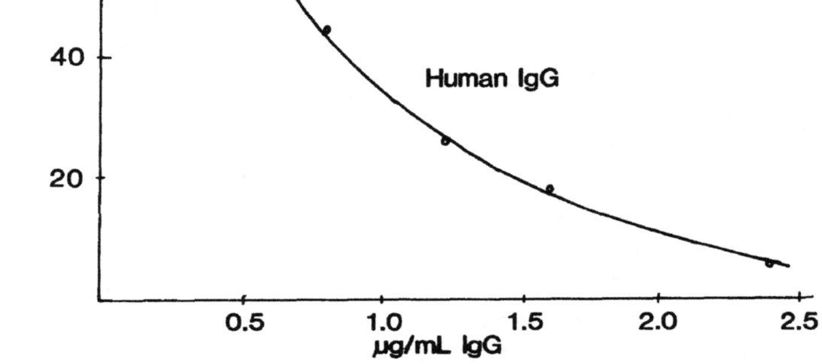

Fig. 3. Calibration curve for measurement of human IgG with homogeneous potentiometric enzyme immunoassay.

beads provide sufficient concentration of the enzyme-antibody complex to bring the carbon dioxide reaction product within the sensitivity limits of the probe. The homogeneous method takes advantage of a shift in the pH optimum of the enzyme following its conjugation to the antibody. The fortuitous shift from pH 3 to 6 provided the necessary compatibility between the enzyme and immunochemical reactions. Such phenomena, however, cannot be depended upon when designing analytical techniques. Besides pH compatibility within the assay system, several other parameters require attention during development of an EIA with the carbon dioxide probe detector.

REQUIREMENTS FOR A CARBON DIOXIDE PROBE EIA

The carbon dioxide sensing probe, as previously mentioned, is a very selective analytical tool by virtue of the gas permeable membrane. In addition, the probe does not require extensive upkeep or advanced technical skills by its users. Yet other features of this probe, such as the pH limitation already mentioned, translate into possible drawbacks when it is being considered for use with immunoassays. Three areas of concern are addressed here.

pH

The operating pH range of the sensor, the pH optimum of the immunochemical reaction and the pH for optimal enzyme activity must all be compatible. The carbon dioxide probe functions best in acidic solutions. This is due to the change in the distribution of the CO_2, HCO_3^-, and $CO_3^{=}$ species in solution at different pHs (Orion Research). A pH of five or less is generally recommended for best results (Orion Research; Ross et al., 1973). The probe will function in alkaline solution, but with a noticeably attenuated response. Table I lists the potential output of the probe at pH 5 and 7.

The acidic pH range requirement for the carbon dioxide probe is of some significance when combining this sensor to an immunochemical reaction. Not surprisingly, antigen-antibody interactions are generally strongest near physiological pH -- neutral to slightly alkaline. While the immunochemical bond usually requires a pH less than three or greater than ten before being disrupted (Goding, 1983), at least some dissociation can be expected when operating outside the optimal pH range. Decreased sensitivity results from sub-par conditions. One can see, then, that the sensitivity of an immunoassay with a carbon dioxide probe will suffer regardless of whether one compromises in favor of the measuring device or the immunochemistry.

Finally, the third component of a carbon dioxide probe EIA highly affected by pH is the enzyme. Maximum enzymatic activity is achieved only over a particular pH range which may be broad, but can be quite

TABLE I. Response of Carbon Dioxide Probe at pH 5 and 7

	citrate		phosphate	
$HCO_3^{=}$	pH 5	pH 7	pH 5	pH 7
3×10^{-4}M	48mV	15.2mV	45.9mV	17mV
3×10^{-3}	94	45.2	89.8	52

narrow. To be feasible for use with the carbon dioxide sensor the enzyme should have a pH optimum of approximately five. A casual inspection of a compendium of enzymes (Barman, 1969) reveals barely a handful of decarboxylating enzymes where the pH of the reaction is less than five or six. This would account for the much higher percentage of ammonia-based rather than carbon dioxide-based enzyme electrodes (Freiser, 1980). The mismatch between the pH optimum of the enzyme and probe has not prevented development of some carbon dioxide probe enzyme techniques (Guilbault and Shu, 1972; Seegopaul and Rechnitz, 1983) which indicates that both the probe and the catalyst can be stretched toward their limits, albeit at the cost of sensitivity.

Nonetheless, a tabulation of the more commonly used enzymes in immunoassays [Table II] (Blake and Gould, 1984; Sittampalam and Wilson, 1984) reveals that the majority of the enzymes listed have a pH optimum of six or greater. In addition, none of these enzymes catalyzes reactions that consume or produce carbon dioxide. Though these points are worth noting, the list is not very long and reflects, in part, limited research on the utility of a wide scope of enzymes for immunoassay labels. The immunoinhibition of enzyme activity may be a broad-based phenomenon and if used with suitable decarboxylases would be most effective for homogeneous EIAs with a carbon dioxide sensor.

There is only nominal flexibility when trying to mesh the pH sensitivity of each of the components of an EIA using a carbon dioxide probe. Judicious selection of enzymes and wise compromises will often have to suffice. Should the immunochemistry of interest involve an antigen-antibody couple with a high affinity constant, some of the pressure for pH

TABLE II. Enzyme Labels of Enzyme Immunoassays

Enzyme	pH optimum	Detection Method
Acetylcholinesterase	7-8	potentiometric (H^+, NH_3) spectrophotometric
Adenosine deaminase	7.5-9	potentiometric (NH_3)
Alkaline phosphatase	8-10	amperometric (phenol) fluorometric spectrophotometric
L-amino acid oxidase	6.5	amperometric (H_2O_2) potentiometric (NH_3)
Amine oxidase	7.2	amperometric (H_2O_2) potentiometric (NH_3)
Catalase	6-8	amperometric (O_2) spectrophotometric thermometric
β-galactosidase	6-8	fluorometric spectrophotometric
Glucose oxidase	4-7	amperometric (H_2O_2) spectrophotometric
Glucose-6-phosphate dehydrogenase	7.4-7.8	amperometric (NADH) fluorometric spectrophotometric
Lysozyme	4.5-5.5	spectrophotometric
Malate dehydrogenase	8.5-9.5	spectrophotometric
Peroxidase	5-7	amperometric (H_2O_2) spectrophotometric
Urease	6.5-7.5	potentiometric (NH_3) spectrophotometric

compatibility would be alleviated. This is because the complex should be more stable over a wider range of pHs than complexes with lower affinity. On the other hand, stronger binding immune complexes introduce the disadvantage of requiring harsher conditions for the dissociation of the complex. This is a concern when reversibility of the binding is an integral part of the assay procedure such as when one component of the complex is immobilized and must be recycled for subsequent measurements.

Dissociation of antigens from antibodies is often accomplished by subjecting the immunocomplex to a solution of very low pH. This will not have an adverse effect on the carbon dioxide probe, but a large protein such as an antibody may be denatured. This problem may be circumvented by using certain ions which have been shown to successfully break immunochemical bonds (Dandliker et al., 1967; Ternynck and Avrameas, 1971). These so-called chaotropic ions (e.g. I^-, F^-, SCN^-) ought to provide a useful alternative to the low pH approach.

Measurement Time

Because antigen-antibody reactions are studied indirectly by a secondary process, it is not surprising that proposed methods cover a plethora of instrumentation. The electrochemist developing an immunoassay can also choose from many tools of the trade including amperometric and potentiometric sensors. When a simple dipstick type device is being planned, the carbon dioxide probe in a sense competes with the lower linear range of the ammonia gas-sensing probe (Covington, 1979; Ross et al., 1973), the advantages of amperometric sensors (Pinkerton and Lawson, 1982) and the unique characteristics of a host of modern biosensors (North, 1985). At least one property of the carbon dioxide sensor, its long recovery time, may hinder its application to immunoassays.

Although the response time of the carbon dioxide sensor is on the order of a few minutes, it can take much longer than that for the probe to recover its baseline between measurements. Each of the measurements taken for Table I required nearly an hour recovery time. While shorter times have been reported (Takano et al., 1985), especially in flow systems (Midgley, 1975), the long reequilibration time is recognized as a problem and has been addressed (Keeley and Walters, 1983; Guilbault et al., 1985). The principal attribute of the carbon dioxide probe, its selective membrane, is also a cause of the prohibitive recovery time because of slow diffusion through and equilibration with the probe membrane, the electrolyte film and any immobilized reagent layers. Diffusion is also slow from the internal pH electrode solution. These and other processes limit the speed with which baseline can be attained between each measurement. The individual steps that comprise the analytical result obtained with gas sensors have been delineated and several suggestions made to optimize the probe (Guilbault et al., 1985). One significant improvement came with a small alteration of the probe design. The outer jacket was made with small holes to allow for convenient replacement of the inner filling solution between each measurement. The recovery period was thus reduced by half an hour to about one minute. While the investigators used an ammonia probe, their observations and conclusions should apply equally well to the carbon dioxide probe. Other workers have suggested use of a reconditioning buffer between each test (Keeley and Walters, 1983). Use of a 0.1M pH 10 phosphate buffer with the carbon dioxide probe decreased recovery time to ten minutes.

Relative to immunotechniques needing long incubation times, the total measurement time of the carbon dioxide probe is not a major concern, but when compared to other electrochemical sensors, it falls short in this category. For example, experiments equivalent to the carbon dioxide measurements shown in Table I were carried out for peroxide on an amperometric probe. The turn-around time in this case, with few exceptions, was under ten minutes without any special efforts to optimize between-run equilibration. Potentiometric probes continue to be popular partially because of advances in field effect transistors (Pinkerton and Lawson, 1982), but amperometric probes have also garnered attention as immunoelectrochemical sensors (Pinkerton and Lawson, 1982; Uditha de Alwis and Wilson, 1985; Ngo and Lenhoff, 1983; Broyles and Rechnitz, 1986). (An additional attraction of amperometric probes is the commercial availability of horseradish peroxidase and glucose oxidase conjugated immunochemicals, both of which catalyze reactions that can be readily coupled to amperometric probes. The question arises whether researchers tend to design methods with commercially available reagents or if suppliers feed the need and desires of the researchers. Currently, very few enzyme labels are routinely conjugated to immunochemicals for general sale.)

Immobilization

Quantitative immunoassays may or may not require a separation step, but to date, the great majority fall in the first category. In these heterogeneous assays, one element of the immunocomplex (or other reagent) is labeled for later detection. The labeled and unlabeled species must be separated before the final measurement step. Homogeneous immunoassays require no separation of components and offer increased speed and ease of operation while also eliminating the imprecision and inaccuracy of the separation step. Homogeneous immunoassays have been generally unsatisfactory for determining large molecules, but are well-utilized in therapeutic drug monitoring.

The separation step in heterogeneous immunoassays is often facilitated by immobilizing one or more of the reagents. Attachment of reaction constituents to a support medium is not a new field of science, but a broad and continually expanding one. The advantages and disadvantages of the possible modes for immobilizing proteins have been discussed (Silman and Katchalski, 1966; Weetall, 1975; Zaborsky, 1973) and should be evaluated based on the particular application at hand. The immobilization phase of heterogeneous EIAs has taken on many faces, but homogeneous EIAs have employed immobilization less frequently though they could benefit from its wise application. The advantages of a separation-free assay lure researchers toward developing such methods. Immunoelectrochemical sensors with immobilized reactants offer the appealing prospects of simple dipstick methodology.

Immunoelectrochemical probes seem destined to carry immobilized antigens, antibodies or enzyme-immunochemical conjugates. The membrane with the immobilized species may be attached to the probe after preliminary steps in the reaction sequence (Boitieux et al., 1984) or the reagent membrane may be immobilized on the probe originally, remaining there throughout the assay (Aizawa et al., 1980). The former methodology is always heterogeneous, but the latter can be homogeneous if the probe with the immobilized reagent does not have to be transferred between solutions at any time during the assay. An electrochemical homogeneous EIA that includes immobilization of reactants directly on the detector is often categorized as a chemically modified electrode (Janata, 1975; Yamamoto et al., 1980). Further development of these and a host of other amperometric and potentiometric probes should follow. The immobilization need not be a necessary part of the method, but as an integral part of the technique, can be of considerable merit because immobilization provides simplification in the testing procedure, conservation of reagents and the option to measure a steady state output rather than a rate. Of course, the immobilization adds still another dimension in the successful development of an immunoelectrode.

Enzymes have been held on the tip of electrodes in a variety of ways, including the popular dialysis membrane. Because substrates, cofactors and products are of much smaller size than enzymes, dialysis membrane presents no barrier for easy passage of solution components to and from the enzyme layer. In fact, when combined with cross-linking agents, dialysis membrane has proven reliable in this regard. Entrapment of enzyme, antibody or enzyme conjugate behind dialysis membrane is unlikely to be feasible in those immunoassays where the enzyme conjugate on one side of the membrane and the antibody on the other must come in contact during the assay. The molecular weight of these large proteins and the molecular weight cutoff of dialysis membranes complicates the assay design. [R. Frost: "Before I build a wall I'd ask to know what I was walling in or walling out" . . .]

Cellulosic materials play a dominant role in immobilization for immunological techniques. Immunochemists have been using nitrocellulose and its derivatives on a large scale since the introduction of protein blotting techniques (Gershoni and Palade, 1983). Electrochemical sensors have been described where cellulose acetates and similar materials serve as the immobilizing medium (Aizawa et al., 1980; Seiyama et al., 1983). The cellulosic material is impregnated with the immunochemical by mixing the components in solution and making sheets of the immunomembrane. Simple adsorption of immunochemicals on nitrocellulose acetates has been effective for ELISA, but preliminary work with adsorption of enzymes to cellulose acetate disks and collagen sheets which were then placed on the tip of electrochemical probes, has met with limited success because of poor lifetimes (Broyles, 1985). Stronger binding to these two materials via chemical crosslinking has shown more promise.

Immobilization causes a change in structure of the immobilized substance and inevitably the activity of that substance is altered. When immobilizing an enzyme-immunochemical conjugate, both enzymatic and immunochemical activity must be preserved at sufficient levels to yield a sensitive and useful probe. Adapting past immobilization procedures to current interests is necessary. Cross-linking techniques on cellulose derivatives and collagen are being pursued as barrier-free, flexible membranes for electrodes. Finding other suitable and convenient chemicals to act as immobilization material continues to be a field deserving considerable attention.

PROSPECTS FOR EXPANDED UTILITY OF PROBE

The carbon dioxide gas sensor can have a niche as a valuable detector in immunoassays if one takes advantage of the positive characteristics of the probe such as its selectivity. Moreover, the electrode is stable and not easily contaminated by proteins and serum constituents as are certain other electrochemical sensors (notably platinum and carbon surfaces).

It is important to examine the limitations of a system in order to understand where improvements can be made and what research would be most helpful. Suggestions have been given in the areas previously discussed. Understanding and patience in selecting and testing suitable enzymes is important. General investigations into immobilization techniques could prove beneficial for the carbon dioxide sensor.

Furthermore, improvements in the carbon dioxide probe itself should be pursued with the intention of reducing recovery time and extending the linear response range. The bulkiness of the current probe confines it to certain applications, prohibiting its use in assays requiring a small sample size and for in-vivo measurements. Scattered reports of micro-carbon dioxide probes (Sohtell and Karlmark, 1976; Pui et al., 1978) are encouraging though such miniaturized sensors have not yet been established.

Expansion of the carbon dioxide probe EIA measurement system into other areas of immunological importance would be feasible once some of the above-mentioned hurdles are successfully handled. In cytotoxicity studies, cell-mediated cytotoxic responses are monitored by observing the reaction between sensitized T lymphocytes and cell surface antigens on labeled target cells (Grabstein, 1980; Schlager and Adams, 1983). The lymphocytes cause the target cells to lyse, releasing the label into solution. The released label, usually a radioisotope, is measured and is directly proportional to the percentage of target cells killed. If target cells can be labeled with a substrate or cofactor of an enzyme immobilized on a carbon dioxide probe, that probe, when placed in the cell solution after lysis, would determine the amount of carbon dioxide produced due to the release of the reactant. This approach could also be extended to quantification of complement fixation. This principle has been applied in lysis of electroactive marker-loaded liposomes (Shiba et al., 1982) and sheep red blood cells (D'Orazio and Rechnitz, 1979) used to measure antibodies following a complement fixation step.

There is certainly room for improvement in the design of current carbon dioxide gas-sensing probes and past experience shows that efforts to upgrade the probe are fruitful. However, because of the plethora of alternative enzyme immunoassays, these improvements might be attempted only when necessitated by a clear need for the carbon dioxide sensor in a particular application. In the meantime, the advantageous features of the sensor remain available for those interested in pursuing its use in immunochemical assays.

REFERENCES

Aizawa, M., Morioka, A. and Susuki, S., 1980, An enzyme immunosensor for the electrochemical determination of the tumor antigen α-fetoprotein, Anal. Chim. Acta 115:61-67.

Alexander, P. W. and Maltra, C., 1982, Enzyme-linked immunoassay of human immunoglobulin G with the fluoride ion selective electrode, Anal. Chem. 54:68-71.

Analytical Chemistry Fundamental Reviews (see, e.g., April 1984, April 1982).

Barman, T. E., 1969, "Enzyme Handbook," Springer-Verlag, New York.

Blake, C. and Gould, B. J., 1984, Use of enzymes in immunoassay techniques. A review, Analyst 109:533-547.

Boitieux, J. L., Thomas D. and Desmet, G., 1984, Oxygen electrode-based enzyme immunoassay for the amperometric determination of hepatitis B-surface antigen, Anal. Chim. Acta 163:309-313.

Boitieux, J. L., Desmet, G. and Thomas, D., 1979, An "antibody electrode," preliminary report on a new approach in enzyme immunoassay, Clin. Chem. 25:318-321.

Brontman, S. B. and Meyerhoff, M. E., 1984, Homogeneous enzyme-linked assays mediated by enzyme antibodies; a new approach to electrode-based immunoassays, Anal. Chim. Acta 162:363-367.

Broyles, C. A. and Rechnitz, G. A., 1986, Drug antibody measurement by homogeneous enzyme immunoassay with amperometric detection, Anal. Chem., in press.

Broyles, C. A., 1985, unpublished results.

Cais, M., 1983, Metalloimmunoassay: principles and practice, Meth. Enzymol. 92:445-458.
Covington, A. K., ed., 1979, "Ion-Selective Electrode Methodology," CRC Press, Inc., Boca Raton, Florida.
Dandliker, W. B., Alonso, R., deSaussure, V. A., Kierszenbaum, F., Levison, S. A. and Schapiro, H. C., 1967, The effect of chaotropic ions on the dissociation of antigen-antibody complexes, Biochem. 6:1460-1467.
D'Orazio, P. and Rechnitz, G. A., 1979, Potentiometric electrode measurement of serum antibodies based on the complement fixation text, Anal. Chim. Acta 109:25-31.
Doyle, M. J., Halsall, H. B. and Heineman, W. R., 1982, Heterogeneous immunoassay for serum proteins by differential pulse anodic stripping voltammetry, Anal. Chem. 54:2318-2322.
Eggers, H. M., Halsall, H. B. and Heineman, W. R., 1982, Enzyme immunoassay with flow-amperometric detection of NADH, Clin. Chem. 28:1848-1851.
Fonong, T. and Rechnitz, G. A., 1984, Homogeneous potentiometric enzyme immunoassay for human immunoglobulin G, Anal. Chem. 56:2586-2590.
Freiser, H., ed., 1978, 1980, "Ion Selective Electrodes in Analytical Chemistry," Vol. I, Vol. II, Plenum Press, New York.
Gebauer, C. R. and Rechnitz, G. A., 1981, Immunoassay Studies using adenosine deaminase enzyme with potentiometric rate measurement, Anal. Lett. 14:97-109.
Gershoni, J. M. and Palade, G. E., 1983, Protein blotting: principles and applications, Anal. Bioch. 131:1-15.
Goding, J. W., 1983, "Monoclonal Antibodies: Principles and Practice," Academic Press, New York.
Grabstein, K., 1980, Cell-mediated cytolytic responses, in: "Selected Methods in Cellular Immunology," Mishell, B. B. and Shiigi, S. M., eds., W. H. Freeman and Co., San Francisco.
Guilbault, G. G., Czarnecki, J. P. and Nabi Rahni, M. A., 1985, Performance improvements of gas-diffusion ion-selective and enzyme electrodes, Anal. Chem. 57:2110-2116.
Guilbault, G. G., 1983, Immobilised biological and immuno sensors, Anal. Proc. 20:550-552.
Guilbault, G. G. and Shu, F. R., 1972, Enzyme electrodes based on the use of a carbon dioxide sensor. Urea and L-tyrosine electrodes, Anal. Chem. 44:2161-2166.
Jackson, C. J., Neuberger, C. and Taylor, M., 1981, Applicability and cost effectiveness of ion chromatographic and ion-selective electrode techniques as applied to environmental monitoring by the health and safety executive, Anal. Proc. 18:201-204.
Janata, J., 1975, An immunoelectrode, J. Amer. Chem. Soc. 97:2914-2916.
Keating, M. Y. and Rechnitz, G. A., 1985, Potentiometric enzyme immunoassay for digoxin using polystyrene beads, Anal. Lett. 18:1-10.
Keating, M. Y. and Rechnitz, G. A., 1984, Potentiometric digoxin antibody measurements with antigen-ionophore based membrane electrodes, Anal. Chem. 56:801-806.
Keeley, D. F. and Walters, F. H., 1983, Use of a conditioning buffer to regenerate gas sensing ion selective electrodes, Anal. Lett. 16:1581-1584.
Kost, G. J., Chow, J. L. and Kenny, M. A., 1983, Transcutaneous carbon dioxide for short-term monitoring of neonates, Clin. Chem. 29:1534-1536.
Mascini, M., Zolesi, F. and Palleschi, G., 1982, pH electrode-based enzyme immunoassay for the determination of human chorionic gonadotropin, Anal. Lett. 15:101-113.
Midgley, E., 1975, Investigations into the use of gas-sensing membrane electrodes for the determination of carbon dioxide in power station waters, Analyst 100:386-399.

Monroe, D., 1984, Enzyme immunoassay, Anal. Chem. 56:920A-931A.
Ngo, T. T. and Lenhoff, H. M., 1983, Antibody-induced conformational restriction as basis for new separation-free enzyme immunoassay, Bioch. Biophys. Res. Comm. 114:1097-1103.
North, J. R., 1985, Immunosensors: antibody-based biosensors, Trends Biotech. 3:180-186.
Orion Research Instruction Manual; Carbon dioxide electrode model 95-02.
Pinkerton, T. C. and Lawson, B. L., 1982, Analytical problems facing the development of electrochemical transducers for in vivo drug monitoring, Clin. Chem. 28:1946-1955.
Pui, C. P., Rechnitz, G. A. and Miller, R. F., 1978, Micro-size potentiometric probes for gas and substrate sensing, Anal. Chem. 50:330-333.
Ross. J., W., Riseman, J. H. and Kruegar, J. A., 1973, Potentiometric gas sensing electrodes, Pure Appl. Chem. 36:473-487.
Schlager, S. I. and Adams, A. C., 1983, Use of dyes and radioisotopic markers in cytotoxicity tests, Meth. Enzymol. 93:233-245.
Seegopaul, P. and Rechnitz, G. A., 1983, Enzymatic determination of thiamine pyrophosphate with a pCO_2 membrane electrode, Anal. Chem. 55:1929-1933.
Seiyama, T., Fueki, K., Shiokawa, J. and Suzuki, S., 1983, "Chemical Sensors," Elsevier, New York.
Severinghaus, J. W. and Bradley, A. F., 1958, Electrodes for blood pO_2 and pCO_2 determination, J. Appl. Physiol. 13:515-520.
Shiba, K., Umezawa, Y., Watanabe, T., Ogawa, S. and Fujiwara, S., 1982, Thin-layer potentiometric analysis of lipid antigen-antibody reaction by tetrapentylammonium (TPA^+) ion loaded liposomes and TPA^+ ion selective electrode, Anal. Chem. 52:1610-1613.
Silman, I. H. and Katchalski, E., 1966, Water insoluble derivatives of enzymes, antigens, and antibodies, Ann. Rev. Biochem. 35:873-908.
Sittampalam, G. S. and Wilson, G. S., 1984, Enzyme immunoassays with electrochemical detection, Trends Anal. Chem. 3:96-99.
Sohtell, M. and Karlmark, B., 1976, In vivo micropuncture P_{CO_2} measurements, Pflugers Arch. 363:179-180.
Stow, R. W., Baer, R. F. and Randall, B. F., 1957, Rapid measurement of the tension of carbon dioxide in blood, Arch. Phys. Med. Rehab. 38:646-650.
Takano, S., Kondoh, Y. and Ohtsuka, H., 1985, Determination of carbonates in detergents by a carbon dioxide gas selective electrode, Anal. Chem. 57:1523-1526.
Ternynck, T. and Avrameas, S., 1971, Effect of electrolytes and of distilled water on antigen-antibody complexes, Biochem. J. 125:297-302.
Uditha de Alwis, W. and Wilson, G. S., 1985, Rapid sub-picomole electrochemical enzyme immunoassay for immunoglobulin G, Anal. Chem. 57:2754-2756.
Weetall, H. H., ed., 1975, "Immobilized Enzymes, Antigens, Antibodies and Peptides. Preparation and Characterization," Marcel Dekker, Inc. New York.
Wehmeyer, K., Doyle, M. J., Halsall, H. B. and Heineman, W. R., 1983, Immunoassay by electrochemical techniques, Meth. Enzymol. 92:432-444.
Wimberley, P. D., Pedersen, K. G., Thode, J., Fogh-Andersen, N., Sorensen, A. M. and Siggard-Anderson, O., 1983, Transcutaneous and capillary P_{CO_2} and P_{O_2} measurements in healthy adults, Clin. Chem. 29:1471-1473.
Yamamoto, N., Nagasawa, Y., Shuto, S., Tsubomuro, H., Sawai, M. and Okumura, H., 1980, Antigen-antibody reaction investigated with use of a chemically modified electrode, Clin. Chem. 26:1569-1572.
Zaborsky, O. R., 1973, "Immobilized Enzymes," CRC Press, Cleveland.

ENZYME-LINKED IMMUNOSORBENT ASSAYS USING OXYGEN-SENSING ELECTRODE

Masuo Aizawa

Department of Bioengineering, Faculty of Engineering,
Tokyo Institute of Technology,
Ookayama, Meguro-ku, Tokyo 152, Japan

INTRODUCTION

Quantitative radio-immunoassay (RIA) techniques are now standard practice (Yalow, 1971). However, the drawbacks of radio-isotopic handling such as high costs, health and environmental hazards limited variety of usable isotopes, short half-life of the radio-labeled antigens and difficult introduction of the radio-isotopes onto the substances to be determined, have stimulated the research for non-isotopic methods in immunological assays. Of the many non-isotopic immunoassays, enzyme immunoassay (EIA) has been proved as an alternative to RIA for determining antigens and antibodies in clinical medicine (Engvall, 1971; Wisdom, 1976; Monroe, 1984), as it can claim easy preparation, simplified handling and relatively low cost. Enzyme-linked immunosorbent assay (ELISA) has also been used to describe an immunoassay using enzyme-labeled antigens, antibodies, or haptens.

Enzymes make suitable labels because their catalytic properties allow them to act as amplifiers. The enzyme labels are usually measured by visible or ultraviolet spectrophotometry. In some cases, luminescence detection systems have been used in highly sensitive assays for the labels such as peroxidase. The combination of electrochemical detection with enzyme immunoassay results also in extremely sensitive assays (Doyle, 1984; Ngo, 1985).

In the last decade biosubstances such as enzymes and antibodies have been used in conjunction with electrochemical sensoring devices to form biosensors, which include enzyme sensors, microbial sensors, and immunosensors (Aizawa, 1983; Suzuki, 1984). An immunosensor depends its selectivity on immunochemical affinity of an antigen to the corresponding antibody. Several immunosensors use the generation of a transmembrane potential across an antibody (or antigen)-bound membrane after immunochemical reaction with antigen (or antibody) in solution (Aizawa, 1977), and use the potentiometric response of an antigen-ion carrier conjugated in a ligand membrane after binding of an antibody. An antibody-bound electrode also works as immunosensor (Yamamoto, 1978). These immunosensors may provide ultimately high sensitivity. One of the problems remained unsolved, however, is to enhance the sensitivity of the immunosensors for the trace analysis of a specific substance.

An enzyme immunosensor is an analytical device which is dependent on the immunochemical affinity for selectivity and on the chemical amplifications of a labeling enzyme for sensitivity. In case that catalase, which catalyzes the evolution of oxygen from hydrogen peroxide,

$$H_2O_2 \longrightarrow H_2O + 1/2\ O_2 \qquad (1)$$

is a labeling enzyme for an antigen, the enzyme immunosensor is constructed by assembling an antibody-bound membrane and an oxygen sensing electrode. In heterogeneous enzyme immunoassay, the labeling enzyme is measured by amperometry with the oxygen sensing device. The enzyme immunosensor requires an extremely short time for measuring the labeling enzyme. Consequently, rapid and highly sensitive enzyme immunoassay may be accomplished with the enzyme immunosensor(Aizawa et al., 1976; 1978; 1979; 1980). Two different types of enzyme immunosensors for α-fetoprotein (AFP) and ochratoxin A (OTA) (Hongyo et al., 1986) are described in this chapter.

PRINCIPLE OF ENZYME IMMUNOSENSOR

The principle of an enzyme immunosensor with an oxygen electrode is schematically represented in Fig. 1. The principle is illustrated for case of competitive enzyme immunoassay. In order to determine an antigen, the corresponding antibody is immobilized on the membrane matrix. The immobilized antibody can then be bound the antigen specifically on the membrane surface. If the antigen to be determined reacts with the immobilized antibody in the presence of a labeled antigen, the binding sites of the membrane will be occupied by the unlabeled and labeled forms in a ratio appropriate to their concentrations. When catalase is used as label, the antigen to be determined and the catalase-labeled antigen react competitively with the membrane-bound antibody. Since catalase decomposes hydrogen peroxide to release oxygen, the catalase-labeled antigen bound to the membrane-bound antibody can be measured from the rate of oxygen generation in the presence of a known amount of hydrogen peroxide. The generation of oxygen is inhibited by an increase in the concentration of the antigen to be determined.

The enzyme immunosensor is prepared by attaching the antibody-bound membrane to a Clark-type oxygen electrode, which has an oxygen permeable plastic (e.g. Teflon) membrane on the cathode surface and responds sensitively and rapidly to oxygen.

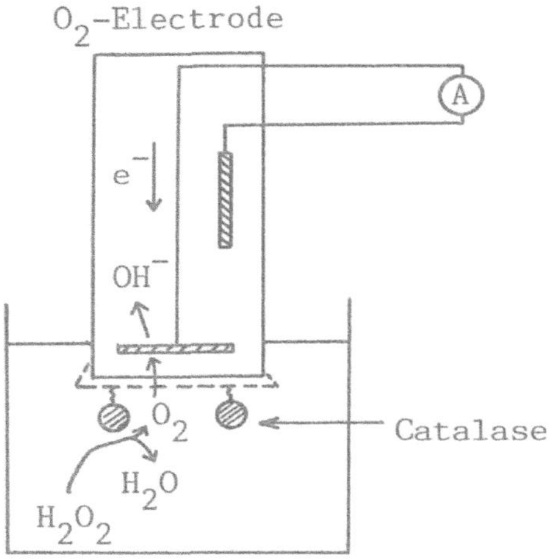

Fig. 1. Principle of enzyme immunosensor with oxygen electrode.

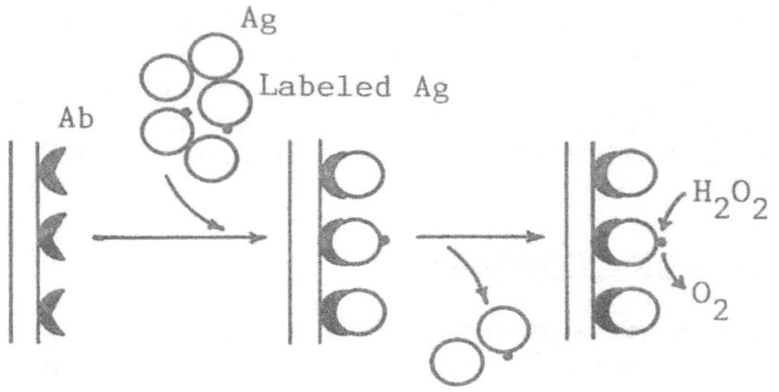

Fig. 2. Enzyme immunosensor based on competitive enzyme immunoassay.

Figure 2 illustrates another type of enzyme immunosensor based on competitive enzyme immunoassay. An antigen to be determined is immobilized to the membrane matrix which is attached to an oxygen electrode. The corresponding antibody is labeled with catalase. A known amount of catalase-labeled antibody is added to a sample solution containing the antigen to be determined. Catalase-labeled antibody competitively reacts with membrane-bound antigen and free antigen in solution. After rinsing the membrane, the labeling catalase is measured by amperometry.

Sandwitch type of enzyme immunoassay may also be performed with an enzyme immunosensor as shown Fig. 3.

EXPERIMENTAL

Preparation of the Antibody Membrane

An antibody-bound membrane was prepared for an enzyme immunosensor for AFP. Cellulose triacetate (250 mg) was dissolved in 5 ml of dichloromethane and 200 μl of 50 % glutaraldehyde was added followed by 1 ml of 1,8-diamino-4-aminomethyloctane. This was mixed to a homogeneous paste which was cast on a glass plate. The preparation was allowed to stand for 2 days at room temperature to complete intermolecular cross-linking of 1,8-diamino-4-aminomethyloctane via glutaraldehyde. The membrane was then peeled off and cut into 1.5×1.5 cm^2 pieces, which were washed with 0.05 M phosphate buffer, pH 7.5.

The membranes were placed in a 0.1 % (w/v) solution of glutaraldehyde (pH 8.0) at 30°C for 2 h. After prolonged washing with water, they were

Fig. 3. Enzyme immunosensor based on sandwitch type of enzyme immunoassay.

then immersed in 10 ml of 0.05 M phosphate buffer containing the anti-AFP antibody at 5 °C for 20 h. To reduce the unreacted aldehyde groups and the Schiff's base, the resultant membranes were immersed in 0.1 M sodium tetrahydroborate at pH 7.5 for 30 min. The antibody-containing membranes were then washed with water and stored in 0.05 M phosphate buffer, pH 7.0, at 5 °C.

Preparation of the Antigen Membrane

An antigen-bound membrane was prepared for an enzyme immunosensor for ochratoxin A (OTA). Anti-OTA monoclonal antibody was produced in ascites fluid of mice bearing hybridoma secreting anti-OTA monoclonal antibody. OTA (5 mg) was dissolved in the mixture of ethanol (0.1 ml) and 3 ml of 0.1 M phosphate buffer (pH 7). Bovine serum albumin (BSA) (50 mg) in 10 ml of PBS was added and then stirred with 36 mg of 1-ethyl-3-dimethylaminopropyl-carbodiimide for 24 h at 20°C in the dark. In order to remove the unreacted low molecular substances, it was dialyzed against the distilled water of 2 ℓ for one day at 4 °C. A piece of polypropylene (P.P.) membrane was immersed into the solution, where OTA-BSA concentration was 14 $\mu g \cdot ml^{-1}$, at 4 °C over night for the fixation of OTA-BSA on it.

Preparation of Catalase-labeled AFP

AFP (6 mg) and catalase (20 mg, 40,000 units) were dissolved in 3 ml of 0.05 M phosphate buffer, pH 8.0. To the solution was added 3 µl of 25 % glutaraldehyde, and the solution was allowed to stand at 25°C for 50 min. It was then applied to a Sepharose CL-6B column and eluted with 0.05 M phosphate buffer, pH 7.0. The catalase-AFP conjugates were fractionated. The catalase-AFP conjugates were assayed for catalase activity and protein content and used as catalase-labeled AFP.

Preparation of Catalase-Labeled Antibody

Catalase (1 mg) in 0.5 ml of PBS (pH 7) was incubated with 0.5 ml of 2.5 % glutaraldehyde for 40 min at 4°C. The solution of 1 ml containing 645 µg of monoclonal antibody was added to the catalase solution and then incubated for 2 h. The obtained mixture was chromatographed on a Sepharose CL-6B column and eluted with PBS (pH 7). Each 2 ml was fractionated and assayed for the contents of catalase and protein by monitoring absorbances at 407 and 208 nm. The fraction pattern is shown in Fig. 4. Fraction 8 was collected as catalase-labeled monoclonal antibody.

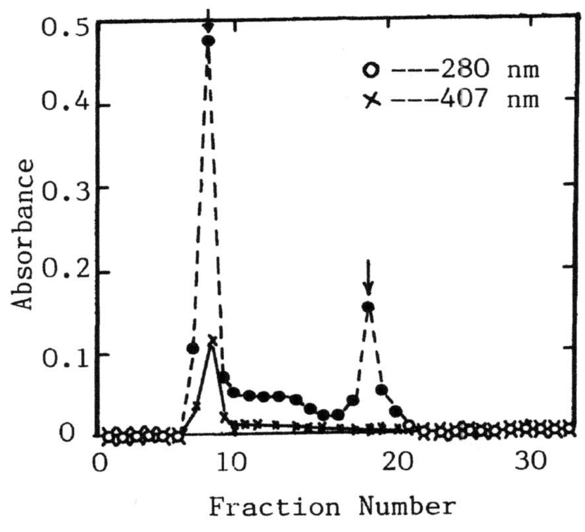

Fig. 4. Fraction pattern of catalase-labeled antibody (see the text).

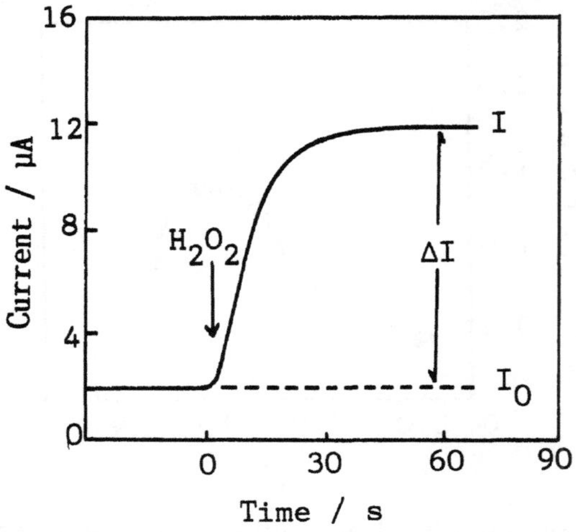

Fig. 5. Response of an enzyme immunosensor for AFP.

Enzyme Immunosensor

A Clark-type galvanic oxygen electrode, consisting of a platinum cathode, lead anode, alkaline electrode and oxygen-permeable Teflon membrane, was used. The antibody membrane was tightly fixed on the Teflon membrane surface; special care was taken to eliminate air bubbles from the cathode surface and the Teflon-antibody membrane ‿ertaces. The output current was displayed on a cnart recorder.

RESULTS

Characterization of Enzyme Immunosensor for AFP

A sheet of antibody membrane, attached to the sensor was placed in contact with 260 "catalase units" of catalase-labeled AFP at pH 7 and 37°C for 2 h. After thorough washing, the sensor was placed in phosphate buffer. When it gave a steady state current from dissolved oxygen, 100 1 of 3 % hydrogen peroxide was injected. The sensor responded very

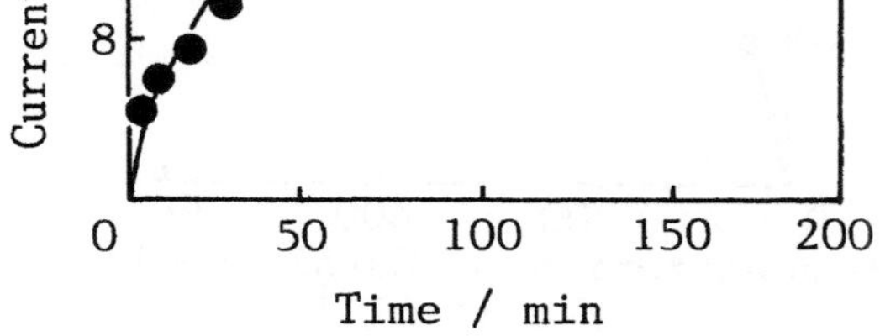

Fig. 6. Time course of the immunochemical reaction of membrane-bound antibody with catalase-labeled AFP.

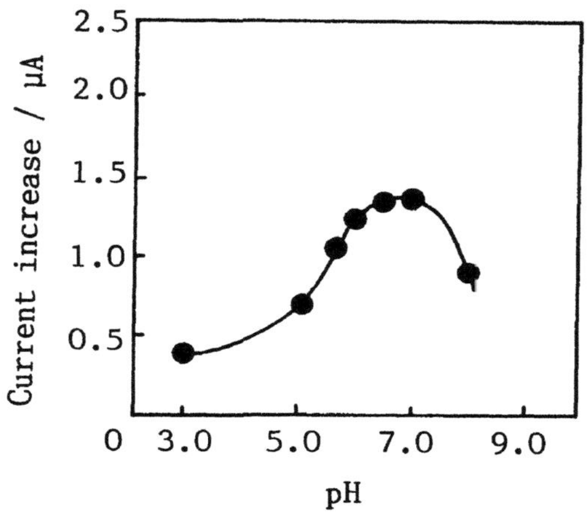

Fig. 7. Effect of pH on the immunochemical reaction of membrane-bound antibody.

rapidly to the generation of oxygen, as shown in Fig. 5. A steady current was obtained within 30 s.

Figure 6 shows the time course of the immunochemical reaction of the membrane-bound antibody with catalase-labeled AFP. Several sheets of antibody membrane were treated with catalase-labeled AFP at 37°C. Each membrane was removed after particular time, washed with 0.5 M sodium chloride and then attached to the sensor. The amount of catalase-labled AFP complexed with membrane-bound antibody was estimated by measuring the response current of the sensor in the presence of hydrogen peroxide. The current increase is proportional to the amount of catalase-labeled AFP bound by the membrane. The extent of the immunochemical reaction increased with time, reaching a steady state after 2 h. Thus the immunochemical reaction time was fixed at 2 h in all further experiments.

As shown in Fig. 7, the immunochemical reaction is greatly influenced by pH. Membrane-bound antibody was reacted with catalase-labeled AFP at different pH values for 2 h. Each membrane was assayed for the amount of antibody-antigen complex as described above. The pH range 6.0 - 7.5 was optimum for the immunochemical reaction, and all further experiments were conducted at pH 7.0.

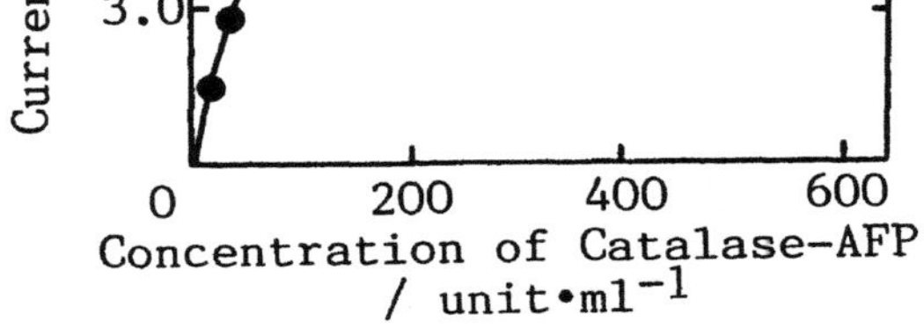

Fig. 8. Binding capacity of membrane-bound antibody for catalase-labeled AFP.

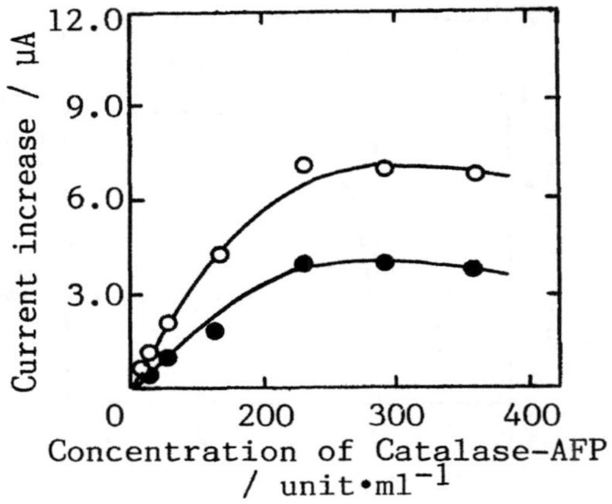

Fig. 9. Inhibition effects of AFP on the binding of catalase-labeled AFP by membrane-bound antibody. (○) 1 ng AFP added, (●) 5 ng AFP added.

Figure 8 established the binding capacity of the membrane-bound antibody for catalase-labeled AFP. The amount of antigen-antibody complex markedly increased with an increase in AFP concentration, reaching a maximum at 200 "catalase units"• ml^{-1}. Excess of catalase-labeled AFP appeared to cause some dissociation of the complex.

Determination of AFP with the Enzyme Immunosensor

Competitive enzyme immunoassay was employed. The AFP to be determined reacted with the antibody membrane in competition with the catalase-labeled AFP.

Figure 9 shows the inhibition effects of AFP on the binding of catalase-labeled AFP by the antibody membrane. The two curves were obtained with a different concentration of AFP, but show a similar dependence on the concentration of catalase-labeled AFP to that without AFP. However, it is evident that the presence of AFP markedly retards the binding of catalase-labeled AFP. This is probably because the antigen-

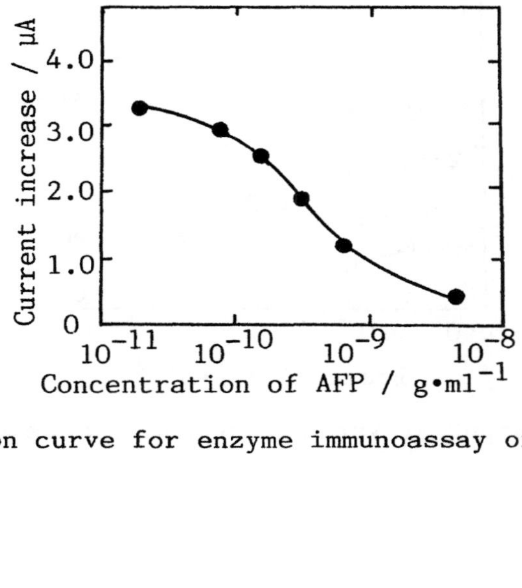

Fig. 10 Calibration curve for enzyme immunoassay of AFP with the sensor.

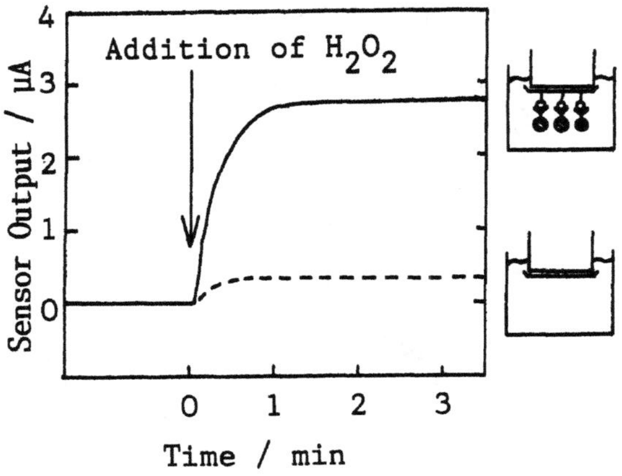

Fig. 11 Response of enzyme immunosensor for OTA.

binding sites of the antibody membrane are partly occupied by AFP in place of catalase-labeled AFP. It should be emphasized that the presence of 1 ng of AFP effectively prevented binding of the catalase-labeled AFP.

A calibration curve is presented in Fig. 10. It shows that AFP can be determined in the range 5×10^{-11} - 5×10^{-8} g•ml^{-1}. The standard deviation for 25 assays of 10^{-9} g of AFP was 15 %. All the assays were performed with different antibody membranes.

Competitive Enzyme Immunoassay of OTA with the Sensor

In order to evaluate the response of the OTA sensor the sensor was in contact with a solution of catalase-labeled antibody in the absence of OTA at 35°C for 1 h. After thorough rinsing with a phosphate buffer saline (PBS), the sensor was set in the PBS (pH 7.2) at 30°C. The time course of the change in sensor output upon H_2O_2 addition is illustrated in Fig. 11. The sensor output reached a steady state value within 100 s.

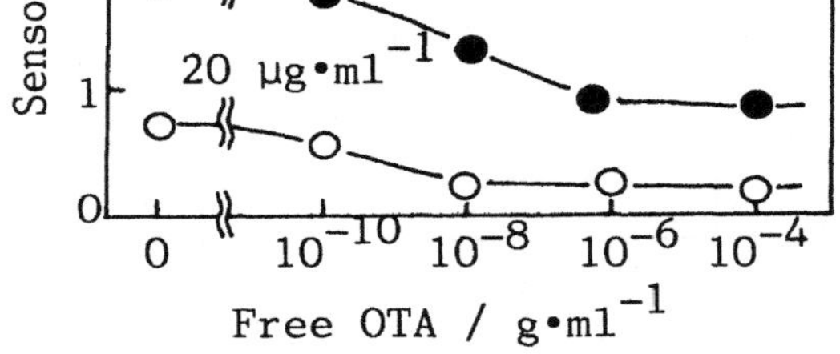

Fig. 12. Sensor output at various concentrations of catalase-labeled antibody.

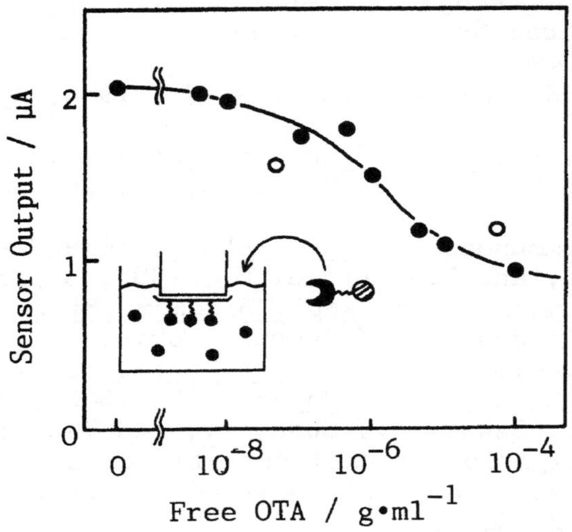

Fig. 13. Standard curve for enzyme immunoassay of OTA with the sensor.

The concentration of catalase-labeled antibody linearly related with the change in sensor output as shown in Fig. 12. The OTA-bound membrane was changed for each measurement. The replacement of the membrane resulted in no serious effects on the reproducibility. The concentration of catalase-labeled antibody was varied up to 100 $\mu g \cdot ml^{-1}$. No saturation in the sensor output was observed in the antibody concentration range investigated.

Competitive enzyme immunoassay of OTA was carried out using the enzyme immunosensor. The concentration of catalase-labeled antibody was fixed at 20, 50, and 100 $\mu g \cdot ml^{-1}$ depending on the OTA concentration range to be covered. Sensor response was obtained for each concentration of free OTA, where the measurement was carried out in the PBS at 35°C. The change of sensor output decreased with the addition of free OTA to be determined. A standard curve at a constant labeled antibody concentration of 50 $\mu g \cdot ml^{-1}$ is shown in Fig. 13. Under the concentration of 50 $\mu g \cdot ml^{-1}$ labeled antibody, OTA was determined in the concentration range of 10^{-6} to 10^{-8} $g \cdot ml^{-1}$.

The standard curves shifted to the lower concentration range with a decrease in labeled antibody concentration. The minimum detection limit was in the order of 10^{-10} $g \cdot ml^{-1}$, when the concentration of labeled antibody was fixed at 20 $\mu g \cdot ml^{-1}$.

REFERENCES

Aizawa, M., Morioka, A., Matsuoka, H., Suzuki, S., Nagamura, Y., Shinohara, R., and Ishiguro, I., 1976, An enzyme immunosensor for IgG, J. Solid-Phase Biochem., 1:319-326.

Aizawa, M., Kato, S., and Suzuki, S., 1977, Immunoresponsive membrane, J. Membrane Sci., 2:125-132.

Aizawa, M., Morioka, A., and Suzuki, S., 1978, Enzyme immunosensor II. Electrochemical determination of IgG with an antibody-bound membrane, 3:251-258.

Aizawa, M., Morioka, A., Suzuki, S., and Nagamura, Y., 1979, Enzyme immunosensor III., Anal. Biochem., 94:22-28.

Aizawa, M., Morioka, A., and Suzuki, S., 1980, An enzyme immunosensor for the electrochemical determination of the tumor antigen α-fetoprotein, Anal. Chim. Acta., 115:61-67.

Aizawa, M., 1983, Molecular recognition and chemical amplification of biosensors, in "Chemical Sensors," Seiyama, T., et al., eds., Kodansha-Elsevier, Tokyo, p.683-692.

Doyle, U. J., Halsall, H. B., and Heineman, W. R., 1984, Anal. Chem., 54:2318-2322.

Engvall, E., and Perlmann, P., 1971, Immunochemistry, 8:871-879.

Hongyo, K., Uda, T.. Ueno, H., Aizawa, M., Sano, S., and Shinohara, H., (submitted)

Monroe, D., 1984, Enzyme immunoassay, Anal. Chem., 56:920A-931A.

Ngo, T. T., Bovaird, J. H., and Lenhoff, H. M., 1985, Separation-free amperometric enzyme immunoassay, Appl. Biochem. Biotechnol., 11:63-70.

Suzuki, S., ed., 1984, "Biosensors," Kodansha, Tokyo.

Wisdom, G. B., 1976, Enzyme-immunoassay, Clin. Chem., 22:1243-1255.

Yalow, R. S., and Berson, S. A., 1959, Nature, 184:1648-1653.

Yamamoto, N., Nagasawa, Y., Sawai, M., Suda, T., and Tsubomura, H.,1978, Potentiometric investigation of antigen-antibody and enzyme-enzyme inhibitor reactions using chemical modified metal electrode, J. Immunol. Methods, 22:309-315.

BIOAFFINITY ELECTROCHEMICAL SENSOR WITH

PREFORMED METASTABLE LIGAND-RECEPTOR COMPLEX

Masuo Aizawa

Department of Bioengineering, Faculty of Engineering,
Tokyo Institute of Technology, Ookayama, Meguro-ku, Tokyo 152

INTRODUCTION

Molecular recognition and signal transduction are amalgamated in a biosensor in various manners. Molecular recognition is responsible for the selectivity of a biosensor. Many a biomolecule has an intrinsic nature to recognize its counterpart; an enzyme recognizes its substrate, and an antibody selectively binds its corresponding antigen. These biomolecules have been utilized as sensor material to develop such biosensors as an enzyme sensor and an immunosensor (Aizawa et al., 1983).

A new biosensor has been designed on the basis of bioaffinity difference between two ligands, i.e., one a determinant and the other an analogue compound in a given binding reaction. In general an analogue compound shows lower affinity to the binding protein than a determinant does. Therefore, one can expect the following displacement reaction when a membrane-bound analogue compound complexed with its binding protein is exposed to a determinant molecule. The binding protein is displaced from the membrane-bound analogue molecule as schematically shown in Fig. 1. The displacement is supposed to depend on the bioaffinity difference an analogue molecule and a determinant as well as the determinant concentration. The determinant concentration may be easily measured by detecting the residual molecular complex which remains on the membrane surface. High sensitivity can be attained by chemical amplification by the usage of an enzyme catalyst as a label. A biosensor based on the above principle may be termed "Bioaffinity sensor".

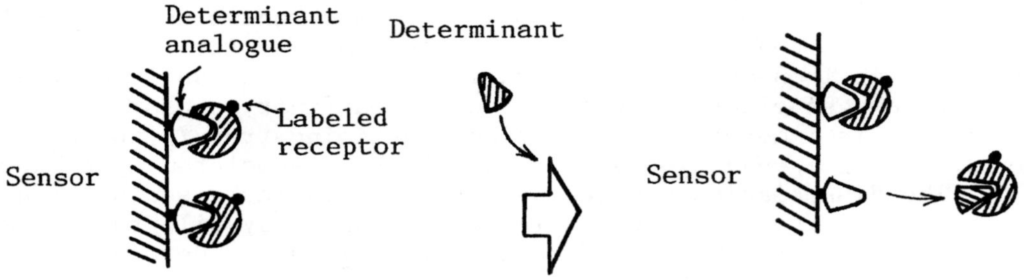

Fig. 1 Principle of bioaffinity sensor.

Table 1. A few applications of bioaffinity sensor

Ligand(L)	Determinant(D)	Binding(B) Protein	Measurable range($g{\cdot}ml^{-1}$)	Detection
HABA*	Biotin	Avidin	10^{-9} - 10^{-7}	Electrochemical
	Dethiobiotin	Avidin	10^{-9} - 10^{-7}	Electrochemical
Lipoic acid	Biotin	Avidin	$5x10^{-10}$ - 10^{-8}	Electrochemical
Thyroxine (modified)	Thyroxine	Antibody	10^{-8} - 10^{-5}	Electrochemical
Insulin (porcine,modified)	Bovine insulin	Antibody**	10^{-7} - 10^{-4}	Optoelectronic
Insulin (bovine,modified)	Bovine insulin	Antibody**	10^{-8} - 10^{-6}	Optoelectronic

* 2-(4'-Hydroxylphenylazo) benzoic acid, ** Anti bovine insulin antibody

There have been reported several bioaffinity sensors (Table 1). Schultz et al. (1979) suggested an affinity sensor for glucose using fluorescein-labeled dextran complexed with Con-A. The Con-A-immobilized fiber optic sensor is a convenient device for metabolites which are present in the 10^{-3} M range. On the other hand thyroxine was sensitively determined, in our study with membrane-bound thyroxine complexed with catalase-labeled antibody, in the concentration range 10^{-8} to 10^{-6} $g{\cdot}ml^{-1}$ (Ikariyama et al., 1982). A bioaffinity sensor was also assembled to determine biotin (vitamin H) in the concentration range from 10^{-9} to 10^{-7} $g{\cdot}ml^{-1}$ (Ikariyama et al., 1985). Insulin was determined in the concentration range from 10^{-8} to 10^{-4} $g{\cdot}ml^{-1}$ with a bioaffinity sensor which incorporated different sources of insulin as determinant analogue (Ikariyama et al., 1983).

PRINCIPLE OF BIOAFFINITY SENSORS

Bioaffinity Sensor for Thyroxine

The working principle of a bioaffinity sensor for thyroxine (3,5,3',5'-tetraiodothyroxine: T_4), a thyroid hormone, is diagrammatically shown in Fig. 2. The sensor is composed of membrane-bound T_4 and enzyme-labeled antibody, i.e., thyroxine is chemically immobilized on a membrane and the membrane-bound T_4 is undergone immunochemical reaction with enzyme-labeled anti-T_4 antibody to form an immunocomplex. Membrane-bound T_4 has less affinity to the antibody than free T_4 in solution. Attachment of the membrane where antigen-antibody complex is formed on the surface of an galvanic-type oxygen electrode results in a bioaffinity sensor for T_4. When the sensor is immersed in a solution containing free T_4 to be determined, the antigen-antibody complex will be dissociated upon exposure to T_4. The dissociation may be enhanced with an increase in the analyte concentration. The released enzyme-labeled antibody will then form a stable complex with free T_4. Consequently, T_4 can be determined by measuring the enzyme activity of enzyme-labeled antibody retained on the T_4 membrane.

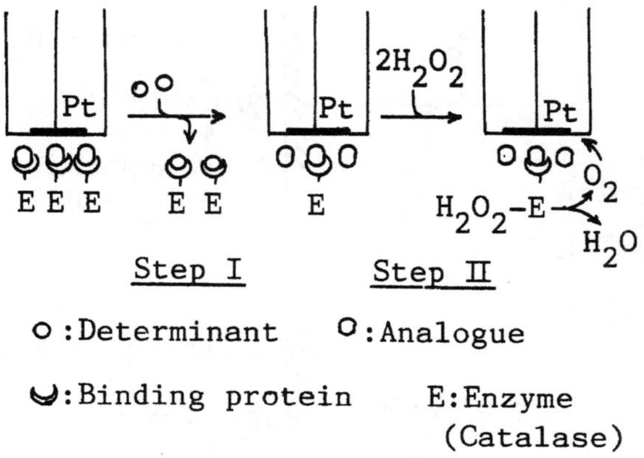

Fig. 2. Schematic representation of bioaffinity sensor. Molecular complex of low affinity is employed as a receptor. The first step is the biosensing of determinant, and the remaining molecular complex is sensitively detected by enzyme amplification.

Bioaffinity Sensor for Biotin

Avidin, an egg white protein, forms a very stable complex with biotin (vitamin H). The protein can also bind analogue compounds of biotin such as 2-[(4-hydroxyphenyl) azo] benzoic acid [HABA] and lipoic acid to form metastable complexes (Green, 1965; 1966; 1970). These ligand-avidin complexes dissociate upon exposure to biotin to form a very stable biotin-avidin complex. Figure 3 shows the association constants of a few biotin-related compounds in a ligand-avidin binding reaction.

The sensor is fabricated from a membrane, upon which a molecular complex between an analogue compound of biotin and enzyme-labeled avidin is prepared, and a galvanic oxygen electrode in a similar manner as shown in Fig. 2. HABA and lipoic acid are employed as the analogue compounds, whereas biotin and dethiobiotin are the determinants. Sensitivity is to be attained by an enzyme amplification technique.

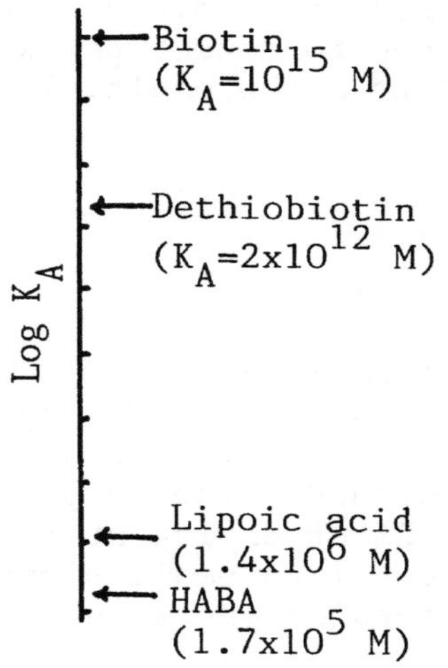

Fig. 3. Association constant of complex between avidin and biotin related compounds.

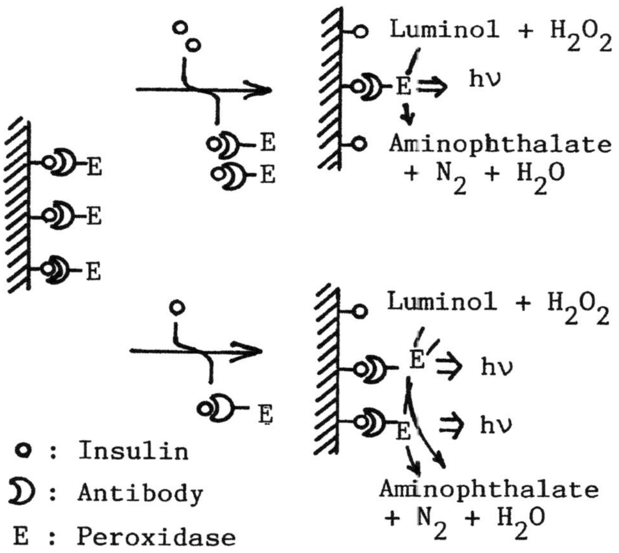

Fig. 4. Principle of bioaffinity sensor for insulin with luminescent detection.

$$[\text{Avidin}] + 4[\text{HABA}] \rightleftharpoons [\text{Avidin}][\text{HABA}]_4 \xrightarrow{4[\text{Biotin}]} [\text{Avidin}][\text{Biotin}]_4 + 4[\text{HABA}] \quad (1)$$

Bioaffinity Sensor for Insulin

The working principle of a bioaffinity sensor for insulin is schematically shown in Fig. 4. In case of a bioaffinity sensor for bovine insulin, porcine insulin is used as the analogue compound. A porcine insulin-bound plate is undergone immunoreaction with peroxidase-labeled anti bovine insulin antibody to form an immunocomplex. When the plate on which surface the immunocomplex is formed is immersed in a solution containing free bovine insulin to be determined, the complex may be dissociated. The dissociated enzyme-labeled antibody then forms a stable complex with bovine insulin in a solution. Insulin is thus determined by measuring peroxidase retained on an immunoplate. The luminol-H_2O_2 system is employed to detect peroxidase activity . The emitted light is transferred to a photomultiplier through an optofiber.

EXPERIMENTAL

Preparation of Membrane-Bound Ligand

4-(Aminomethyl)-1,8-octanediamine (2 ml) and 50 % glutaraldehyde (400 µl) were added to cellulose triacetate (500 mg) dissolved in 5 ml of dichloromethane, and then cast on a glass plate. The aldehyde cross-linked 4-(aminomethyl)-1,8-octanediamine to polymerize throughout the stick of cellulose triacetate. After drying at room temperature for a few days, the pink membrane was cut into small pieces, peeled off, and incubated for 1 h in 0.1 % glutaraldehyde solution buffered with 0.1 M phosphate of pH 7 for 24 h.

The membranes were used as matrices for immobilizing T_4 and HABA.

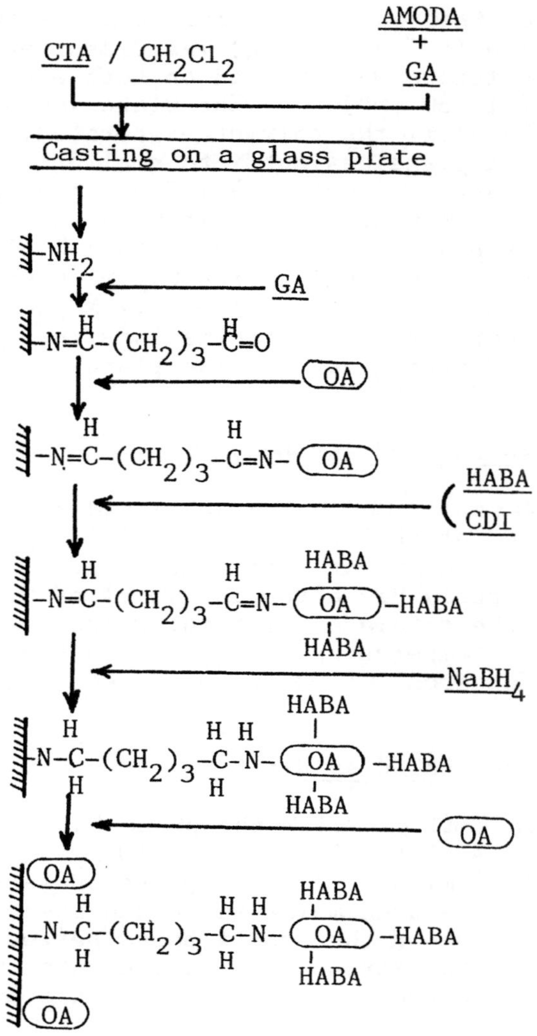

Fig. 5. Preparation of HABA-immobilized membrane: CTA, cellulose triacetate; AMODA, 4-(aminomethyl)-1,8-octanediamine; GA, glutaraldehyde; HABA, 2-[(4-hydroxyphenyl)azo] benzoic acid; CID, water-soluble carbodiimide; OA, ovalbumin.

The membrane was immersed in a 1 % bovine serum albumin (BSA) solution of pH 7.0 which was followed by the treatment with 1 % glutaraldehyde solution for 2 h at pH 7.0 which was followed by the treatment with 1 % glutaraldehyde solution for 2 h at pH 7.0. The chemically activated membrane was then placed in a T_4 solution (500 $mg \cdot ml^{-1}$) of pH 8.0 and was incubated overnight. The membrane was reduced in a solution of 0.1 M sodium borohydride, and was in a solution of 1 % BSA.

To immobilize HABA (or lipoic acid), the membrane was incubated in 1 % ovalbumin solution for 1 h. It is a well-known fact biotinyl enzymes where biotin is bound to the ε-amino groups of the enzymes are inhibited by avidin. Therefore, HABA (or lipoic acid) was immobilized with the method of carbodiimide conjugation to the membrane via the ε-amino group of the albumin to decrease steric hindrance as little as possible. During coupling reaction the pH was controlled at 4.5. Finally, the polymer membranes were reduced with sodium borohydride under pH control at 7.0 with NaH_2PO_4. The reaction scheme is presented in Fig. 5.

Insulin was immobilized on the surface of a transparent plastic plate. Polyvinylbenzyl chloride (PVBC) (500 mg) was dissolved in 20 ml of toluene. Several drops of the solution were poured on both sides of a polyvinyl chloride (PVC) plate (21×30 cm^2). The plate was allowed to stand at room temperature (20 °C) until the solvent evaporated, and was then placed in a desiccator. The plate was cut into smaller pieces (6×150 mm^2), which were placed in 5 % 4-(aminomethyl)-1,8-octanediamine for a few days at 20°C, and then were thoroughly washed with phosphate buffer (0.1 M, pH 7.0). The pieces of plate were reacted with 5 % glutaraldehyde at 20°C overnight. After thorough washing with 0.1 M phosphate buffer (pH 7.0), the pieces of plate were placed in 10 ml of 0.1 M phosphate buffer (pH 7.0) containing 5 mg of either bovine or porcine insulin. Insulin was covalently bound at 20 °C overnight. The plate was then stored in a solution containing 1 % BSA at 4°C.

Preparation of Catalase-Labeled Antibody

Catalase (Tokyo Kasei, 2,000 units•mg^{-1}) (20 mg) was dissolved in 2 ml of 0.1M phosphate buffer of pH 7.0 containing 0.1 % glutaraldehyde, and stirred magnetically for 10 min at room temperature. Fifteen milligram of anti-T_4 antibody (Cappel) dissolved in 0.5 ml of 0.1 M phosphate buffer (pH 7.0) was added to the catalase solution, which was followed by incubation for 30 min at room temperature. The resulting solution was chromatographed on a Sepharose CL-6B column to separate catalase-labeled antibody.

Preparation of Catalase-Labeled Avidin

Catalase (4 mg) in 1 ml of 0.1 M phosphate buffer (pH 7.0) was incubated with 50 μl of 2 % glutaraldehyde for 10 min. Avidin (1.5 mg) previously dissolved in 1 ml of 0.1 M phosphate buffer (pH 7.0) was added to the catalase solution in a unimolar ratio and then incubated for another 20 min. After reduction of the Schiff bases between catalase and avidin with sodium borohydride, the mixture was concentrated with Sunwet IM-300 of Sanyo Kasei (Kyoto) and then chromatographed on a Sepharose CL-6B column.

Preparation of Peroxidase-Labeled Antibody

Anti bovine insulin antibody was labeled by peroxidase in a similar manner as the catalase-labeled antibody.

DETERMINATION OF THYROXINE WITH A BIOAFFINITY SENSOR

Figure 6 represents the Sepharose Cl-6B chromatogram of the catalase-conjugated antibody. Each fraction (2 cm^3) was monitored at 410 nm for catalase and at 280 nm for both catalase and antibody. The first peak was ascribed to a catalase-conjugated antibody. The conjugate fraction 9 was spectrophotometrically estimated to have one to two enzymes per IgG. Fraction 9 was used in the hereafter investigation and is referred to as catalase-conjugated antibody in the following description.

The membrane-bound T_4 was reacted with catalase-labeled anti-T_4 antibody to form an immunocomplex at pH 7.0 and 37°C for 1 h in the presence of 1 % BSA. A galvanic oxygen electrode was mounted with the resulting membrane to yield a bioaffinity sensor for T_4.

After the immunoreaction with free T_4 in solution, the biosensor was transferred to a phosphate buffer (0.1 M, pH 7.0). The output current of the sensor soon became steady when the solution was magnetically stirred. Then hydrogen peroxide was added to make the final concentra

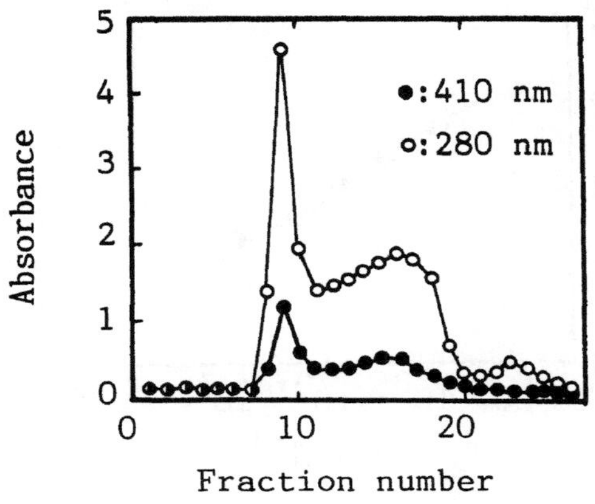

Fig. 6. Fractionation of catalase-conjugated antibody on Sepharode CL-6B column.

tion 3 mM. As Figure 7 shows the output current, oxygen-reducing current, of the sensor increased instantaneously upon addition of hydrogen peroxide. The sharp increase was followed by a gradual increase to reach another steady state within 1 min. The oxygen electrode generates current when dissolved oxygen molecules penetrate an oxygen-permeable Teflon membrane and is reduced on the Pt cathode in the manner shown in equation (2). The oxygen-reducing current increased due to an increase in the concentration of dissolved oxygen in the vicinity of the antibody-bound membrane where catalase-conjugated antibody degraded hydrogen peroxide to H_2O and O_2.

$$\text{Catode (Pt): } O_2 + 2H_2O + 4e^- \longrightarrow 4OH^- \qquad (2)$$

$$\text{Anode (Pb): } Pb + 4OH^- \longrightarrow PbO_2^{2-} + 2H_2O + 2e^- \qquad (3)$$

$$2H_2O_2 \xrightarrow{\text{Catalase}} 2H_2O + O_2 \qquad (4)$$

Figure 8 shows the relationship between T_4 concentration and output current of the sensor. Change in current caused by the residual catalase-conjugated antibody in the presence of hydrogen peroxide was remarkable in the concentration range from 10^{-7} to 10^{-5} $g{\cdot}ml^{-1}$. Consequently, T_4 was coverable in this concentration with the present bioaffinity sensor.

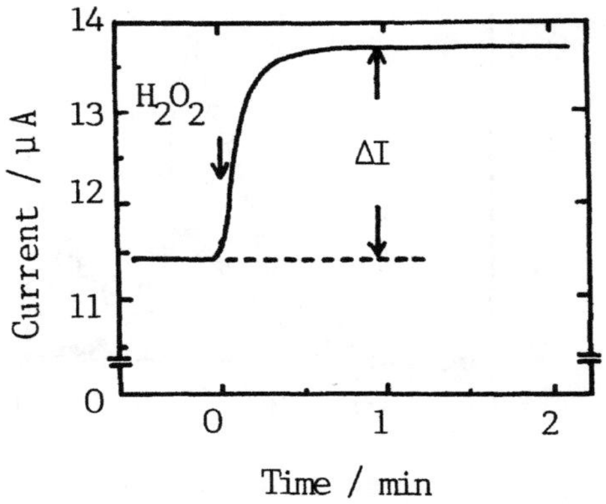

Fig. 7. Typical response of bioaffinity sensor for T_4. The sensor was immersed in a solution containing 210 $ng{\cdot}ml^{-1}$.

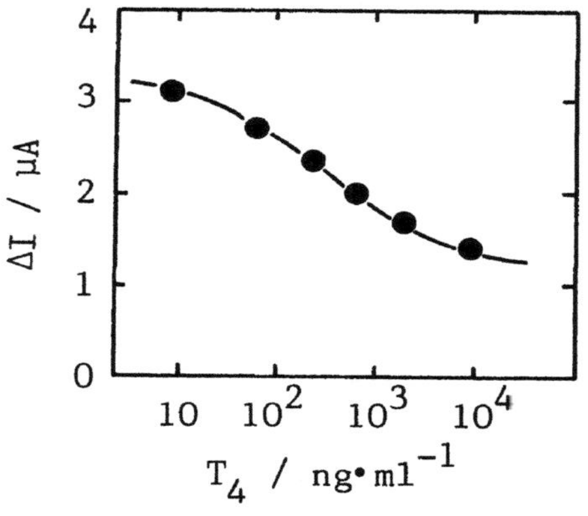

Fig. 8. Standard curve of T_4 with bioaffinity sensor.

FABRICATION OF A BIOAFFINITY SENSOR FOR BIOTIN

A cellulose triacetate membrane blended with a copolymer of 4(aminomethyl)-1,8-octanediamine and glutaraldehyde adsorbed a great deal of HABA; however, most of the adsorbed HABA was removed with thorough washings with 0.1 M carbonate pH 10. Spectroscopic study in the UV region showed that 3 mol of HABA was covalently immobilized to a membrane though the ovalbumin bound to the membrane surface.

Figure 9 illustrates the Sepharose CL-6B column chromatogram of the catalase-labeled avidin. Every fraction (2 ml) was monitored at 280 nm for the protein moieties of catalase and avidin and 406 nm for the heme moiety of catalase. The first and second peaks in the chromatogram were ascribed to catalase-labeled avidin. The second peak (from fraction 16 to 25) was concentrated to 4 ml with Sunwet IM-300. The catalase/avidin molar ratio of the conjugate was estimated for the second peak from the absorbance at 280 nm and 406 nm on the assumption that E_{280} of catalase and avidin, and E_{406} of catalase were 14.6, 15.5, and 18.7, respectively (Yamakawa, 1980). The approximate catalase/avidin molar ratio was 1.2.

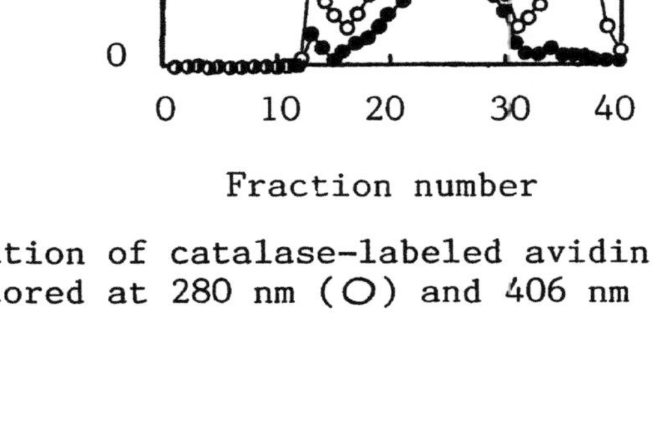

Fig. 9. Fractionation of catalase-labeled avidin. Every fraction (2 ml) was monitored at 280 nm (○) and 406 nm (●).

The binding capacity of catalase-labeled avidin was studied with HABA. A new absorption band at 500 nm was observed when the labeled avidin (0 mg•ml^{-1}) was mixed with free HABA (14.6×10^{-9} mol). The new absorption band at 500 nm was compared with that of native avidin complexed with free HABA. The estimated binding capacity of the avidin moiety was 20 % of native avidin. Considerable binding capacity was lost after conjugation. However, approximately 80 % of the labeled avidin seemed to bind HABA immobilized on the membrane, since avidin is a tetrameric protein. The absorption band disappeared immediately when free biotin was added to the avidin-HABA complex. Also the catalase activity in the conjugate estimated to be 40 % of free catalase with a spectroscopic method (Beers, 1952).

The membrane-bound HABA was incubated at 37°C in a solution containing 0.66 mg of catalase-labeled avidin and 10 mg of ovalbumin per 1 ml. The complex formation, i.e., receptor preparation, terminated within 60 min. The HABA-immobilized membrane on which the molecular complex of low affinity was formed was tightly attached to a galvanic oxygen electrode through a Teflon membrane.

CHARACTERIZATION OF A BIOAFFINITY SENSOR FOR BIOTIN

The time course of molecular recognition with the proposed bioaffinity sensor was studied. The sensor was immersed in a biotin solution (10^{-5} g•ml^{-1}) in the presence of 1 % ovalbumin at 37°C to observe the outlook of molecular recognition. Either HABA or lipoic acid was used as an analogue compound to be immobilized to the membrane. After the biosensing, the sensor was transferred to the measuring medium. The change in sensor output upon addition of hydrogen peroxide was recorded.

Figure 10 illustrates the relation between the incubation time and the change in sensor output. The current change was caused by the molecular complex receptor that remained on the sensor. Biotin recognition with the receptor finished within 10 min after the start of biosensing. The measurement of the remained receptor finished within 1 min as shown in Figure 10. Similar outlook of biosensing was observed with membrane-bound lipoate complexed with the labeled avidin. The catalase-labeled

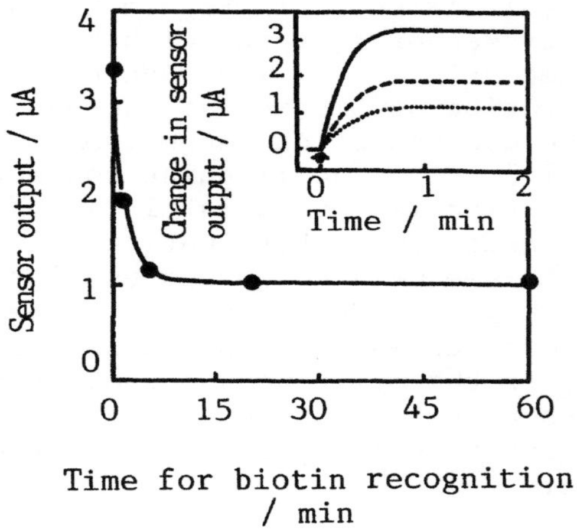

Fig. 10. Time course of molecular recognition of bovine with membrane-bound HABA complexed with catalase-labeled avidin. Change in sensor output H_2O_2 addition in shown in the inset figure (step II of Figure 2). Sensing times were 0 min (——), 2 min(----), 5 min (·······). Hydrogen peroxide was added at the time indicate with an arrow.

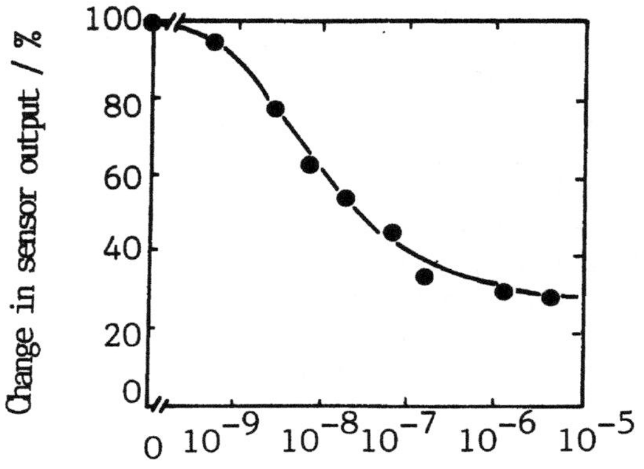

Fig. 11. Standard curve for biotin with bioaffinity sensor. The molecular complex of low affinity, membrane-bound HABA complexed with catalase-labeled avidin, was used as a receptor.

avidin adsorbed was not fully dissociated even in the presence of excess biotin. Approximately half of the receptor was undissociated. The HABA-immobilized membrane binds the labeled avidins by complex formation (specific binding) and adsorbs the labeled avidins on its nonspecific binding sites. The labeled avidin, specifically bound to membrane-immobilized HABA (or lipoic acid), is expected to easily dissociate in the presence of excess biotin, whereas the nonspecifically adsorbed avidin is not. The response time of the proposed bioaffinity sensor was no longer than 10 min, although biosensing time was here set to 60 min.

After biosensing at 37°C for 60 min, the sensor was washed three times with 0.1 M phosphate buffer of pH 7. The sensor was transferred to a magnetically stirred medium buffered with 0.1 M phosphate of pH 7. After the oxygen-reducing current reached a steady state, 5 ml of 30 mM hydrogen peroxide prepared in 0.1 M phosphate buffer of pH 7 was quickly injected to make the remained molecular complex receptor detected. The change in sensor output promptly increased and reached a constant value within 1 min. The result is shown in Figure 11 when biotin was determined with the bioaffinity sensor equipped with the membrane-bound HABA complexed with the labeled avidin. The change in sensor output decreased as the concentration of biotin was increased. The determinable biotin was in the concentration range from 10^{-9} (90 % response) to 10^{-7} (10 % response) $g \cdot ml^{-1}$. The midpoint concentration was 10^{-8} $g \cdot ml^{-1}$. The standard deviation is approximately 11 % at the midpoint concentiation (n = 10). After the determination, the bioaffinity sensor was reimmersed in the catalase-avidin conjugate solution to prepare the molecular complex in the way mentioned above. The sensor was repeatedly used when the molecular complex was reprepared in the conjugate solution. It was not the stability of the membrane but that of the conjugate which limited the life of the sensor. So long as the catalase-labeled avidin is stable, the membrane was repeatedly used as membrane receptor for the biotin-related molecules.

Figure 12 is also a calibration curve for biotin with the bioaffinity sensor. The receptor was the membrane-bound lipoic acid complexed with the labeled avidin. The coverable range for biotin was from 5×10^{-10} (90 % response) to 5×10^{-8} (10 % response) $g \cdot ml^{-1}$. The midpoint concentration was 2×10^{-9} $g \cdot ml^{-1}$.

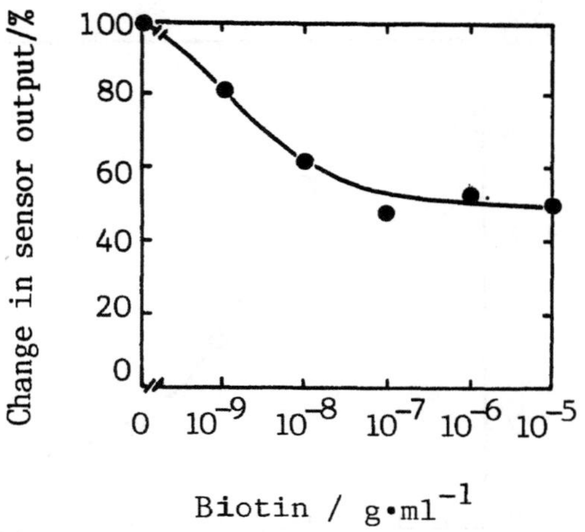

Fig. 12. Standard curve for biotin with bioaffinity sensor. The low affinity molecular complex, membrane-bound lipoate complexed with catalase-labeled avidin, was employed as a receptor for biotin recognition.

DETERMINATION OF INSULIN

The insulin-immobilized plate was undergone immunoreaction with peroxidase-conjugated antibody (7×10^{-5} g•ml^{-1} at 37°C for 1 h in the presence of 1 % BSA, which resulted in the receptor preparation for insulin. The plate receptor was then immersed in an analyte (bovine insulin) solution. After sensing of bovine insulin, the plate receptor was transferred to the luminescent medium of luminol-H_2O_2 system to detect the residual peroxidase activity on the immunoplate. The luminescence generated was led to a photomultiplier through an optofiber. Total luminescence during 1 min was measured at each insulin concentration and compared. Typical calibration curve for bovine insulin by the bio-affinity sensing system is presented in Fig. 13. The ordinate gives the

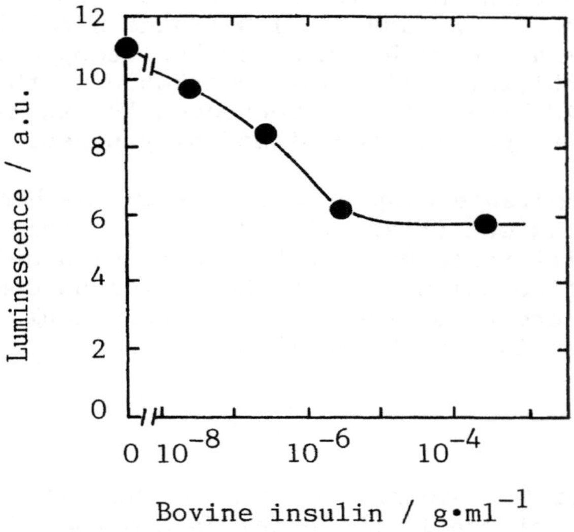

Fig. 13. Standard curve for bovine insulin with bioaffinity sensor (I). Receptor for bovine insulin: Plate-bound bovine insulin complexed with anti bovine insulin antibody.

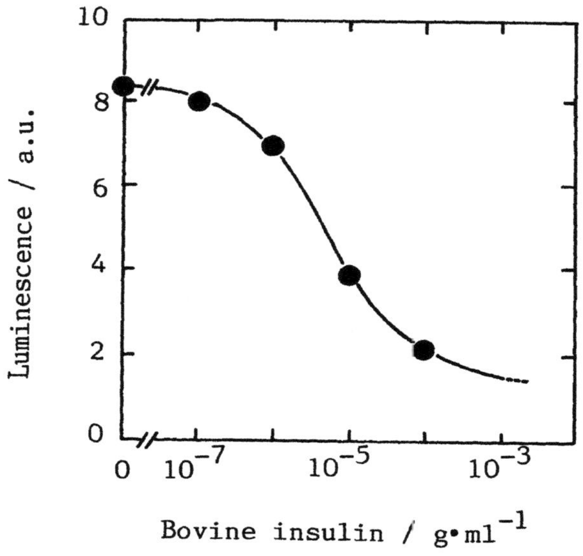

Fig. 14 Standard curve for bovine insulin with bioaffinity sensor (II). Receptor for bovine insulin: Plate-bound porcine insulin complexed with anti bovine insulin antibody.

relative luminescence at a given analyte concentration. The result shows that remarkable dissociation of the immunocomplex on the plastic plate was observed in the insulin concentration ranging from 10^{-8} to 10^{-5} $g \cdot ml^{-1}$. The insulin-antibody complex on the plate recognized the determinant (insulin) to form stable complex in a solution.

Anti bovine insulin antibody reacts with not only bovine insulin but a few insulins of other animals. However, the affinity of anti bovine insulin antibody to an insulin of different animal species is less strong than that to the original one. Thus the authors prepared a porcine insulin-immobilized plate complexed with the peroxidase-conjugated anti bovine insulin antibody. The heterologous immunocomplex, i.e., porcine insulin-anti bovine insulin antibody complex, was applied to the determination of bovine insulin. Figure 14 illustrates a calibration curve for bovine insulin determination with immunocomplex receptor. Change in luminescence (in an arbitrary unit) was remarkable in the concentration ranging from 10^{-7} to 10^{-4} $g \cdot ml^{-1}$. Consequently, insulin was coverable in this concentration range with the bioaffinity sensor.

The ratio of dissociable complex was 50 % in the homologous system and 75 % in the heterologous one (Figs. 13 and 14). This may be the reflection of bioaffinity difference between bovine insulin and porcine insulin in the present binding reaction. the heterologous complex can be easily dissociated upon exposure to bovine insulin since stable immunocomplex is formed with bovine insulin and its antibody.

CONCLUSION

The performance of the bioaffinity sensor was assessed to indicate that the system has considerable potential as a method for sensing of bioactive substances which are present at low concentrations. The characteristics of the bioaffinity sensors investigated are summarized in Table 1. Although a few points remain still to be improved, the present sensitive sensing concept based on bioaffinity difference and chemical amplification is a unique approach to the recognition of many a small molecule of biochemical importance. One further point requires emphasis. Physiologically active substances such as hormones and vitamins, which are susceptible to degradation during chemical modification, are easily determined, since the sensor does not require labeled determinant.

REFERENCES

Aizawa, M., 1983, Molecular recognition and chemical amplification of biosensors, in: "Chemical Sensors," Seiyama T., et al., eds., Kodansha-Elsevier, Tokyo, p.683.

Beers, R. F., Jr., and Sizer, I. W., 1952, A spectrophotometric method for measuring the breakdown of hydrogen peroxide by catalase, J. Biol. Chem., 195: 133.

Green, N.M., 1965, A spectrophotometric assay for avidin and biotin based on binding of dyes by avidin, Biochem. J., 94: 23c.

Green, N.M., 1966, Thermodynamics of the binding of biotin and some analogues by avidin, Biochem. J., 101: 774.

Green, N.M., 1970, Purification of avidin, in: "Methods in Enzymology, vol.18 Part A," McCormich, D. B., and Wright, L. D., eds., Academic Press, New York, p.414.

Ikariyama, Y. and Aizawa, M., 1982, Bioaffinity sensor (Determination of Thyroxine), Proc. 2nd Sensor Symp., p.97.

Ikariyama, Y. and Aizawa, M., 1983, Bioaffinity sensor for insulin with luminescence detection, Proc. 3rd Sensor Symp., p.17.

Ikariyama, Y. and Aizawa, M., 1985, Sensitive Bioaffinity sensor with metastable molecular complex receptor and enzyme amplifier, Anal.Chem., 57: 496.

Schultz, J. S., and Sims, G., 1979, Affinity sensors for individual metabolites,Biotechnol. Bioeng. Symp., 9: 65.

Yamakawa, T., ed., 1980, "Data Book of Biochemistry vol.1," Tokyo Kagaku Dojin, Tokyo, p.94.

ELECTRIC PULSE ACCELERATED IMMUNOASSAY

Isao Karube and Eiichi Tamiya

Research Laboratory of Resources Utilization
Tokyo Institute of Technology, Nagatsuta-cho
Midori-ku, Yokohama, 227, Japan

INTRODUCTION

Various antigens including proteins, peptides, drugs and microorganisms are determined by immunoassay. The immunoassay is based on the specific binding reaction of an antigen to its antibody. Immunoreaction results in the formation of antigen-antibody agglutination. As the amount of agglutination is considerably small, radioisotopes, enzymes, fluorescence dyes or spin labels are incorporated to enhance response detection (Yalow 1976, Johnson et al. 1978). Use of radioisotope (radioimmunoassay;RIA) has proved the most successful and is nowadays routinely performed in many clinical laboratories. RIA requires the provision of specialized and exclusive instruments and facilities, therefore, a simple, rapid, and sensitive method which could replace RIA is intensively required.

Enzyme immunoassay or fluorescence immunoassay can sometimes attain high sensitivity, but they require, however, complicated and delicate procedures. Recently, a simple and rapid method has been developed based on turbidimetry, using antibody bound latex beads (Masson et al. 1981, Collet-Cossart et al. 1981). This method is however, still inadequate, an improvement in its sensitivity is necessary. The promotion of the agglutinating reaction is expected by increase of contact frequency between antigen and antibody. Mechanical stirring may not be effective for a minute amount of reaction solution. Increase of temperature is also not effective for biologically active substances including antibodies. It was formerly observed that conducting or nonconducting particles suspended in various fluids formed linear linkage under electric field (Hu et al. 1975, Schwan et al. 1969, Füredi et al. 1962). These particles are aluminum powder, carbon powder, potato starch, polystyrene particles, red blood cells and any yeast cells. These electric field effects were considered promising for the increase of contact frequency between antigen and antibody. In the case of antigen, with a size of micrometers, the agglutination is composed of two steps : (1) binding of antibody on a particle (antigen) and (2) binding of another particle with the antibody-bound particle. The electric pulse probably accelerates the second step and increases the total reaction rate. Similar effects were expected in the case of small antigens(e.g., protein, peptide) and antibodies immobilized on a large particle such as latex bead.

In this study, electric pulses are applied to the immunoassay for human immunoglobulin (H-IgG) and Candida albicans(Matsuoka et al. 1985). H-IgG is an important indicator of abnormality in protein metabolism, and is used as typical example of serum proteins. The antibody to H-IgG is covalently bound on the surface of latex beads and reacted with H-IgG. C.albicans is a pathogenic yeast and the determination of its concentration has clinical importance.

METHODS AND MATERIALS

Materials

Human immunoglobulin G (H-IgG) and its antibody (IgG fraction of rabbit serum) were purchased from Miles Laboratory Ltd. Latex beads (1.0 µm) were obtained from Japan Synthetic Rubber Co. Ltd. C. albicans was cultured in a medium containing glucose (1%), peptone (1%) and yeast extract (0.5%) for 17 hr at 37°C , pH 6.5. The cells were collected and suspended in distilled water after washing twice with distilled water. Antiserum to C.albicans was obtained from IATORON Company and dialyzed against distilled water for 10 hr before use. All other reagents were commercially available analytical grade.

Preparation of antibody bound latex beads (Ab-L)

Latex beads were suspended in 40 ml of water (1.0 wt%) and 22.6 mg of 1-cyclohexyl-3 (2-morpholinoethyl) carbodiimide was added to the suspension. The suspension was then incubated at room temperature for 3 hr with stirring, the pH being maintained at 4.5 throughout. About 0.16 mg of antibody to H-IgG was added and reacted at 4° C overnight. The latex beads were then washed with phosphate buffer solution and stored at 4°C.

Apparatus for Immunoreaction

Figure 1 shows the reactors used for the immunoreaction. The slit cell was composed of a couple of electrodes (electrode distance, 1 mm ; electrode thickness, 25µm) and a slide glass. The cuvette cell was constructed with glass plates. A couple of electrodes (3 x 1 cm^2) were attached on the inside surface of the cuvette. The distance of the electrode was 1 mm.

Procedure of immunoreaction and agglutination rate

A cuvette type cell as shown in Fig.1 was prepared and used for the immunoreaction. One hundred µl of Ab-L or C.albicans suspension and 40 µl of antibody were mixed in the cuvette cell with a microsyring. Then electric

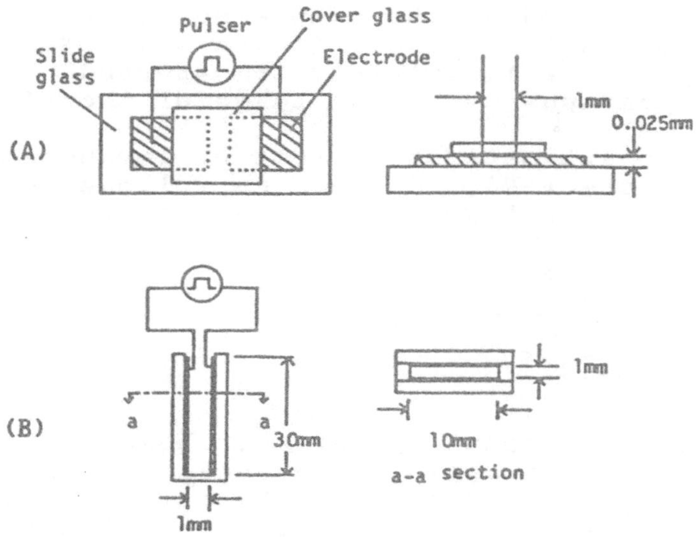

Figure 1. Reactors for pulse immunoassay:(A) slit reactor; (B) cuvette reactor.

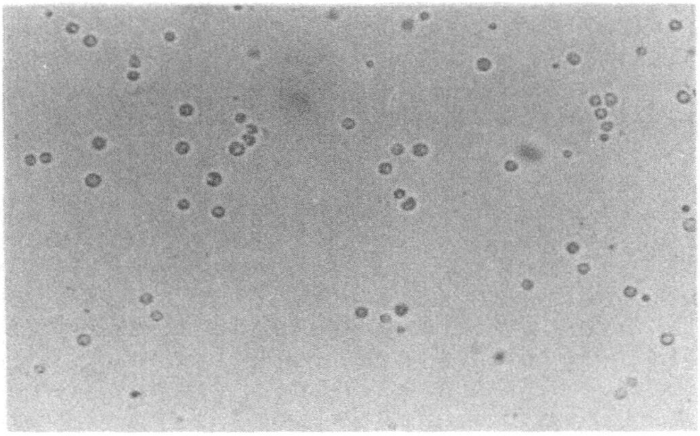

Figure 2. Homogeneously dispersed C.albicans without electric pulse.

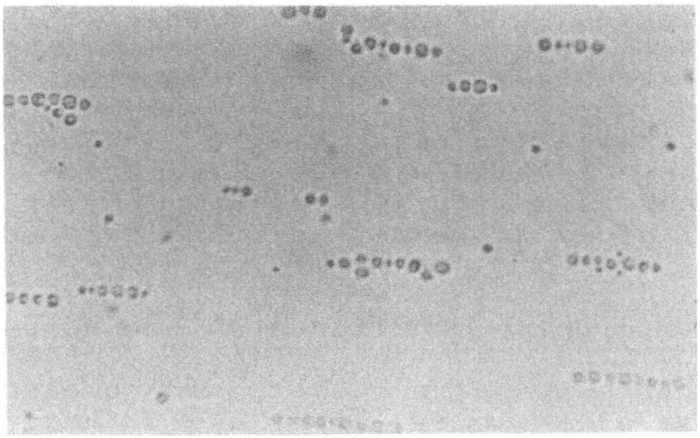

Figure 3. Linear agglutination of C.albicans under electric field. Pulse frequency $1/T$ = 8 kHz Pulse width τ = 20 μs, pulse height H = 100 V

pulses were applied to the sample solutions for a certain period of time, after which, it was stopped and an aliquot of the sample was immediately transferred onto a glass slide. Photographs were then taken of the slide through a microscope.

In order to estimate the extent of immunoreaction, the agglutination rate (AR), defined as follows, was determined by analyzing five photographs for each sample:

$$AR = \frac{\sum_{n=5}^{\infty} Nn}{\sum_{n=1}^{\infty} Nn} \times 100 \ (\%) \qquad (1)$$

where Nn is the total number of beads forming n-bead agglutinations or n-agglutinated cell. As a control, immunoreaction was conducted in the cuvette reactor without electric pulse. As another control, the effect of electric pulse on samples containing no antigen (H-IgG or _C.albicans_) was checked.

RESULTS AND DISCUSSIONS

Effect of electric pulse condition on formation of linear agglutination

A drop of _C.albicans_ cell suspension was put on the slit reactor, covered with a cover glass, and observed under a microscope. Then electric pulses were applied with a pulse generator. Reversible agglutination and dispersion of microbial cells were observed under electric field and its removal. _C.albicans_ cells are homogeneously distributed as shown in Figure 2. When electric pulses were applied to the suspension, linear linkages were formed within several seconds as shown in Figure 3. The extent of linear linkage formation depended on electric pulse conditions as described below. Linear agglutinations, however, dispersed immediately after the removal of the electric pulse. Alternation of pattern, as shown in Figures 2 and 3 was repeatedly observed. Figure 4 shows the effects of electric pulse height on the formation of linear agglutination at 8 kHz. No agglutination was observed below 20 V. Above 20 V, the extent of agglutination increased with increasing pulse height. The effect of the frequency is shown in Figure 5. The formation of linear agglutination increased with increasing frequency. Similar agglutinations of latex beads was observed under electric pulse.

Effect of electric pulse conditions on immunoreaction

When the cell suspension was exposed to electric pulses in the presence of antibody, some agglutinations remained even after the removal of electric pulse as shown in Figure 6. In Figure 6A concentration of antibody was 2.3 mg of protein/ ml, while it was 2.9 mg of protein/ml in Figure 6B. AR in Fig.6B was obtained by counting the number of agglutinations and the number of cells in each agglutina-

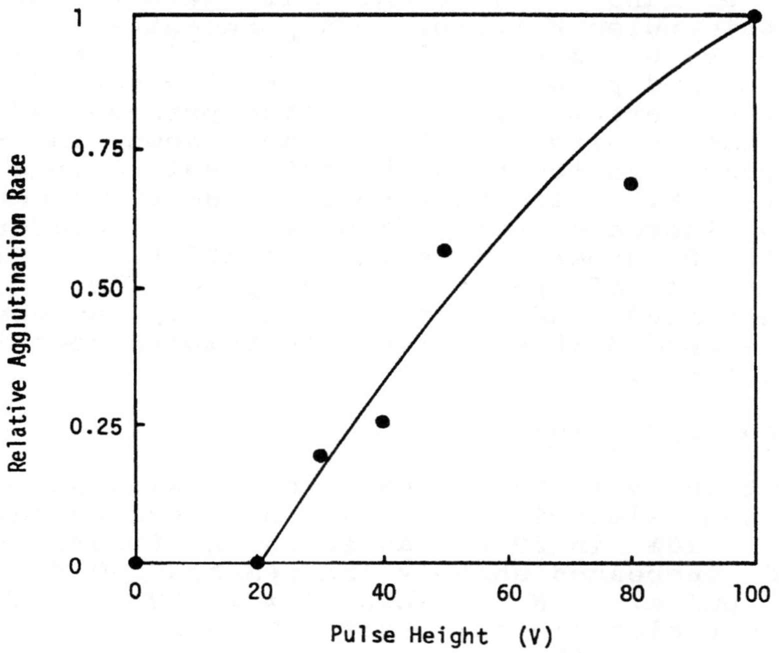

Figure 4. Effect of pulse height on the formation of linear agglutination. 1/T = 8 kHz, τ = 20 μs.

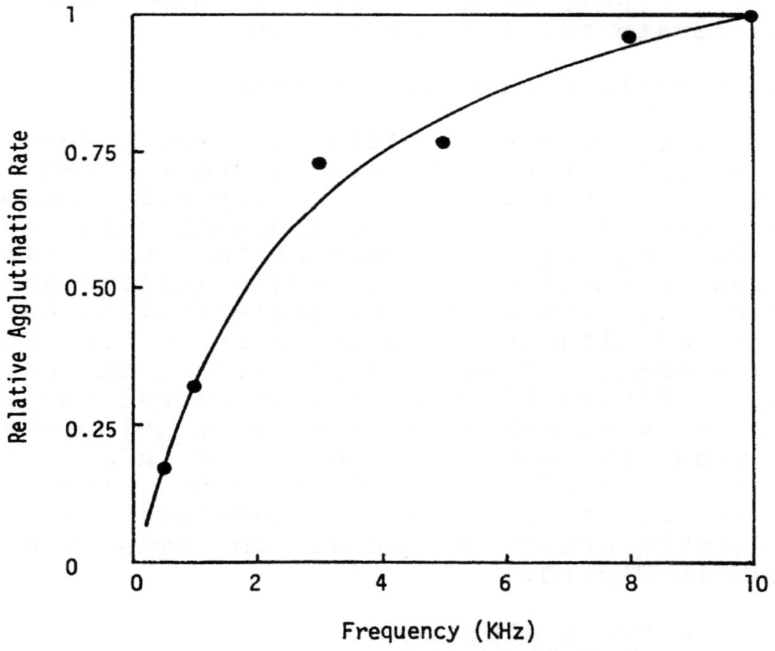

Figure 5. Effect of pulse frequency on the formation of linear agglutination. τ = 20 μs. H = 100 V.

tion. Adapting this result to eq 1, AR was estimated as 60 %. Similarly, AR was 15 % for Figure 6A.

On the other hand, Ab-L and H-IgG were reacted under varying electric pulse heights. The concentrations of Ab-L and H-IgG were 0.3 % and 6.7 x 10^{-6} g/ml^{-1} respectively. Pulse frequency and pulse application period were fixed at 8 kHz and 10 min respectively. In the presence of H-IgG, maximum AR was obtained at 200 V, as shown in Fig. 7. Further increase in the pulse height resulted in a slight decrease of the AR. In the absence of H-IgG, however, the AR began to increase above 200 V. Therefore, the difference in AR showed a maximum at 200 V. Figure 8 shows the effect of pulse frequency on AR. As the frequency increased, AR increased gradually and reached a steady level around 8-10 kHz, for both samples containing H-IgG and without H-IgG.

Time course for agglutination

Immunoreaction was performed with or without electric pulse. Without electric pulse, AR increased gradually and reached about 10% in 20 min as shown in Figure 9. In contrast, AR increased sharply and reached 50 % in 5 min with electric pulse. AR was about 10% in 20 min, though it increased under electric pulse without antibody.

On the other hand, the time course of immunoreaction for IgG was shown in Figure 10. When electric pulse was applied to Ab-L without H-IgG, the AR increased slightly and reached 20 % at 20 min. When H-IgG was reacted with Ab-L in the absence of electric pulse, a simple result was obtained. In sharp contrast, when Ab-L was reacted with H-IgG under electric pulse, the AR increased drastically and reached 50 % at 10 min. These results clearly show that electric pulse can accelerate the immunoreaction.

Effect of other protein on agglutination

In order to check the specificity, the effect of human serum albumin(HSA) or bovine serum albumin (BSA) on the response was investigated as a typical example of contaminating protein. When C.albicans cell suspension containing 2.9 mg of protein/ml of HSA was exposed to electric pulses, a small amount of agglutination appeared as shown in Figure 11. The extent of agglutination was similar to that obtained without antibody as shown in Figure 9. Therefore the presence of HSA did not affect the nonspecific agglutination. Figure 12 shows a time course for AR, when Ab-L was reacted with BSA with or without electric pulse. BSA concentration was adjusted to 6.7 x 10^{-6} g/ml. The AR was found to be less than 20 % even after 20 min, independent of electric pulse. Therefore BSA did not present appreciable affect on the AR, in comparison with the controls shown in Fig.10.

Calibration curve for C.albicans

Reaction rate depends considerably on the relative concentration of antibody and antigen. Therefore effect of antibody concentration on AR was examined. Antigen(C.albicans) concentration was fixed to 6.0 x 10^{7} cells/ml. Without electric pulse, no agglutination was

Figure 6. Irreversible agglutination caused by immunoreaction under electric field: (A) *C.albicans* 6×10^7 cells/ml, antibody 2.3 mg of protein/ml;(B)*C. albicans* 6×10^7 cells/ml,antibody 2.9 mg of protein/ml.

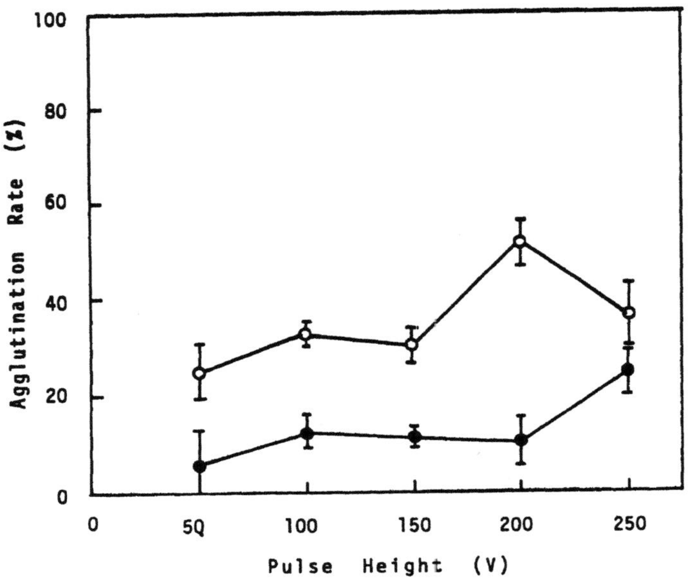

Figure 7. Effect of pulse height.
(●) Pulse, (○) Pulse + H-IgG.
Reaction condition: Latex concentration 0.3%, H-IgG concentration 6.7 x 10^{-6} g/ml, Pulse frequency 8 kHz, Reaction time 10 min, Temperature 20°C.

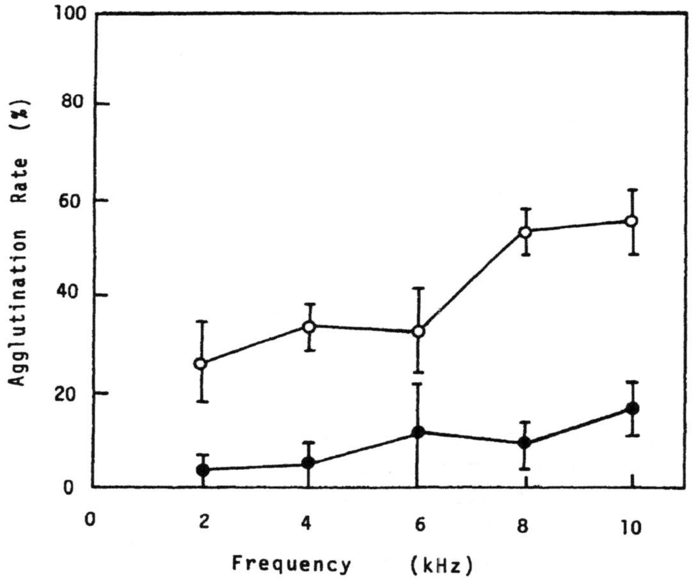

Figure 8. Effect of pulse frequency.
(●) Pulse, (○) Pulse + H-IgG, Pulse height 200 V.

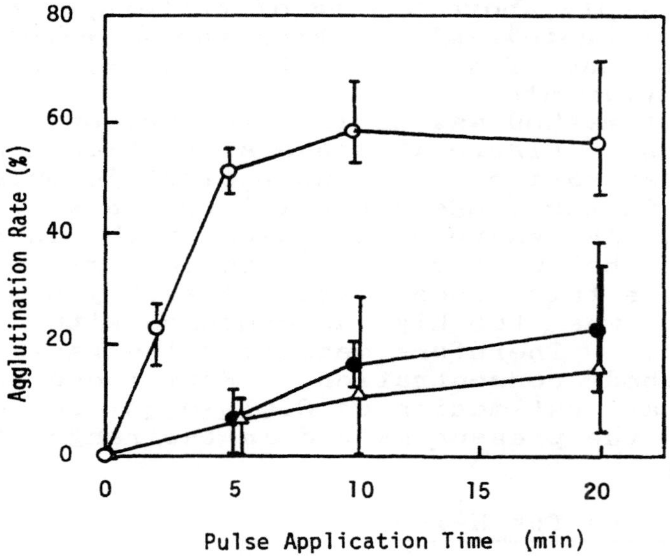

Figure 9. Time courses of agglutination.
(O) Antibody(2.9 mg of protein/ml) with electric pulse, (△)Antibody(2.9 mg of protein/ml) without electric pulse, (●)No antibody with electric pulse. Electric pulse conditions were 1/T = 8 kHz, τ = 20 μs, H = 100 V.

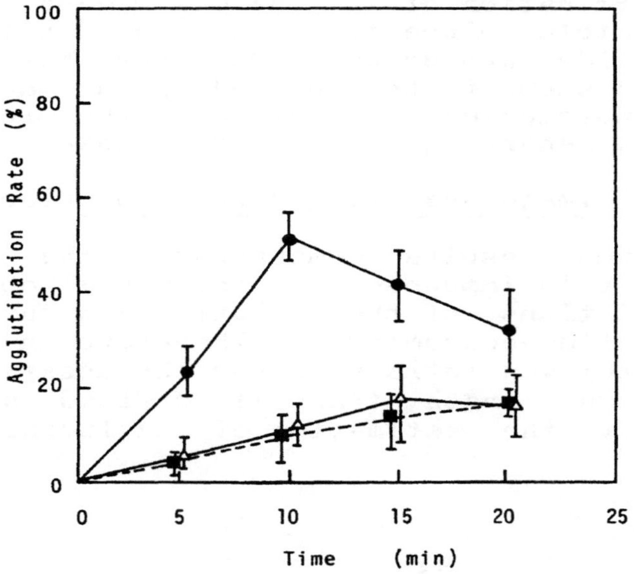

Figure 10. Time course for agglutination.
(●) H-IgG + pulse, (△) H-IgG, (■) Pulse.
Reaction conditions: Pulse height 200 V.
Other conditions were same as Fig.7 except for reaction time.

observed below 2.4 mg of protein/ml, but AR increased slightly above this concentration. On the other hand, AR increased sharply above 2.4 mg of protein/ml and reached 50 % at 2.9 mg of protein/ml. From these results, antibody concentration was fixed to 2.9 mg of protein/ml in the following experiments.

The present method was applied to various concentrations of C.albicans. Figure 13 shows a relationship between AR and cell concentration. Without electric pulse, AR was at most 10 % in the range 3.8×10^7 to 3.6×10^8 cells/ml. In contrast, AR sharply increased with increasing cell concentration below 6.0×10^7 cells /ml. However AR decreased above this concentration possibly because the cell concentration was too high as compared with the antibody concentration. Therefore measuring the range might depend upon the antibody concentration. From these results, it is concluded that estimation of C.albicans concentration is possible by the present method in the range of 10^7 to 6×10^7 cells/ml.

Calibration curve for H-IgG

Various concentration of H-IgG were reacted with Ab-L. As shown in Figure 14, AR increased with increase in H-IgG concentration in the range 6.7×10^{-8} - 6.7×10^{-6} g/ml. At the lower concentrations, AR was around 10%, but the AR decreased above 6.7×10^{-6} g/ml. This is probably because the relative amount of Ab-L is too small, resulting in retardation of agglutination formation. Therefore, the measuring range for H-IgG was 6.7×10^{-8} - 6.8×10^{-6} g/ml under the present experimental conditions.

The pulse immunoassay demonstrated here is concerned with the determination of IgG and C.albicans. In systems it would be possible to use many other antigen including serum proteins, peptides and drugs. Moreover, the sensitivity of the assay is shown to be improved by an acceleration of specific immunoreaction. Therefore, the present method is promising as a general method of immunoassay.

Application of image analyzing system to pulse immunoassay

Although such results demonstrated the value of the electric pulses in immunoassay, the visual observations and manual calculations of the agglutination rates were very tedious and time-consuming. Therefore an on-line image analyzing system was introduced for the assay system.

Mean of run length ($\langle x\rangle\langle y\rangle$) defined as below was introduced for the estimation of agglutination with a microcomputer :

$$\langle x\rangle = \frac{\sum_{i=1}^{m} x_i}{m} \qquad \langle y\rangle = \frac{\sum_{j=1}^{n} y_j}{n}$$

where x is the run length of the i-th aducts when scanned in the x direction, and y is the run length of the j- th aducts when scanned in the y direction as depicted in Figure 15. The run length means a unit of aggregated particles. Between the $\langle x\rangle\langle y\rangle$ and AR, a linear correlation was observed as shown in Figure 16. Therefore $\langle x\rangle\langle y\rangle$ was applicable to the on-line estimation of antigen antibody agglutination.

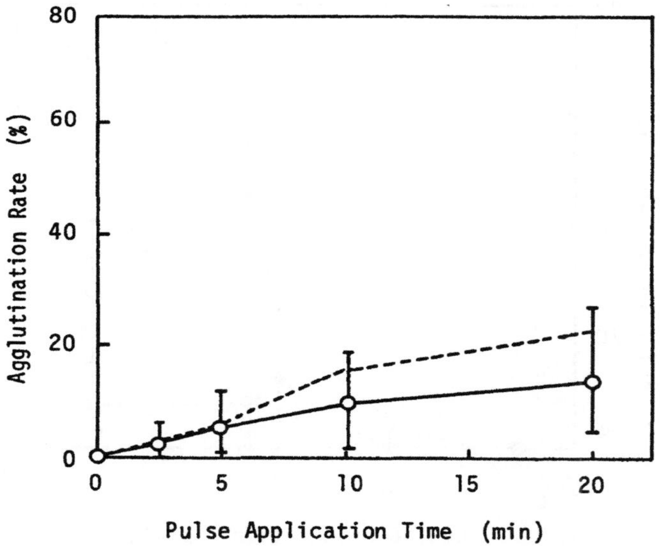

Figure 11. Agglutination caused by human serum albumin(HSA). (-O-) HSA(2.9 mg of protein/ml) with electric pulse, (---) no antibody with electric pulse.

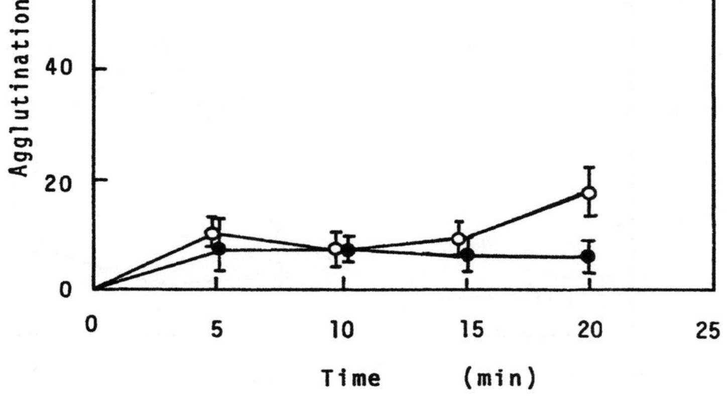

Figure 12. Effect of bovine serum albumin on the agglutination of Ab-L.
(O) BSA + pulse, (●) BSA.
BSA concentration 6.7×10^{-6} g/ml

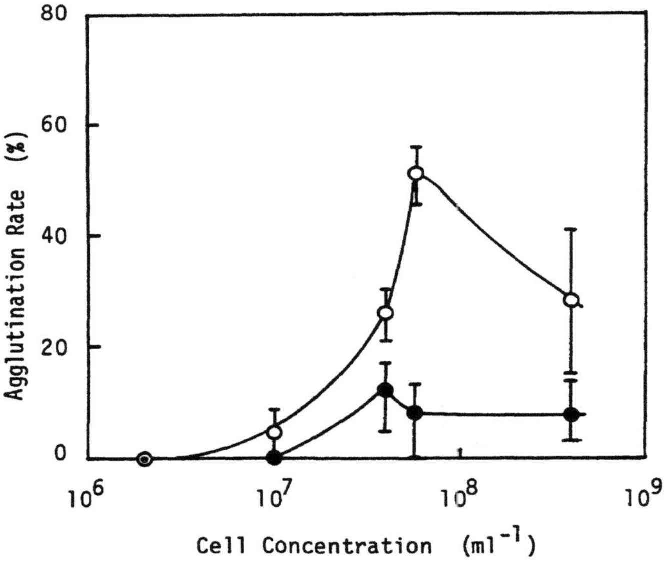

Figure 13. Effect of *C.albicans* concentration on the agglutination rate.
(○) with electric pulse, (●) without electric pulse.

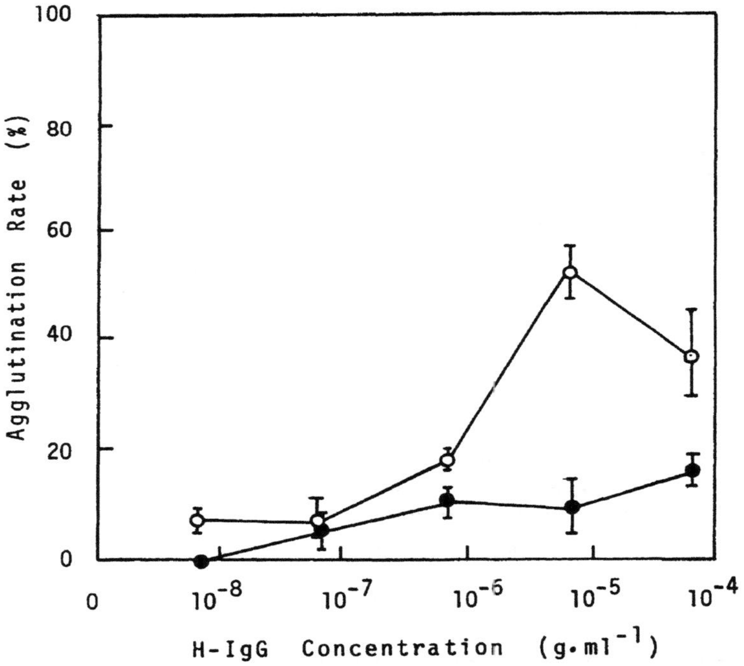

Figure 14. Calibration curve for H-IgG.
(○) H-IgG + pulse, (●) H-IgG.
Reaction time 10 min.

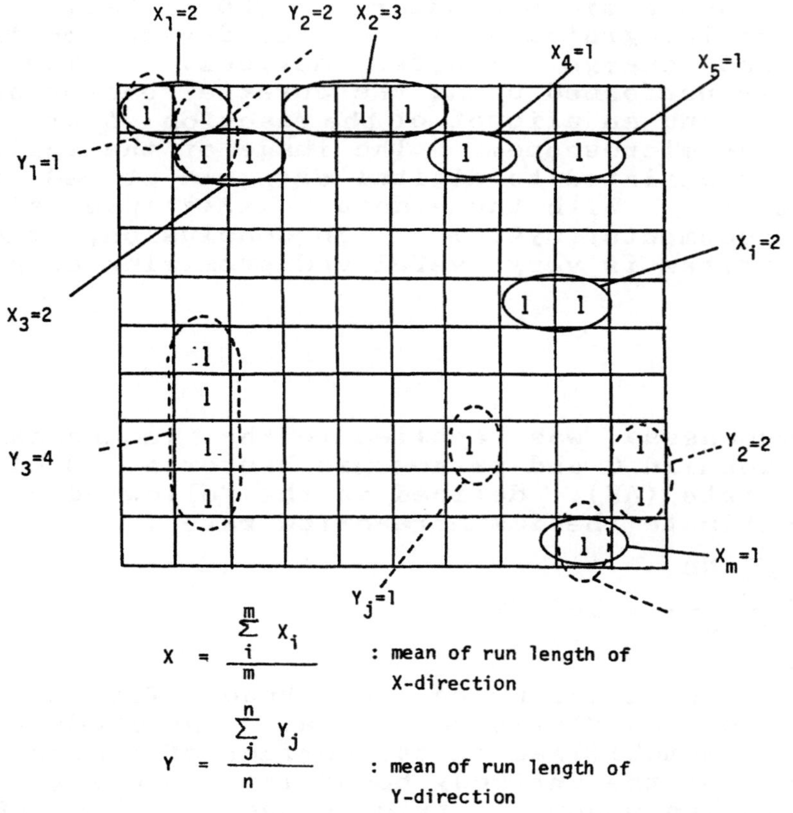

Figure 15. Definition of a new agglutination rate.

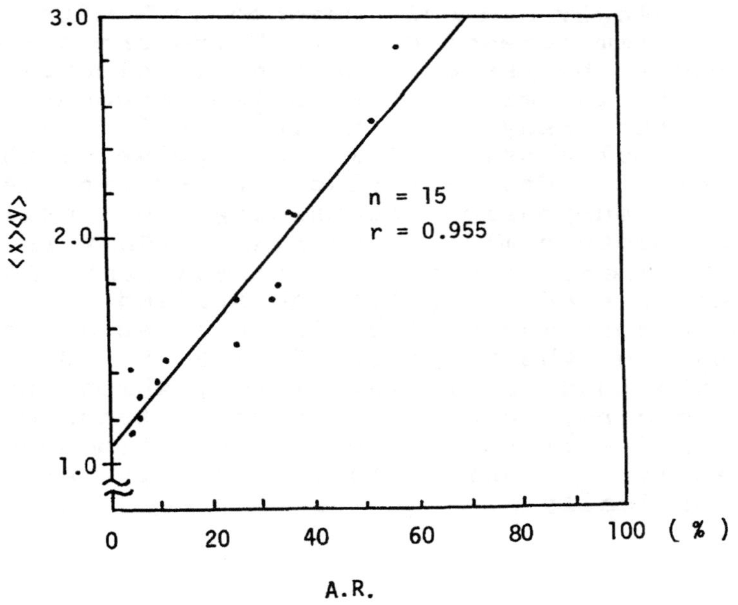

Figure 16. Correlationship between AR and <x><y>.

The image analyzing system was composed of a CCD video camera, a analog-digital(A/D) converter, a transient memory, a microscope and a microcomputer. CCD (Charge Coupled Devices) is an integrated semiconducter device consisted of photoarrays and charge transfer devices. The pulse immunoassay was performed using the cuvette reactor as shown in Figure 1, and an aliquot of the reaction solution was placed under the microscope. The image of the agglutinate adducts was translated to digital data and stored in the transient memory. With these data <x><y> was calculated with the microcomputer system. In conclusion, the pulse immunoassay system is very useful and promising as a rapid immunoassay.

Summary

Pulse immunoassay was applied to the immunoassay for human immunoglobulin G and pathogenic bacteria, C.albicans. Agglutination rate (AR), defined as the following equation, was used to estimate the immunoreaction rate :

$$AR = \frac{\sum_{n=5}^{\infty} Nn}{\sum_{n=1}^{\infty} Nn} \times 100 \ (\%)$$

where Nn is the total number of beads forming n-bead agglutination or n-agglutinated cells. Antibody to H-IgG was covalently immobilized on the surface of latex beads. Immunoreaction of the antibody bound latex beads (Ab-L) and H-IgG was performed under electric pulses. When 140 µl of Ab-L suspension (0.7%) containing H-IgG (6.7 x 10^{-6} g/ml) was reacted under electric pulses (peak height 2 kV/cm, pulse width 20 µsec, pulse frequency 8 kHz), AR increased sharply and reached 50 % after 10 min. In contrast, when the same sample was reacted without electric pulse, AR increased very slowly and only reached 20 % after 20 min. Therefore, the immunoreaction of H-IgG and Ab-L can be performed rapidly by the application of electric pulses. From the AR at 10 min, the H-IgG concentration was determined in the range 6.7 x 10^{-8} - 6.7 x 10^{-6} g/ml. Electric pulses (pulse height 1 kV/cm, pulse width 20 µsec, pulse frequency 8 kHz) were also applied to 140 µl of C.albicans cell suspension containing 10^{7} cells/ml and antibody(4 mg protein/ml). AR became 50 % in 5 min of irradiation with electric pulse. In contrast, AR was less than 10 % when no electric pulse was applied to the system. The pulse immunoassay was applicable to a rapid immunoassay for C.albicans in the range of 10^{7} - 6 x 10^{7} cells/ml. Furthermore, the image analyzing system, which was composed of a CCD video camera, a A/D converter, a transient memory, a microscope and a microcomputer, was introduced for the pulse immunoassay. Using this system, AR could be estimated automatically.

REFERENCES

Collet-Cossart,D., Magnusson,C.G.M., Cambiaso,C.L., Lesne M. and Masson,P.L., 1981, Automated particle-counting immunoassay for digoxin, Clin.Chem., 27, 1205.

Furedi,A.A. and Valentine,R.C., 1962, Factors involved in the orientation of microscopic particles in suspensions influenced by radio frequency fields, Biochim.Biophys.Acta, 56, 33.

Hu,C.J.and Barnes F.S., 1975, A simplified theory of pearl chain effects, Rad. and Environm.Biophys., 12, 71.

Johnson,G.D.,Holborow,E.J.and Dorling,J., 1978, in "Handbook of Experimental Immunology" Weir,D.M.,ed., Blackwell Scientific Publications, chapter 15.

Masson,P.L., Cambiasa,C.L., Collet-Cassart,D., Magnusson,C.G.M., Richards,C.B. and Sindic,C.J.M., 1981, Particle counting immunoassay, in "Methods in Enzymology" vol.74, Academic Press, New York.

Matsuoka,H., Tamiya,E. and Karube,I., 1985, Pulse immunoassay for Candida albicans, Anal.Chem., 57, 1998.

Schwan,H.P. and Sher,L.D., 1969, Alternating-current field-induced forces and their biological implications, J.Electrochem.Soc., 22C.

Yallow,R.S.,1976, "Methods in Radioimmunoassay of Peptide Hormones" North Holland, Amsterdam.

HETEROGENEOUS ENZYME IMMUNOASSAY WITH AMPEROMETRIC DETECTION

Kenneth R. Wehmeyer and Matthew J. Doyle

The Procter and Gamble Company
Miami Valley Laboratories, P. O. Box 39175
Cincinnati, Ohio 45247

H. Brian Halsall and William R. Heineman

Department of Chemistry
University of Cincinnati
Cincinnati, Ohio 45221

INTRODUCTION

Immunoassay is a very selective and sensitive approach to trace metabolite analysis (Thorell, 1978). As a consequence of binding geometries the interaction between antigen (Ag)† and antibody (Ab) molecules may be extremely specific under favorable conditions (Eisen, 1974). The utility of immunoassay in diagnostic medicine is well established with many routine clinical methods already in use for the determination of a wide variety of biologically important substances (Chalt and Ebersole, 1981).

Immunoassay methods can be divided into two general classes. In the competitive binding or saturation analysis assay (Zettner, 1973) a competitive equilibrium is established between excess labeled (Ag*) and unlabeled Ag for a limited amount of highly specific Ab as outlined in Scheme I.

Scheme I. Competitive Binding Immunoassay

$$\begin{matrix} \mathrm{Ag} \\ \mathrm{Ag}^* \end{matrix} + \mathrm{Ab} \rightleftharpoons \begin{matrix} \mathrm{Ag{:}Ab} \\ \mathrm{Ag^*{:}Ab} \end{matrix}$$

Following saturation of the Ab binding sites, the relative amounts of bound and free label are determined and a standard curve plotted. As a result of mass action, the bound-to-free ratio (B/F) of any test solution is inversely proportional to the concentration of Ag originally present in that solution. Both large (e.g. proteins) and small (e.g. drugs/hormones) molecules can be determined in this manner.

† Antigen (Ag) is used throughout this text as a general describer for all ligands, haptens included, capable of binding to a specific Ab.

Scheme II. Sandwich Binding Immunoassay

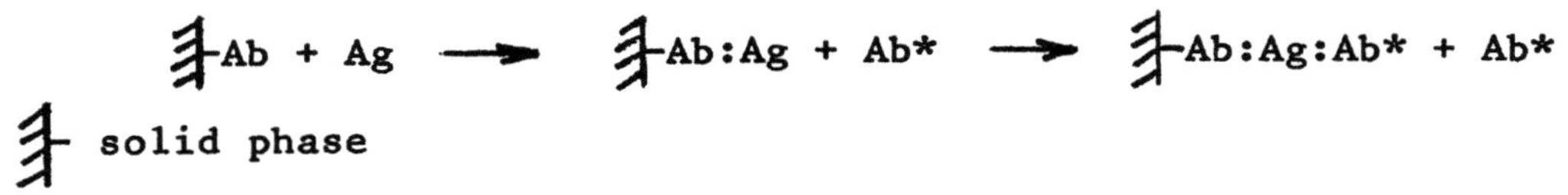

A second form, the sandwich immunoassay, is based upon non-competitive binding (Wisdom, 1976). Typically, Ab is attached to a solid support and exposed to a sample containing the Ag of interest as depicted in Scheme II. The amount of Ag bound to the solid phase-Ab is determined following incubation with a second specific labeled Ab (Ab*). In contrast to a competitive assay, the amount of Ab-bound Ag is directly proportional to its concentration in solution and sandwich immunoassays are limited to the determination of large molecules.

Since the pioneering work of Yalow and Berson (1959), radioisotopes have been employed as the label of choice. However, the drawbacks associated with the use of isotopic labels have prompted the development of immunoassays based on nonisotopic labeling schemes (Charlton, 1979). A number of nonisotopic alternatives have been developed including electron spin resonance for detecting radical labels (Leute et al., 1972), nephelometry (Deaton et al., 1976), fluorescence (O'Donnelly and Suffin, 1979), chemiluminescence (Hersh et al., 1979), and enzyme labels (Voller et al., 1978) among others (Chais et al., 1978; Schall and Tenoso, 1981; Nakamura and Dito, 1980).

Enzyme labels have been the most successful, and both homogeneous and heterogeneous assays are commercially available. Both of these assay formats are based on the inherent signal amplification capabilities of an enzyme label, with quantitation being achieved by measuring the conversion of substrate to product. Some enzymes can catalyze the conversion of 10^1-10^3 mole of substrate to product/mole enzyme/min, enabling the detection of as little as 0.1 fmole/mL of enzyme (Clark and Engvall, 1980).

The rationale for using enzyme labels extends beyond sensitivity considerations. Most enzymes are highly selective for a specific substrate, and rate methods can be employed to minimize error due to competing side reactions. Many are commercially available in high purity and at low cost. Several enzymes are readily adaptable to electrochemical methods and generate an electroactive product from an electroinactive, and therefore noninterfering, substrate.

Homogeneous assays do not require the separation of bound and free Ag* leading to a higher sample throughput. At present, these assays are limited to the determination of low molecular weight compounds that are present in the μg/mL to ng/mL concentration range. Heterogeneous techniques on the other hand, require separation of bound and free Ag* prior to distribution analysis but are applicable to molecules of all sizes. Typical limits of detection using heterogeneous procedures approach the ng/mL-pg/mL range and, in addition, potentially interfering matrix components are removed prior to determination of the enzyme product.

Enzyme products are commonly detected by spectroscopic techniques, however, absorbance measurements lack high sensitivity while fluorescence detection often suffers from endogenous interferences. The wide dynamic range and low detection limits of modern electroanalytical techniques (Kissinger and Heineman, 1984) represent an attractive alternative to spectroscopic methods. In addition, electrochemical procedures are convenient, fast, inexpensive, and selective. Two approaches to heterogeneous enzyme immunoassay employing amperometric detection have been reported. One technique utilizes conventional solid phase separation procedures for the analysis of both large and small molecules (Doyle et al., 1984; Wehmeyer et al., 1985; Wehmeyer et al., 1986; Heineman and Halsall, 1985). Alternatively, de Alwis and Wilson (1985) has applied affinity chromatography to separate bound and free Ag*.

The finite current technique known as liquid chromatography/electrochemistry (LCEC) as applied to conventional competitive and sandwich heterogeneous enzyme immunoassay is the subject of this chapter. Assays for digoxin (a cardiac glycoside), orosomucoid (a common serum glycoprotein), and immunoglobulin G are described.

MATERIALS AND METHODS

Apparatus

A prototype electrochemical immunoassay system (Bioanalytical Systems, Inc., West Lafayette, IN) was used for liquid chromatography with hydrodynamic amperometric detection. The thin-layer electrochemical cell consisted of a paraffin oil-based carbon paste working electrode, a silver/silver chloride (Ag/AgCl) with KCl (3 M) reference electrode, and a stainless-steel block auxiliary electrode that constituted one-half of the cell.

A slurry-packed 5 cm x 2 mm precolumn with 10-μm irregularly shaped RSiL material (Alltech Associates, Deerfield, IL) was used to separate phenol from the components of the assay buffer. A 12 cm x 4 mm Knauer column dry-packed with 37 to 44 μm pellicular C_{18} packing material (Alltech Asociates) was placed between the pump and the injection valve to saturate the mobile phase. All hydrodynamic analyses were done using phosphate buffer (0.1 M, pH 7.0) containing 4% methanol. A flow rate of 1.0 to 1.6 mL/min, an applied potential of +870 mV vs Ag/AgCl and a fixed 20-μL injection loop were used.

Buffers

All buffer solutions were prepared from distilled/deionized H_2O of at least 10^6 Ω resistivity using ACS grade chemicals obtained from MCB (Norwood, OH) or Fisher Scientific (Cincinnati, OH). The 0.15 M phosphate buffered saline (PBS, pH 7.4) and 0.05 M carbonate (pH 9.6) buffers were prepared from the appropriate sodium salt. Tween 20 solutions (Fisher) were prepared as 0.5 mL of Tween 20 in 1 L of 0.15 M PBS.

Enzyme Substrate

Alkaline phosphatase (AP) substrate solutions were prepared from 0.0146 g of phenyl phosphate (Calbiochem-Behring Corp., La Jolla, CA) and 0.0055 g of $MgCl_2 \cdot 6H_2O$ (Baker, Phillipsburg, NJ) in 25 mL of 0.05 M carbonate, pH 9.6. Substrate solutions were prepared just prior to use in order to prevent nonenzymatic hydrolysis.

Procedures

Digoxin Assay. Digoxin (Sigma Chemical Co. St. Louis, MO) standards (0 ng/mL - 5 ng/mL) were prepared in phosphate buffer or pooled human plasma (University Hospital, Cincinnati, OH). An antibody coating solution of 10 μg/mL was prepared by dissolving the appropriate amount of digoxin specific Ab (Center for Disease

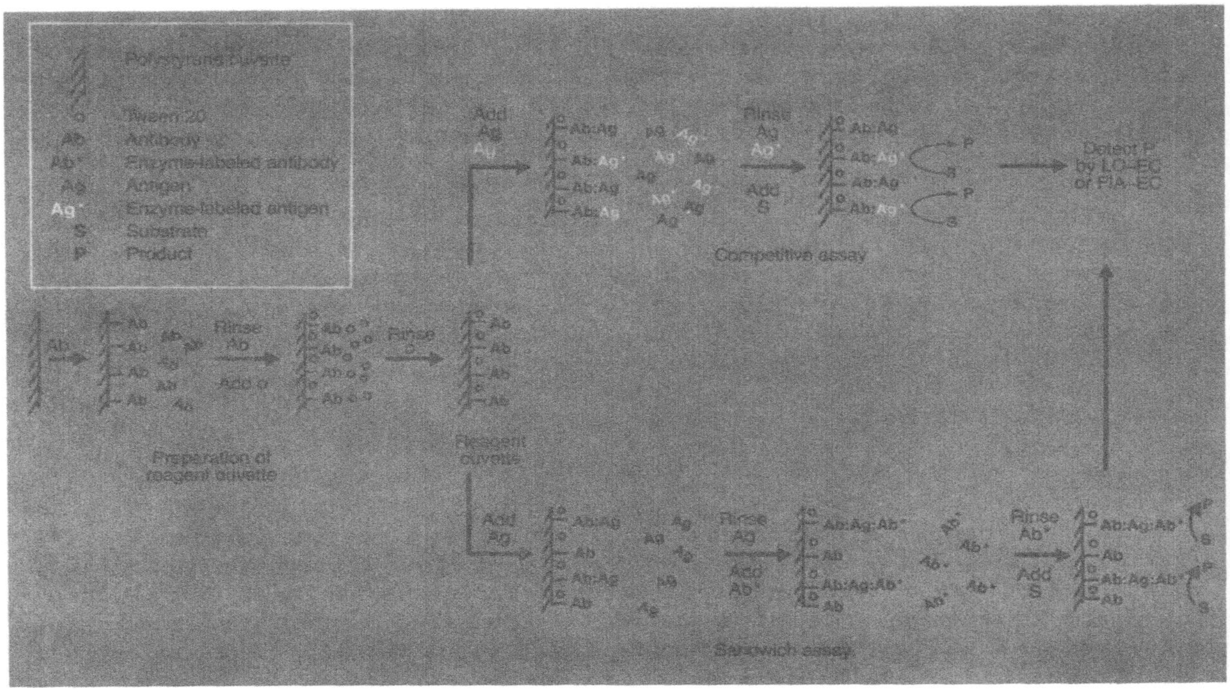

Figure 1. General protocol for heterogeneous enzyme immunoassay. Preparation of reagent cuvettes by coating with Ab, competitive assay format, and sandwich assay format. (Reprinted with permission from Heineman and Halsall. Copyright 1985, American Chemical Society.)

Control, Atlanta, GA) in carbonate buffer containing 0.02% sodium azide. Digoxin-alkaline phosphatase conjugate (Immunotech Corp., Cambridge, MA) was diluted with an appropriate amount of PBS-Tween. For analysis of patient samples, both the samples and standards were diluted with pooled human plasma.

The format for a competitive enzyme immunoassay is outlined in Figure 1. Polystyrene cuvettes (Gilford Instruments, Cleveland, OH) were coated with digoxin-specific Ab by passive adsorption as previously reported (Parsons, 1981). Following Ab coating, the cuvettes were rinsed with PBS/Tween to minimize nonspecific adsorption by coating unoccupied sites on the support. Then, 375 μL of the digoxin standard or sample and 25 μ L of a digoxin-alkaline phosphatase conjugate dilution were added simultaneously to the antibody-coated cuvettes. The assay solutions were incubated at room temperature for 4 h. The contents of the cuvettes were aspirated and washed consecutively, twice with PBS Tween, once with PBS Tween (5 min) and twice with carbonate buffer. Following the washing step, 300 μL of the enzyme substrate solution were added to each cuvette and incubated at room temperature for a timed interval. The enzyme reaction was then stopped by the addition of 25 μL of 5.5 M NaOH. Immediately before chromatographic injection, 25 μL of 5.5 M HCl were added to each sample. This step reduced a slowly decaying capacitive current observed on the injection of the more alkaline solution. Following the HCl addition, 20 μL of the substrate solution were injected into the chromatograph, and the AP generated phenol was quantitated by oxidative hydrodynamic amperometry.

Human Orosomucoid Assay. Human orosomucoid (OMD, α_1-acid glycoprotein) and bovine intestinal AP (Type VII) were obtained from Sigma Chemical Co. OMD and AP migrated as single bands during electrophoresis and were used without further purification. Coupling of OMD to AP, as well as the purification and characterization of the resulting conjugate, have been discussed in detail previously (Doyle et al., 1984). All standard OMD and enzyme conjugate solutions were diluted with PBS/Tween to minimize nonspecific interactions. Ab coating solutions were prepared in 0.05 M carbonate buffer by adding the appropriate amount of antiserum.

The competitive enzyme immunoassay procedure employed is outlined in Figure 1. OMD specific Ab was passively sorbed on the walls of 10 mm x 5.5 mm polystyrene cuvettes. This solid-phase Ab matrix was then allowed to equilibrate with the test solution. Typically, 0.6 mL of OMD antiserum solution (1.0 μg/mL) or nonspecific IgG (10.0 μg/mL) were placed in each cuvette and incubated at 37° C for 24 h. Each well was then aspirated and rinsed twice with PBS/Tween for 15 min and once rapidly.

An OMD containing solution, either an unknown or a standard (0.4 mL), was mixed with 0.1 mL of a 1/100 dilution of the OMD-enzyme conjugate and allowed to equilibrate with the Ab solid phase. Following a 12 h incubation period, the cuvettes were aspirated, rinsed twice with PBS/Tween for 15 min, once with PBS/Tween quickly, and once using 0.05 M carbonate. The wells were aspirated, and 0.5 mL of substrate solution were added to each. Cuvettes were allowed to stand at room temperature for 1 h, and the enzymatic reaction was stopped by transferring the substrate solutions from each well into corresponding polypropylene microbeakers. The product of the enzymatic reaction was determined amperometrically. LCEC samples were diluted as follows: 10 μL of each solution were mixed with 1.0 mL of 0.05 M carbonate, and 20 μL of this diluted sample were injected into the chromatograph. The amount of AP generated phenol was determined amperometrically as described above.

Rabbit Immunoglobulin G (IgG) Assay. Rabbit IgG (Sigma Chemical Co.) standard solutions were prepared from a 10 mg/mL stock solution by appropriate dilution with PBS Tween. Goat antiserum to rabbit IgG (Sigma Chemical Co), received lyophilized in individual vials, was reconstituted with 2.0 mL of carbonate buffer and diluted 1000-fold with carbonate buffer. Goat (anti-rabbit IgG) Ab-alkaline phosphatase conjugate (Sigma Chemical Co.) was diluted 1000-fold with PBS-Tween.

As shown in Figure 1, polystyrene cuvettes were coated with the goat Ab by passive adsorption from 500 μL of carbonate buffer. After coating, the cuvettes were washed three times with PBS Tween, allowing the PBS Tween to remain in the cuvettes for 10 min during each wash. Then 400 μL of the rabbit IgG standards were added to the cuvettes and incubated at room temperature. Following incubation, the cuvettes were aspirated and washed three times with PBS Tween. After washing, 375 μL of the enzyme-labeled goat antibody solution were added to each cuvette, again incubated at room temperature, and the cuvettes washed consecutively with PBS Tween (twice) and carbonate buffer (twice). Following this step, 300 μL of the enzyme substrate solution were added to the cuvettes, incubated at room temperature and the substrate reaction stopped by the addition of 25 μL of 5.5 M NaOH. Immediately before chromatographic injection, 25 μL of 5.5 M HCl were added to the sample and 20 μL of the subsequent reaction mixture were injected into the chromatograph. The peak heights (i_p) obtained for the oxidation of phenol were used to construct a standard curve.

RESULTS AND DISCUSSION

LCEC Detection of Alkaline Phosphatase Product

A number of enzyme systems have been utilized as labels for immunoassay (Jarvis, 1979; Scharpe et al., 1976). Several generate electroactive product from electroinactive, therefore non-interfering, substrate and are ideally suited for use in electrochemical immunoassay (Wehmeyer et al., 1983). Alkaline phosphatase (AP) was used as the enzyme label for these studies. AP catalyzes the conversion of an electroinactive substrate (phenylphosphate) to an electroactive product (phenol):

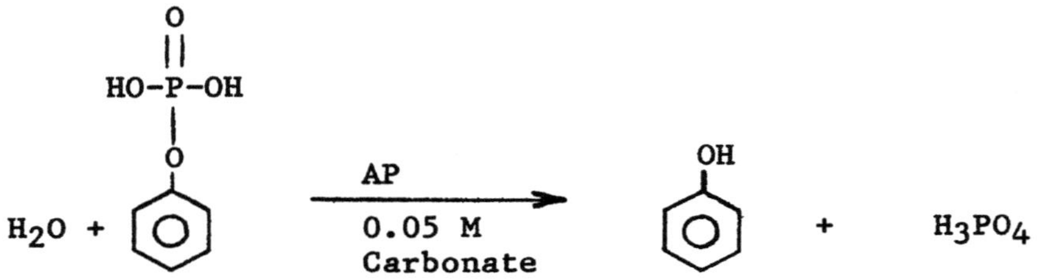

A cyclic voltammogram for phenol at a carbon paste electrode is shown in Figure 2. The irreversible oxidation wave of phenol exhibits a peak potential at +670 mV vs Ag/AgCl in a 0.1 M carbonate buffer. Repetitive scans indicated that electropolymerization of the phenolic radical resulted in fouling of the electrode surface at this relatively high concentration of phenol. A cyclic voltammogram of phenylphosphate showed the substrate to be electroinactive over the potential range in which phenol is oxidized. Therefore, phenol is easily detected in the presence of substrate by electrochemical oxidation.

LCEC was used to determine phenol generated during the immunoassays. A diagrammatic representation of the LCEC system components appears in Figure 3. A precolumn was employed to separate phenol from other assay constituents. Even though the components of the assay buffer are electroinactive, its injection gives rise to a capacitive current, as shown by the chromatogram in Figure 4. This is a result of the ionic difference between the substrate solution and carbonate mobile phase, which contains no phenylphosphate or $MgCl_2$. A hydrodynamic voltammogram for phenol (Figure 5) indicated that maximum sensitivity at a carbon paste electrode is obtained at potentials greater than +850 mV vs. Ag/AgCl. A potential of +870 mV was chosen for subsequent studies. Under these conditions, the LCEC detection of phenol was linear between 9.0×10^{-9}M to 9.6×10^{-6}M (slope = 0.57 nA/nmole, b = -0.30 nA, r = 0.999). This approach represents a simple, sensitive and precise method for the detection of AP generated phenol. No evidence of electrode fouling was observed even at micromolar levels. The LCEC system could be operated continuously for up to 3 months before column replacement or electrode resurfacing was necessary.

Competitive Electrochemical Immunoassay

In its most common form, a competitive solid phase immunoassay involves the competition between labeled and unlabeled ligand for a limited amount of matrix bound antibody (Figure 1). During the development of a competitive immunoassay it is critical to ascertain the optimum conditions for each stage of the procedure. The heterogeneous enzyme assay requires optimization of the antigen/antibody incubation interval, the antibody coating concentration, the enzyme-conjugate dilution and the substrate incubation interval.

Optimizations are routinely performed in a "checkerboard" fashion, where one parameter is varied while the others are held constant. Alternatively, simplex optimization algorithms are available which can greatly facilitate this process (Morgan and Deming, 1974).

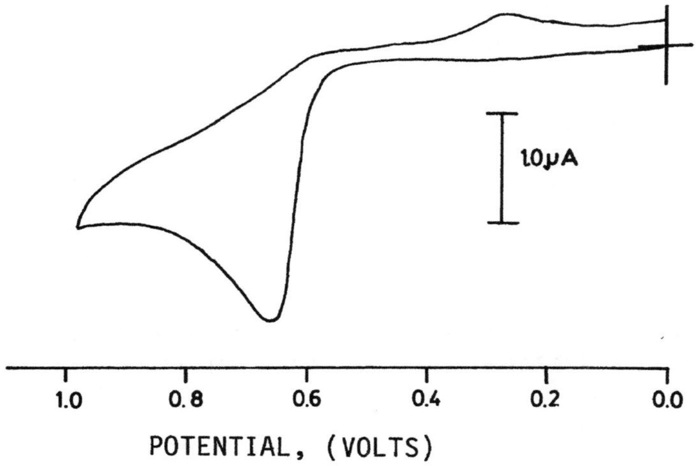

Figure 2. Cyclic voltammogram of 1.5 x 10^{-4} M phenol in 0.05 M carbonate buffer (pH = 9.6); carbon paste working electrode, Ag/AgCl reference electrode, scan rate = 10 mV/s. (From Wehmeyer et al., 1983, with permission.)

In general, ideal immunoassay curves should exhibit the greatest sensitivity (Δ response/ Δ concentration) over the concentration range of interest as well as the lowest limit of detection which is a function of the standard deviation for replicate B_0 analyses. Several theoretical treatments of immunoassay optimization have appeared in the literature (Yalow and Berson, 1964; Yalow and Berson, 1971; Ekins and Newnan, 1970). Assay sensitivity can be tailored to cover a specific dynamic range, limited in the upper range by the amount of Ab which can be sorbed onto the solid support.

Digoxin. A large number of clinically important substances (therapeutic drugs, hormones, steroids) have molecular weights less than 5000 amu. As an example of the application of electrochemical competitive immunoassay to small molecular weight species, digoxin was chosen as a model compound. Digoxin, a steroidal cardiac glycoside, is used in the treatment of chronic heart disease to enhance the force and frequency of contractions. Precise control of plasma digoxin levels is important as this drug exhibits a relatively narrow therapeutic range (0.5 to 2.0 ng/mL).

Representative LCEC assay chromatograms for digoxin standards in plasma are shown in Figure 6. The peak current is proportional to the amount of phenol produced by the antibody-bound digoxin-enzyme conjugate, which is inversely proportional to the amount of digoxin present in the standard. The LCEC method in conjunction with the

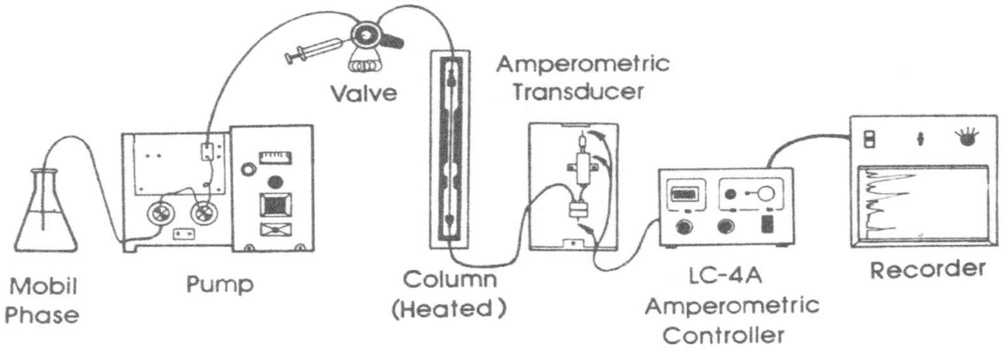

Figure 3. Diagram of liquid chromatography/electrochemistry apparatus. (From Bioanalytical Systems, Inc., with permission.)

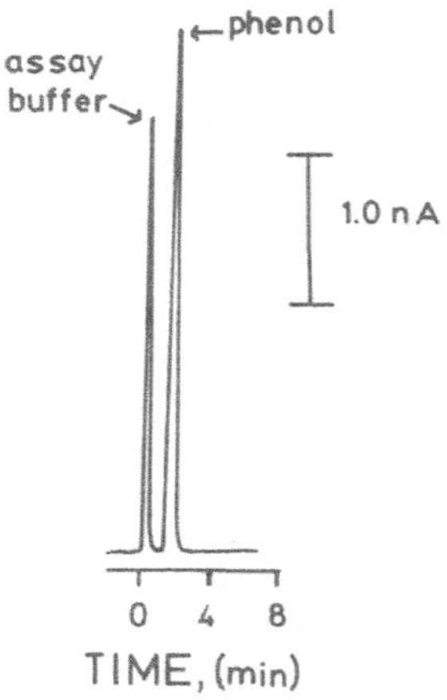

Figure 4. LCEC chromatogram of a 20 μL injection of 5.3 x 10^{-7} M phenol in 0.05 M carbonate buffer. Flow rate, 1.2 mL/min; eluent, 0.1 M phosphate buffer (pH = 7.0). (From Wehmeyer et al., 1983, with permission).

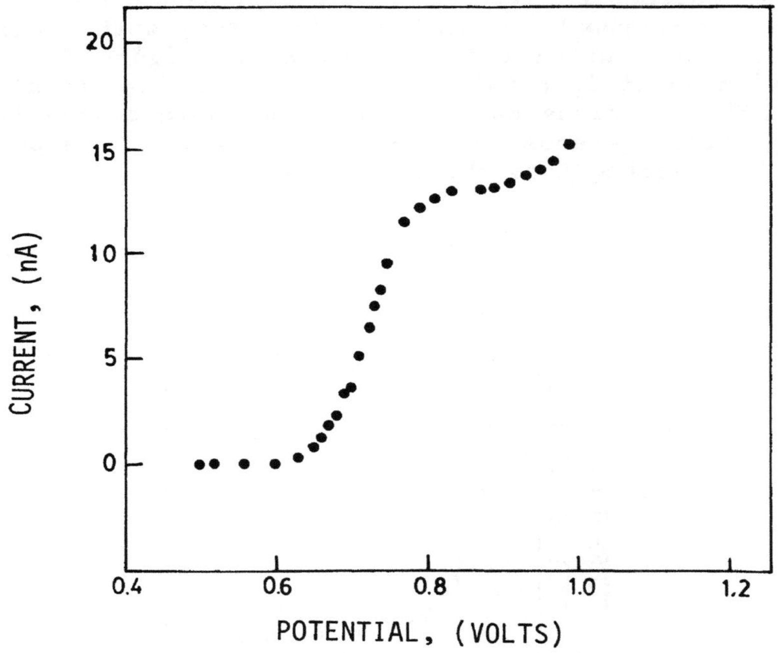

Figure 5. Hydrodynamic voltammogram for a 7.0 x 10^{-7} M phenol solution. See conditions for Figure 4. (From Wehmeyer et al., 1983, with permission).

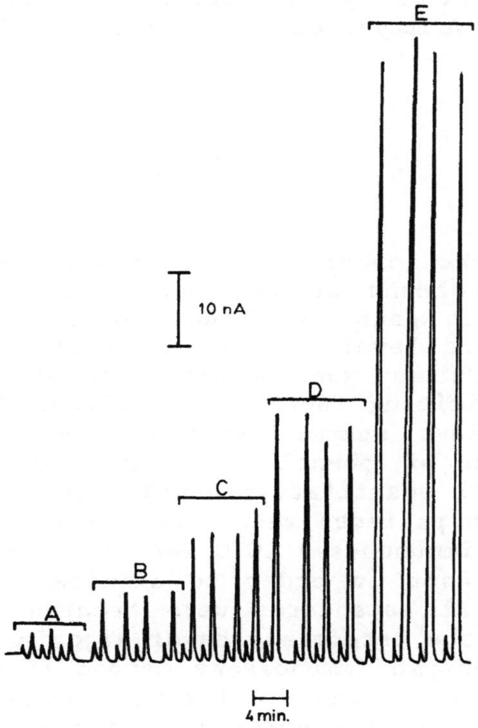

Figure 6. Heterogeneous enzyme immunoassay with LCEC detection for a series of digoxin standards in plasma solutions. Concentration of digoxin in plasma samples was A) 5.0, B) 2.0, C) 1.0, D) 0.5, and E) 0.0 ng/mL. (Reprinted with permission from Wehmeyer et al. Copyright 1986, American Chemical Society.)

optimal assay parameters resulted in a very sensitive immunoassay for plasma digoxin throughout its therapeutic range, with a detection limit of 50 pg/mL. A standard curve is shown in Figure 7. The relative standard deviation of i_p obtained for any given digoxin standard ranged from 4% to 12% on various days and is comparable to results ordinarily obtained by heterogeneous enzyme immunoassays. Approximately 20 samples can be injected into the LC per hour.

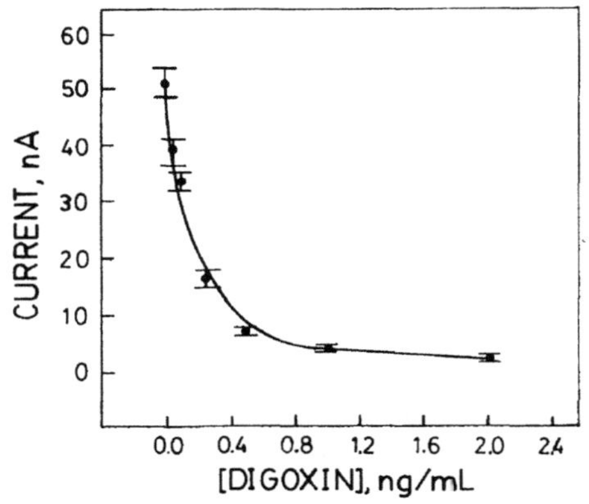

Figure 7. Typical standard curve for the determination of digoxin standards in human plasma by electrochemical immunoassay. (Reprinted with permission from Wehmeyer et al. Copyright 1986, American Chemical Society.)

As a result of the inverse binding relationship, sufficient enzyme labeled material was bound at the lower digoxin concentrations (0.5 ng/mL - 0.0 ng/mL) to generate phenol concentrations in the 10^{-6} M range. Phenol at this level is well above the demonstrated detection limit of the LCEC technique and therefore the measurement of the enzyme activity was not a limiting factor. For higher digoxin concentrations (0.5 - 2.0 ng/mL) less enzyme label was bound and consequently a smaller concentration of phenol (low 10^{-7} M) was generated, which required the low level quantitative capabilities of the LCEC detection system. Samples from patients receiving digoxin therapy were analyzed by the heterogeneous immunoassay LCEC method. Samples were diluted 5/1 with pooled human plasma in order to eliminate an observed antibody matrix effect. Digoxin standards were treated in a similar manner. Results obtained for the determination of digoxin in patient samples by the electrochemical enzyme immunoassay method (EIA-LCEC) were compared to those obtained by RIA. The results for the 54 samples analyzed are presented in Figure 8. A good correlation was obtained between the two methods (r = 0.93).

Orosomucoid. In addition to small M.W. compounds, there are a significant number of large M.W. (> 5000 amu) molecules such as TSH, insulin and immunoglobulins, which are of diagnostic interest. The antigen that was chosen for these model studies was human serum

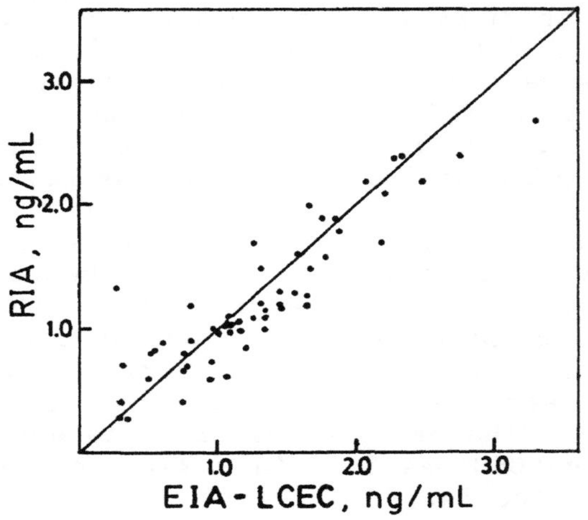

Figure 8. Correlation between digoxin concentration in patients by RIA and competitive heterogeneous enzyme immunoassay liquid chromatography/electrochemistry (EIA-LCEC). A perfect correlation is represented by the solid line. (Reprinted with permission from Wehmeyer et al. Copyright 1986, American Chemical Society.)

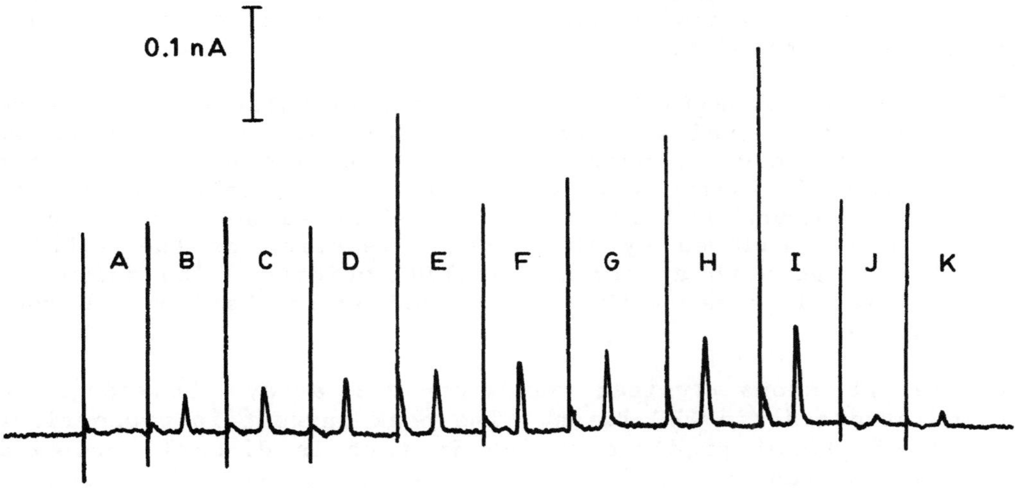

Figure 9. Electrochemical immunoassay of a series of OMD standard solutions and controls: (A) 0.05 M carbonate alone; (B) 200, (C) 100, (D) 60, (E) 10, (F) 5.0, (G) 2.5, (H) 1.0, and (I) 0.75 ng/mL OMD incarbonate buffer; the addition of AP alone (J) or OMD-AP (K) to cuvettes coated with nonspecific IgG. (Reprinted with permission from Doyle et al.. Copyright 1984, American Chemical Society.)

orosomucoid (OMD, α_1-acid glycoprotein). The glycoprotein OMD (41,000 daltons) is an acute phase reactant during inflammation (Jamieson et al., 1972), pregnancy (Adams and Wacher, 1968), and cancer (Winzler, 1955). Serum OMD levels have been shown to vary dramatically during the course of certain malignancies (Gendler et al., 1982; Rashid et al., 1982). Altered OMD-drug binding affinities correlate directly with metastatic progression (Abramson, 1982; Jackson et al., 1982), and the diagnostic value of OMD as a tumoral marker has been suggested (Vea-Martinez et al., 1982). Quantitative methods for the determination of OMD are quite tedious and often insensitive.

Chromatograms for a series of OMD standards appear in Figure 9A-I. The concentration of phenol, and hence peak current, is inversely proportional to the concentration of OMD in the test solution. Nonspecific adsorption was not apparent under these test conditions (Figure 9 J,K). A plot of peak current vs. OMD concentration indicated that the assay is most sensitive in the region between 1 and 10 ng/mL (Figure 10). The percent relative standard deviation for a 40 ng/mL OMD standard by LCEC was 3.7% (n = 9), much improved over conventional immunoassay protocols.

A plot of i_p vs. substrate incubation time for a 5 ng/mL OMD standard following the assay was linear with a slope of 0.0228 nA/min. The substrate incubation period may be shortened to 10-20 min without a loss in sensitivity, since a detectable amount of phenol was reproducibly generated in that time span.

Sandwich Electrochemical Immunoassay

The sandwich immunoassay is non-competitive and involves two separate ligand/antibody interactions (Figure 1). Sandwich assays require a ligand to contain at least 2 distinct antigenic sites and as a result are only applicable to large molecular weight compounds. Optimization of the sandwich assay parameters is also commonly done by a "checkerboard" approach.

Rabbit Immunoglobulin G. Rabbit Immunoglobulin G was selected as a model compound to evaluate the electrochemical sandwich immunoassay. The detection of serum immunoglobulins to specific pathogens can serve as an indicator of disease states in humans and animals (Voller et al., 1978). In the sandwich immunoassay, solid phase antibody reacts with IgG and this is followed by the further reaction of the solid phase antibody bound IgG with an enzyme labeled antibody. The amount of IgG bound to the solid phase is then determined by incubation with the substrate solution.

Figure 11 shows typical sandwich-type assay chromatograms for rabbit IgG standards in PBS Tween. The peak current is proportional to the amount of phenol produced, which in turn is directly proportional to the amount of rabbit IgG in the standard solution. The results of a typical IgG sandwich assay, carried out with 3 h antigen/antibody incubation intervals and allowing 20 min for reaction with substrate, are shown in Figure 12. Concentrations of rabbit IgG determined ranged from 0 to 250 ng/mL with a detection limit of 100 pg/mL. The CV for i_p for 10 repetitions of the 10 ng/mL standard was 3.8%. Increasing the substrate incubation interval to 60 min reduced the detection limit to 50 pg/mL (Table 1). The detection limit was further improved to 10 pg/mL by increasing the antigen/antibody incubation interval to 10 h, with a 60 min substrate reaction interval (Table 1). Table 1 also shows results of a 30 min antigen/antibody incubation interval in conjunction with a 60 min substrate reaction interval. In this rapid assay the detection limit was 100 pg/mL.

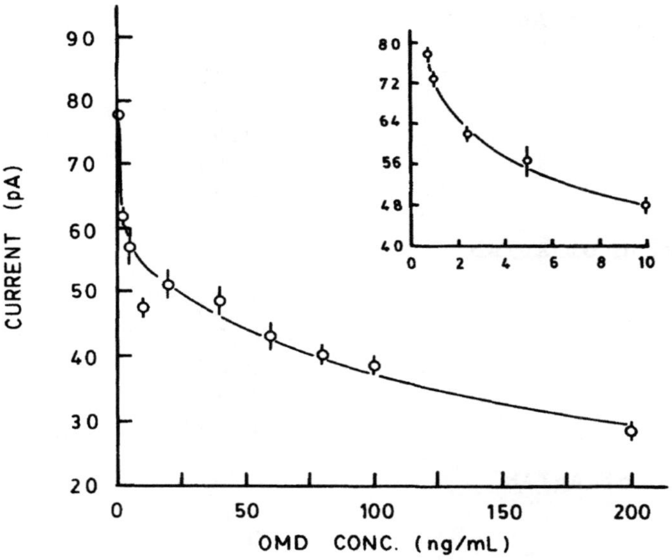

Figure 10. Typical standard curve (expanded scale insert) for the determination of a series of OMD solutions by electrochemical immunoassay. (Reprinted with permission from Doyle et al. Copyright 1984, American Chemical Society.).

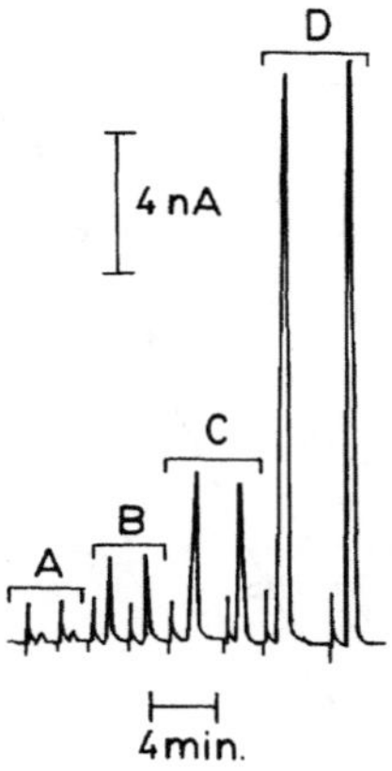

Figure 11. Electrochemical immunoassay for the analysis of rabbit IgG standards in PBS Tween: (A) 0, (B) 1.0, (C) 2.0, and (D) 6.0 ng/mL. (From Wehmeyer et al., 1985, with permission.)

Table 1. Results of Sandwich-Type Immunoassay for Rabbit IgG

Rabbit IgG, pg/mL	i_p, nA Mean	SD
3 h incubation of Ag/Ab		
0	0.18	0.01
50	0.34	0.01
100	0.59	0.07
500	2.96	0.21
1000	8.35	0.17
30 min incubation of Ag/Ab		
0	0.23	0.01
100	0.33	0.04
500	0.87	0.06
1000	1.78	0.06
2000	11.36	--
10 h incubation of Ag/Ab		
0	0.27	0.01
10	0.32	0.02
50	0.61	0.03
100	1.28	0.01

Assays included 60 min incubation of assay mixture with substrate solution. i_p, peak current; nA, nanoamperes.

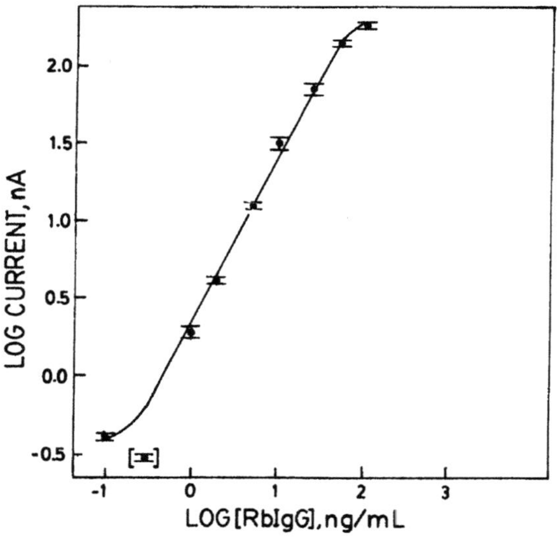

Figure 12. Log/log plot of LCEC peak current vs. rabbit IgG concentration (n = 5) in PBS Tween. [ ] denote the 0 ng/mL standard. (From Wehmeyer et al., 1985, with permission.)

Both under optimum and rapid assay conditions, the labeled antibody bound by the lowest detectable rabbit IgG standard produced a phenol concentration in the 10^{-7} to 10^{-8} M range. Phenol concentrations in these ranges are easily detected by LCEC. For both assays the detection limit for IgG was limited by the extent of nonspecific adsorption (i_p for the 0 ng/mL standard). This nonspecific binding corresponded to a phenol concentration in the low 10^{-8} M region.

CONCLUSIONS

The LCEC detection of phenol provides a relatively rapid, simple and extremely sensitive method of quantitating alkaline phosphatase activity. We have successfully demonstrated the application of LCEC detection to heterogeneous immunoassay methodology utilizing alkaline phosphatase as a label. Competitive and sandwich assays have been developed for both small and large molecular weight compounds. The precision and limits of detection achieved using amperometric based assays are equivalent to or better than conventional radioimmunoassay techniques for similar antigens. Detection limits in the ng/mL - pg/mL range, achieved using heterogeneous amperometric immunoassay techniques, are far superior to the μg/mL levels obtainable by present homogeneous enzyme based assays.

A point should be made regarding the theoretical detection limits of competitive and sandwich immunoassay methods. The lowest detectable amount of analyte achievable via a competitive format is limited by the ratio of the experimental error to the magnitude of the antibody binding constant (Ekins and Newnan, 1970). Assuming a constant level of experimental error, the quality of the antiserum and not the activity of the label will determine the ultimate detection limit. Conversely, an increase in the specific activity of the label will, in general, result in lower detection limits for the sandwich assay. As one approaches zero nonspecific binding and infinite label specific activity, the detection limit of the sandwich assay should approach 1 molecule of antigen (Ekins and Newnan, 1970). Therefore, the signal amplification advantages of enzyme labels can be more fully exploited when the sandwich immunoassay format is employed.

ACKNOWLEDGMENT

Financial support provided by NSF Grants CHE-7911872 and CHE-82 17045 and NIH Grant HD 13207 is gratefully acknowledged.

REFERENCES

Abramson, F. P., 1982, Methadone plasma protein binding: alterations in cancer and displacement from α_1-acid glycoprotein, Clin. Pharmacol. Ther. 32:652-658.

Adams, J. B. and Wacher, A., 1968, Specific changes in the glycoprotein components of serumucoid in pregnancy, Clin Chim. Acta 21:155-157.

Chais, M., Slovin, E. and Snarsky, L., 1978, Metalloimmunoassay: II. iron-metallohaptens from estrogen steroids, J. Organomet. Chem. 160:223-230.

Chalt, E. M. and Ebersole, R. C., 1981, Clinical analysis: a perspective on chromatographic and immunoassay technology, Anal. Chem. 53:680A-692A.

Charlton, J. C., 1979, Overcoming the radiological and legislative obstacles in radioimmunoassay, Antibiot. Chemother. 26:27-37.

Clark, B. R. and Engvall, E., 1980, Enzyme-linked immunosorbent assay (ELISA), in: "Enzyme Immunoassay," E. T. Maggio, ed., Boca Raton, pp. 167-180.

de Alwis, W. U. and Wilson, G. S., 1985, Rapid sub-picomole electrochemical enzyme immunoassay for immunoglobulin G, Anal. Chem. 57:2754-2756.

Deaton, O. D., Maxwell, K. W., Smith, R. S. and Creveling, R. L., 1976, Use of laser nephelometry in the measurement of serum proteins, Clin. Chem. 22:1465-1471.

Doyle, M. J., Halsall, H. B. and Heineman, W. R., 1984, Enzyme-linked immunoadsorbent assay with electrochemical detection for α_1-acid glycoprotein, Anal. Chem. 56:2355-2360.

Eisen, H. N., 1974, Immunology: an introduction to molecular and cellular principles of immune response, in: "Microbiology," 2nd ed., Harper and Row, New York, p. 547.

Ekins, R. P. and Newnan, G. B., 1970, Theoretical aspects of saturation analysis, Acta Endocrinol. 64(147):11-36.

Gendler, S. J., Dermer, A. B., Silverman, L. M. and Tokes, Z. A., 1982, Synthesis of α_1-antichymotrypsin and α_1-acid glycoprotein by human breast epithelial cells, Cancer Res. 42:4567-4573.

Heineman, W. R. and Halsall, H. B., 1985, Strategies for electrochemical immunoassay, Anal. Chem. 57:1321A-1331A.

Hersh, L. S., Vann, W. P. and Wilhelm, S. A., 1979, A luminol-assisted, competitive-binding immunoassay of human immunoglobulin G, Anal. Biochem. 93:267-271.

Jackson, P. R., Tucker, G. T. and Woods, H. F., 1982, Altered plasma drug binding in cancer: role of α_1-acid glycoprotein and albumin, Clin. Pharmacol. Ther. 32:295-302.

Jamieson, J. C., Ashton, F. E., Frieson, A. D. and Chou, B., 1972, Studies of acute phase proteins of rat serum. II. determination of the contents of α_1-acid glcoproteins, α_2-macroglobulin, and albumin in serum from rats suffering from induced inflammation, Can. J. Biochem. 50:871-880.

Jarvis, R. F., 1979, The future outlook for enzyme immunoassays, Antibiotics Chemother. 26:105-117.

Kissinger, P. T. and Heineman, W. R., eds., 1984, "Laboratory Techniques in Electroanalytical Chemistry," Dekker, New York, p. 751.

Leute, A. K., Ullman, E. F., Goldstein, A erzenberg, L. A., 1972, Spin immunoassay technique for determination of morphine, Nature 236:93.

Morgan, S. L. and Deming, S. N., 1974, Simplex optimization of analytical chemical methods, Anal. Chem. 46:1171-1181.

Nakamura, R. M. and Dito, W. R., 1980, Nonradioisotopic immunoassay for therapeutic drug monitoring, Lab. Med. 11:807-817.

O'Donnelly, C. M. and Suffin, S. C., 1979, Fluorescence immunoassays, Anal. Chem. 51:33A-40A.

Parsons, G. H., 1981, Antibody-coated plastic tubes in radioimmunoassay, in: "Methods of Enzymology," J. J. Langone and H. Van Vunakis, ed., Vol. 73, Academic Press, New York, 1224-1239.

Rashid, S. A., O'Quigley, J., Axon, A. T. R. and Cooper, E. H., 1982, Plasma protein profiles and prognosis in gastric cancer, Br. J. Cancer 45:390-394.

Schall, R. F. and Tenoso, H. J., 1981, Alternatives to radioimmunoassay: labels and methods, Clin. Chem. 27:1157-1164.

Scharpe, S. I., Cooreman, W. M., Bloome, W. J. and Leakeman, G. M., 1976, Quantitative enzyme immunoassay: current status, Clin. Chem. 22:733-738.

Thorell, J. I. and Larsond, S. M., 1978, "Radioimmunoassay and Related Techniques," C. V. Mosby Company, St. Louis, p. 298.

Vea-Martinez, A, Gatell, J. M., Segura, F., Heiman, C., Elena, M., Ballesta, A. M. and Mundo, M. R., 1982, Diagnostic value of tumoral markers in serous effusions, Cancer 50:1783-1788.
Voller, A., Bartlett, A. and Bidwell, P. E., 1978, Enzyme immunoassays with special reference to ELISA techniques, J. Clin. Pathol. 31:507-520.
Wehmeyer, K. R., Doyle, M. J., Wright, D. S., Eggers, H. M., Halsall, H. B. and Heineman, W. R., 1983, Liquid chromatography with electrochemical detection of phenol and NADH for enzyme immunoassay, J. Liq. Chromatogr. 6(12):2141-2156.
Wehmeyer, K. R., Halsall, H. B. and Heineman, W. R., 1985, Heterogeneous enzyme immunoassay with electrochemical detection: competitive and "sandwich"-type immunoassay, Clin. Chem. 31:1546-1549.
Wehmeyer, K. R. Halsall, H. B. and Heineman, W. R., 1986, Competitive heterogeneous enzyme immunoassay for digoxin with electrochemical detection, Anal. Chem. 58:135-139.
Winzler, R. J., 1955, Determination of serum glycoproteins, in: "Methods of Biochemical Analysis," D. Glick, ed., Interscience Publishers, New York, 2:279-311.
Wisdom, G. B., 1976, Enzyme immunoassay, Clin. Chem. 22:1243-1253.
Yalow, R. S. and Berson, S. A., 1959, Assay of plasma insulin in human subjects by immunological chemistry, Nature 184:1648-1649.
Yalow, R. S. and Berson, S. A., 1964, in: "The Hormones," G. Pincus and K. V. Thimann, eds., Academic Press, New York.
Yalow. R. S. and Berson, S. A., 1971, Problems of validation of radioimmunoassay, in: "Principles of Competitive Protein-Binding Assays," W. D. Odell and W. H. Daugherty, eds., J. B. Lippincot Company, Philadelphia, Chapter 1, pp. 1-24.
Zettner, A., 1973, Principles of competitive binding assay (saturation analysis). I. equilibrium techniques, Clin. Chem. 19:699-705.

ENZYME IMMUNOASSAY IN FLOW SYSTEMS WITH ELECTROCHEMICAL AND OTHER DETECTORS

B. Mattiasson

Department of Biotechnology,
Chemical Center, P.O.B. 124, S-221 00 Lund, Sweden

INTRODUCTION

Heterogeneous immunoassays are in their general performance characterized by a sequence of experimental steps. In a rapid ELISA (enzyme linked immunosorbent assay) all the individual steps were initially performed manually. This resulted in a variation between assays of approximately 10%. A critical step was the binding reaction. To reduce the variation in this step, binding was allowed to come to equilibrium. The washing steps as well as the development with enzyme substrate have been automated using specially developed washing apparati. A way to improve the speed of an assay without loosing in accuracy is to expose antigen to antibody under well controlled conditions. Such a goal is achieved by applying enzyme linked immunosorbent assays in continuous flow systems. Flow systems may offer certain advantages since washing, development and reading may also be performed under well controlled conditions.
In the reaction

$$Ag + Ab \rightleftharpoons AbAg$$

complex formation is very much dependent upon the affinity of the antibody for its antigen. Binding strength and binding kinetics may be of great importance.

When applying a continuous flow system with a preparation of immobilized antibodies, the pulse containing antigen will be exposed to the antibodies for only a very short time period (seconds to minutes). In some cases this exposure may be sufficient for approaching equilibrium, in other cases insufficient.

This paper summarizes results of ELISA in combination with continuous flow systems and various detector systems; calorimeter, polarographic oxygen electrode and spectrophotometer.

MATERIALS AND METHODS

The solid support used for immobilizing the antibodies was in most cases Sepharose CL-4B from Pharmacia AB, Uppsala, Sweden. When dealing

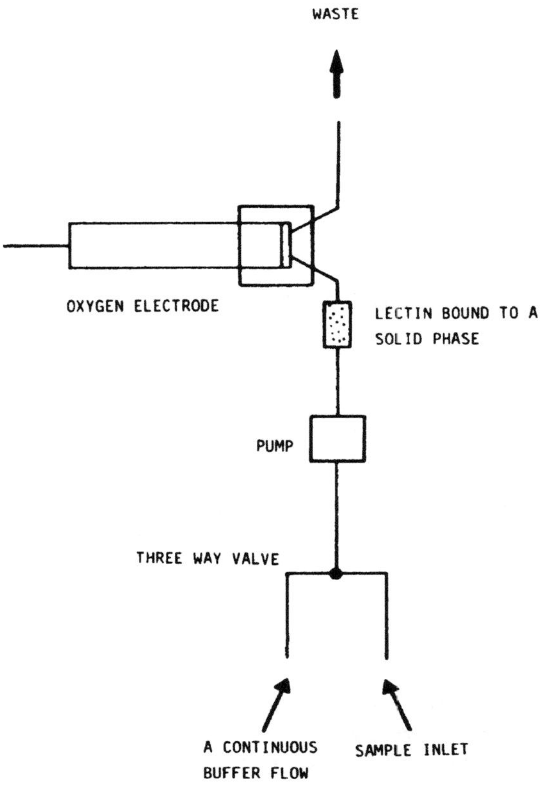

Fig. 1 A schematic representation of the experimental set-up using a column with immobilized lectin. (Reproduced, with permission from Borrebaeck and Mattiasson (1980)).

with polarographic oxygen electrodes a nylon net covering the tip of the electrode was used.

The Sepharose was activated following established procedures using CNBr (Axén, Porath and Ernback 1967). Antibodies were bound to the nylon net using an alkylating method with dimethylsulphate (Hornby, Campbell, Inman and Morris 1974). The flow system used is schematically shown in fig 1.

ASSAY SYSTEMS USED

Calorimetric: The calorimeter used was a simple flow calorimeter usually called "enzyme thermistor" (Mattiasson, Danielsson and Mosbach, 1980; Danielsson, Mattiasson and Mosbach 1981). It consisted of a small thermally insulated column in which the immobilized antibodies were placed. The continuous flow passed a heat exchanger before entering the column from the bottom. At the outlet, which was a thin gold capillary, was placed a thermistor in close contact with the gold. This thermistor was used to register the temperature of the eluate from the antibody column.

Polarographic oxygen electrode: The tip of the electrode was covered by a nylon net carrying the antibodies. To make this useful in a flow system, a small flow chamber was built in such a way that the inlet flushed the central part of the electrode and the outlet was placed to inhibit the development of air bubbles in the flow chamber. Alternatively, a small column was placed in the flow stream and the electrode used to read the oxygen content in the effluent from the column.

Spectrophotometry: A small flow-through cell (80 µl, 10 mm) was used.

Assay cycle:

One characteristic of this kind of assay is the need for reuse of the antibody preparation. If no reuse was possible time would be lost in equilibrations between the assays. This fact is especially clear in the calorimetric assay system. The prerequisite of the antibody reuse raises demands for gentle and efficient dissociation methods.

The general assay cycle is thus: a sample containing a fixed amount of labelled antigen is injected into the continuous flow of buffer with which the system has been equilibrated. The sample pulse is followed by a washing pulse to remove any nonspecifically bound material. After return to equilibration buffer, substrate for the marker enzyme is introduced in a pulse followed by equilibration buffer before the dissociating medium is introduced. After a subsequent reconditioning the system is ready for a new cycle. An assay cycle is schematically illustrated in fig 2.

It is important to investigate the duration of each of these phases of the assay cycle in order to obtain an assay that fulfills the demands concerning specificity, accuracy and reproducibility. Optimization of the steps in order to minimize the duration is of importance since the shorter time for an assay cycle, the more useful is the analysis.

A typical assay

A typical assay consists of first establishing a calibration curve by measuring a series of standard concentrations. The sample to be analyzed is introduced and the concentration of free antigen is read from the calibration curve.

RESULTS

The thermometric detectorsystem was the first to be studied. The thermometric-ELISA, called TELISA, was developed on human serum albumin (Mattiasson, Borrebaeck, Sanfridson and Mosbach 1977). Rabbit antihuman serum albumin antibodies were immobilized to Sepharose CL-4B using the CNBr-coupling method. The column size used was 250-1000 µl

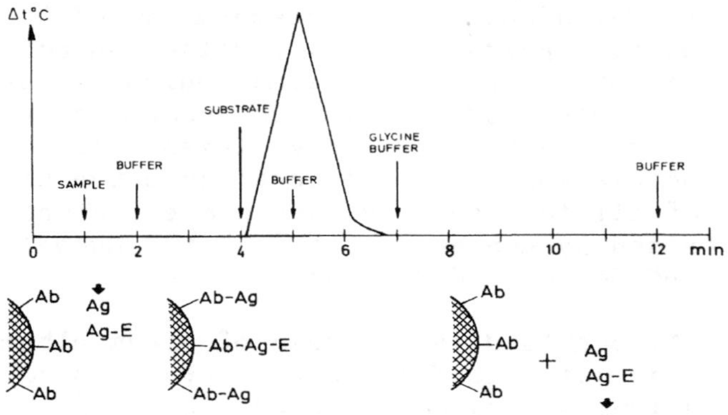

Fig. 2 Diagram of a reaction cycle. The arrows indicate changes in perfusing medium. The schematic peak illustrates the response from the bound enzyme. (Reproduced with permission from Mattiasson, Borrebaeck, Sanfridson and Mosbach 1977).

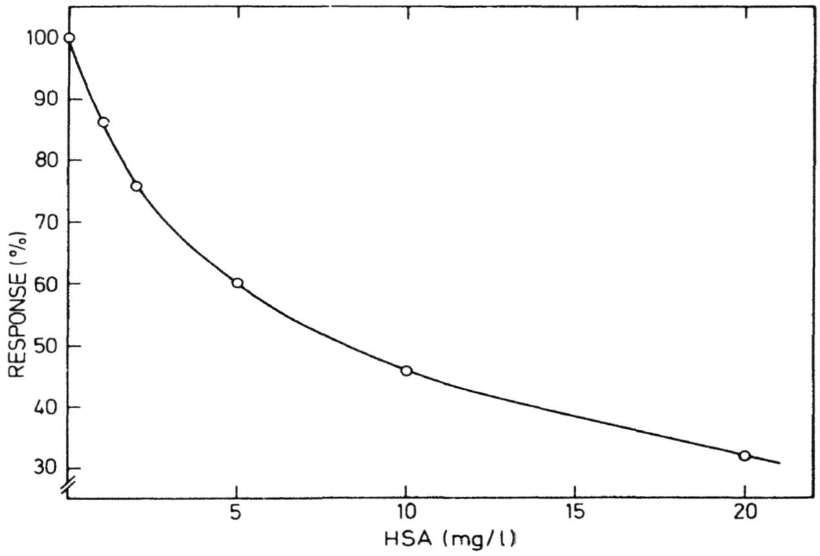

Fig. 3 Calibration curve for HSA using the TELISA-procedure. A 0.4ml column of anti-HSA immobilized to Sepharose was used. The enzyme-antigen conjugate was composed of HSA coupled to catalase using glutaraldehyde. Flow rate used: 0.7 ml/min. Pulse length when sample + conjugate was administered: 1 minute. (Reproduced with permission from Borrebaeck, Börjesson and Mattiasson 1978).

bed volume. When working with the enzyme thermistor and enzymes immobilized to controlled porous glass, flowrates of 1 ml/min were used (Mattiasson, Danielsson and Mosbach 1980; Danielsson, Mattiasson and Mosbach 1981). Controlled porous glass was not useful in this application because of its high nonspecific adsorption. Instead, a chromatographic material with a good record from protein purification was selected. At that time, 1977, Sepharose was an obvious choice. Since Sepharose was used, a flowrate of up to 0.8 ml/min could be used.

A polyclonal antiserum was used in the assay. In order to facilitate the dissociation step in the assay cycle, the antibodies were purified in an affinity chromatographic step before immobilization. The antiserum was adsorbed to a preparation of human serum albumin immobilized to Sepharose CL 4B. Elution was achieved by addition of 0.1 M glycin HCl pH 2.2 - the same medium as used in the dissociation step in the assay cycle. By this treatment the antibodies with highest affinity were removed. An efficient elution is a prerequisite for a reproducible assay. A calibration curve of HSA is shown in figure 3. Assays of clinical samples gave a coefficient of correlation to the values obtained by convention of rocket immunoelectrophoresis of 0.72 (Borrebaeck, Börjesson and Mattiasson 1978).

The sensitivity obtained in the case of serum albumin shows that the assay protocol is suitable for measurements down to 10^{-9} mol/L. If lower concentrations are to be measured, this can be done by increasing the time of exposure of the sample to the antibody-column (Borrebaeck, Börjesson and Mattiasson 1978). Using a time of exposure of 15 min resulted in an improved sensitivity of four orders of magnitude. Samples containing 10^{-13} moles/L could be measured.

The time needed for an assay cycle was 12 min. It should be stated in this context that the optimization was carried out without an access to the flow injection equipment available today. Faster assays can thus be foreseen. In a preliminary study on using FIA-equipment a cycle time of 9 minutes was achieved when using a polyclonal antiserum against a serum protein, transferrin (Larsson, Ohlsson and Mattiasson, unpublished). The use of monoclonal antibodies may improve the conditions substantially.

Gentamycin was quantified in this manner and the results obtained agreed with those obtained with conventional microbiological assays (r=0,98) (Mattiasson, Svensson, Borrebaeck, Jonsson and Kronvall 1978).

A crucial point in all these measurements is the stability of the immobilized antibody-preparation. In the starting situation a limiting amount of antibodies is used. The reading from the binding assay is a reflection of the competitive binding between the native and the enzyme labelled antigen.

The value obtained when no native antigen is present is set at 100%, and all other measurements are compared to this value. When denaturation of the immobilized antibodies takes place, the amplitude of the 100%-reading decreases since less enzyme labelled antigen is bound. All comparisons done with other samples are then made with this value, i.e. the amplitude of the signal is compared to that of the actual reference signal.

The calibration curve is constructed as the relation between signals obtained at different concentrations of free antigen in per cent of that with no native antigen present. This leads to the fact that the calibration curve is valid even after denaturation of parts of the population of immobilized antibodies. The only calibration needed, once the curve is taken up, is to measure the 100%-value. In fig 4 is

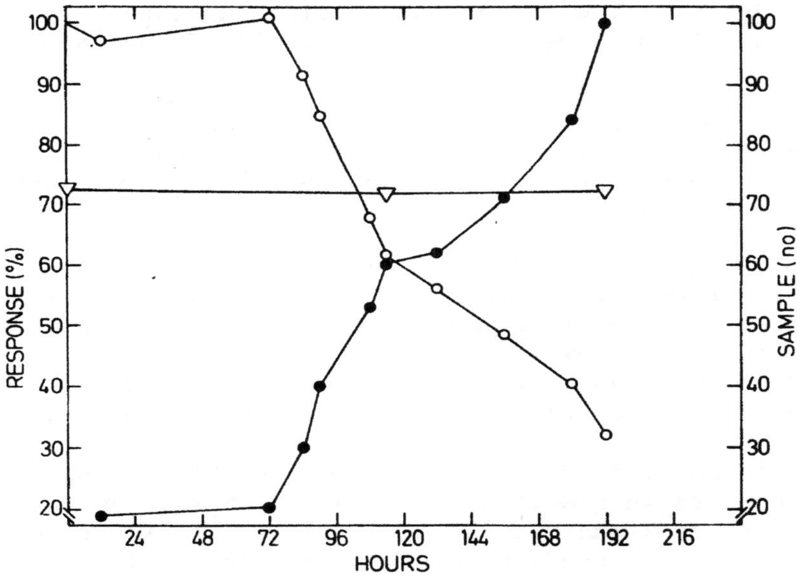

Fig. 4 Reponse stability of an antibody Sepharose CL 4B preparation. (○) peak height obtained when assaying a 100% sample. (●) number of assays run; (▽) peak height in per cent of a preceeding pure aggregate pulse obtained for reference samples containing 40 µg HSA/ml sample. (Reproduced with permission from Borrebaeck, Börjesson and Mattiasson 1978).

shown the outcome from the albumin analyses when run over extended time periods. In this case no specific precautions were taken to keep the antibody- preparation sterile. Others have reported that antibody preparations may be stable for up to 1000 cycles (Dickinson, brochure).

From the figure can be deduced that the binding capacity was reduced to as low as 25% of the starting value without adversely affecting the assay. One way to avoid such dramatic reductions in binding capacity might be to operate with an excess of antibodies in the starting situation. However, this strategy might be successful in enzyme assays but not in binding reactions. With an excess of antibodies no competition will take place and the assay diminishes dramatically in sensitivity. The strategy to follow must be to treat the antibody-preparation in a gentle manner to maintain its binding capacity.

A limiting factor in this context is the dissociation conditions. The use of low pH or chaotropic ions induces partial denaturation of the antibody. If a more gentle procedure is used then a longer operational life time can be anticipated.

This assay strategy of keeping the antibodies in the flow stream and reading the results of the catalytic action of the trapped marker enzymes has also been used in combination with spectrophotometry and polarographic electrodes.

Spectrophotometric assays gave similar results to that obtained with the flow calorimeter. An advantage of using spectrophotometry was that equilibration after an assay was much faster and also that the conventional marker enzymes could be used. Gentamicin analyses were performed with a set up similar to the one discussed above, except for the use of a cheap spectrophotometer and a flow through cuvette (Borrebaeck, Mattiasson and Svensson 1978).

The results from the photometer reading suggests that faster cycling time may be achieved provided suitable FIA and HPLC-equipment is used.

A crucial point in further reducing cycling times may be to select a more rigid column packing allowing higher flow rates. The development in affinity chromatography applying HPLC has led to development of suitable new supports.

When applying the immunoelectrode concept (Mattiasson and Nilsson 1977) the antibodies were either immobilized on a nylon net covering the electrode or on Sepharose packed in a bed preceeding the electrode in the flow stream. The experimental set up is thus very much the same as discussed above. Measurements with a precolumn were more sensitive and also showed a better stability for the antibody preparation.

The continuous flow approach was used in binding assays involving lectin-carbohydrate interactions in conjunction with an oxygen electrode (Borrebaeck and Mattiasson 1980) ($\geqslant$ 15 µl) of concanavalin A bound to Sepharose was used. Glucose oxidase (E.C.1.1.3.4), which is a glycoprotein, was used as an enzyme labelled carbohydrate and allowed to compete with various sugars for the binding sites of Con A. When a fixed concentration of the sugars was used, then plotting the results of the competitive binding against reported binding constants gave a straight line. Such a simple procedure could thus be used to quickly estimate binding constants.

FUTURE APPLICATIONS

The convenient handeling and the fast analytical procedure makes this approach to enzyme immunoassays an attractive candidate when a small volume of samples have to be measured quickly. This may happen in clinical chemistry, in e.g. drug monitoring and in process control. The clinical field is very competitive and already contains very good methods e.g. the homogenous immunoassays for haptens (Ullman and Maggio 1980). The use of immunochemical analyses for process control is just in its infancy (Mattiasson 1984). The potential is great and the speed of analyses is in most cases sufficient. Furthermore, since e.g. fermentation processes may lead to rather high concentrations of products to be monitored an even shorter assay cycle may be used.

Acknowledgements: Parts of this work was supported by The National Swedish Board for Technical Development.

REFERENCES

Axen, R., Porath, J. and Ernback, S., 1967, Chemical Coupling of Peptides and Proteins to Polysaccharides by Means of Cyanogen Halides, Nature 214:1302.

Becton Dickinson Inc., Broschure on ARIA.

Borrebaeck, C., Börjesson, C. and Mattiasson, B., 1978, Thermometric Enzyme Linked Immunosorbent Assay in Continuous Flow System: Optimizaton and Evaluation Using Human Serum Albumin as a Model System. Clin. Chim. Acta 86:267.

Borrebaeck, C. and Mattiasson, B., 1980, Lectin-Carbohydrate Interactions Studied by a Competitive Enzyme Inhibition Assay. Anal. Biochem. 107:446.

Borrebaeck, C., Mattiasson, B. and Svensson, K., 1978, A rapid non-equilibrium enzyme immunoassay for determining serum gentamicin, in: "Enzyme labelled immunoassays of hormons and drugs" S.B. Pal, ed., pp 15-27, Walter de Gruyter & Co., Berlin.

Danielsson, B., Mattiasson, B. and Mosbach, K., 1981, Enzyme Thermistor Devices and Their Analytical Applications, in: "Appl. Biochem. Bioeng." Vol 3, pp 97-143. L.B. Wingard, Jr., E. Katchalski-Katzir and L. Goldstein, eds, Academic Press, New York.

Hornby, W.E., Campbell, J., Inman, D.J. and Morris, D.L., 1974, Preparation of immobilized enzymes for application in automated analysis, in: "Enzyme Engineering " Vol 2, pp 401-407, E.K. Pye and L.B.Wingard,Jr., ed., Plenum Press, New York.

Mattiasson, B., 1984, Immunochemical assays for process control: potentials and limitations. Trends Anal. Chem. 3:245.

Mattiasson, B., Borrebaeck, C.A.K., Sanfridson, B. and Mosbach, K, 1977, Thermometric Enzyme Linked Immunosorbent Assay: TELISA, Biochim. Biophys. Acta 483:221.

Mattiasson, B., Danielsson, B. and Mosbach, K., 1980, Applications of the enzyme thermistor in analysis and process control, in: "Food Process Engineering" vol 2, pp 59-68, P.Linko and J. Larinkari, eds., Appl. Sci. Publ., London.

Mattiasson,B. and Nilsson, H., 1977, An enzyme immunoelectrode, FEBS Lett. 78:251.

Mattiasson, B., Svensson, K., Borrebaeck, C., Jonsson, S. and Kronvall, G., 1978, Non-equilibrium enzyme immunoassay of gentamicin. Clin. Chem. 24:1770.

Ullman, E.F. and Maggio, E.T., 1980, Principles of Homogeneous Enzyme Immunoassay, in: "Enzyme-Immunoassay" pp 105-134, E.T. Maggio, ed., CRC-Press, Boca Raton, Fl. USA.

USE OF IMMUNOCHEMICAL SEPARATION AND ELECTROCHEMICAL DETECTION IN FLOW INJECTION ANALYSIS OF LACTIC DEHYDROGENASE

Shia S. Kuan* and George G. Guilbault

Department of Chemistry
University of New Orleans
New Orleans, Louisiana 70148

INTRODUCTION

Health is one of the most important factors for a high quality of life. People have been trying their best to maintain a healthy condition during their life span. Unfortunately, we sometimes expose ourself to an unfavorable environment that could induce organ disorders and cause diseases. In the living system, enzymes constitute the machinery of the cell, they are sites of functional importance in regulating metabolism. Change in enzyme activity indicate an abnormal metabolism inside the system and can be used to predict an organ disorder. Early detection of such disorder in a screening program could prevent the onset of a large number of diseases.

Some enzymes are known to exist in multiple forms, the so-called isoenzymes, which catalyze the same biochemical reactions. In general, isoenzymes in blood originate from different tissues; hence, they are organ specific. The level of each isoenzyme in blood or serum often provides a useful diagnostic indicator of tissue damage.

Blood lactate dehydrogenase (LD) is a tetramer composed of H and M subunits and exists in five different forms.

LD catalyzes an oxidation of lactate to pyruvate in the presence of NAD^+

$$\text{Lactate} + NAD^+ \underset{\text{low pH}}{\overset{\text{LD: high pH}}{\rightleftharpoons}} \text{Pyruvate} + NADH + H^+ \qquad (1)$$

The fact that elevation of blood LD is associated with myocardial infarction makes it a useful marker for diagnosis by physicians for confirmation of such cases (Galen, 1975).

However, the total LD is not highly specific, and, hence, routinely the "flipped" LD-1 vs. LD-2 ratio, indicating the relative increase in LD-1 against LD-2, is used for diagnostic purposes.

*Food and Drug Administration, 4298 Elysian Fields Avenue, New Orleans, Louisiana 70122

However, this method is not reliable since this symtom occurs usually two or three days following the infarct.

Recently, many papers (Toyoda, et al., 1985) have described an isoenzyme assay of LD-1 or LD-1/total LD ratio in serum which can provide clinically more accurate information for the early detection of acute myocardial infarction.

Methods for separation of LD isoenzymes have utilized differential heat stability (Wilkinson, 1970), substrate analogs, and differential inhibitors (Toyoda, et al., 1985). The most popular instrumental method commonly used in the clinical laboratory is electrophoresis. All methods mentioned above are tedious and time-consuming. An anion-exchange chromatographic method offers some advantages over electrophoresis, but it is time-consuming as well, and inaccurate due to poor resolution. On the other hand, immunochemical methods are very simple, rapid, easy to use and accurate. Several methods, using colorimetric, fluorometric and chemiluminescence detector, have been employed for the determination of LD. Smith and Olson (1975) presented a differential amperometric measurement of LD activity using tubular carbon electrodes and Bundschedler's Green as a mediator, following the current or chromogenicity resulting from the reduction of the dye. In order to handle large number of samples, a continuous measurement in a flow system is a more attractive alternative for large clinical laboratories. A stop-flow measurement of LD in a flow injection system using a spectrophotometer as detector was presented.

Electrochemical detection in a flow system has a number of advantages, including high sensitivity, shorter response time, continuous generation of electrode surface and elimination of problems of a reference electrode. We have combined the benefits from the above methods and developed a novel procedure for the measurement of LD-1 and LD-1/total LD ratio, in serum, using a flow injection system with electrochemical detection after immunochemical separation.

EXPERIMENTAL

1. Apparatus

The flow injection system (Fig. 1) included a peristaltic pump (Model 403, Scientific Industries, Bohemia, NY), an injector (Valco Instruments Co., Houston, TX), teflon tubing (0.038 inch ID and 1/16 inch OD) and a thin-layer transducer (TL-10A, Bioanalytical System, West Lafayette, IN). The three electrode terminals are connected to a polarographic analyzer (PAR 174A, Princeton Applied Research Co., Princeton, NY) and the output of the Y axis is connected to a strip chart recorder (Servograph REC 61, Radiometer, Copenhagen, Denmark). The thin-layer transducer has a thin layer channel with a teflon gasket and contains a glassy carbon electrode, a platinum electrode and Ag/AgCl reference electrode.

Materials and Method

The Isomune-LD kit (Roche Diagnostics, Nutley, NJ) which consists of an goat anti-LD-5 antibody and a suspension of polymer-bound anti-goat immunoglobulin second antibody was used for the separation of LD.

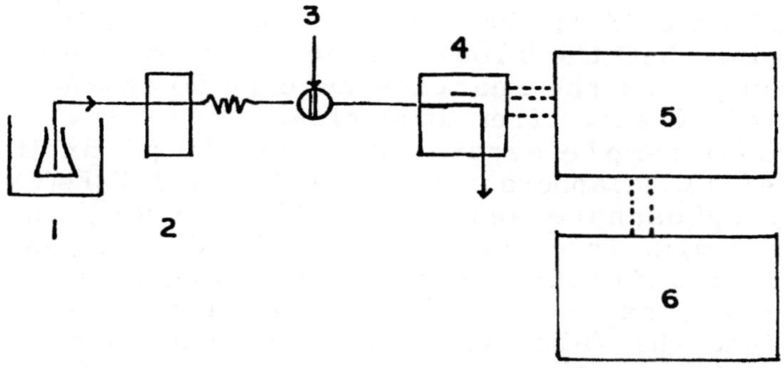

Fig. 1. Flow Injection System with Electrochemical Detector

1. Flow solution in a water bath at 37°C
2. Peristaltic pump
3. Injector
4. Thin-layer transducer
5. Polarographic analyzer, PAR 174 A
6. Chart recorder

Li-lactate, Na-pyruvate, diaphorase, nicotinamide adenine dinucleotide (NAD) and NADH were obtained from Sigma Chemical Co., St. Louis, MO. Other chemicals were reagent-grade quality.

Assay Procedure

Transfer 0.2 ml of diluted LD control (Cardiotrol-LD, Roche Diagnostics) or serum sample into a 10 x 75 mm test tube, add 0.05 ml of anti-LD-5, incubate by mixing on a Volter Genie Mixer (American Hospital Supply Co., Evanston, IL) at room temperature for 5 min. Then add 0.2 ml of the second antibody, mix the solution and allow to incubate for 5 min. Finally, centrifuge all tubes (1000 x g) for 5 min; save the supernatant containing LD-1 for assay.

Measure the activity of LD-1 and total LD by the forward reaction (Li-lactate to pyruvate) with potassium ferricyanide as a mediator, and diaphorase as catalyst. Pretreat the electrode with a potential of first +1.3 V then - 1.3 V, each for two min., in a flow solution of 0.05 M pyrophosphate and 0.05M potassium chloride (pH 9.0). Then fix the potential at 0.07 V; the residual current levels off below 5 nA after 10-20 min. Set the flow rate at 0.8 ml/min using a peristaltic pump. The length of the reaction tube is 50 cm and thus the sample can be detected 30 sec. after injection.

To the LD-1 sample separated above (50 µl or 100 µl), add Li-lactate (1 mg/ml), diaphorase (0.1 mg/ml), and $K_3Fe(CN)_6$ (0.05 mg/ml) in 0.05 M pyrophosphate and 0.05 M KCl, pH 9.0, in a small vial. Incubate for 5 min. in a water bath at 37°C. Add NAD^+ (1 mg/ml) to the solution to initiate the reaction. After two, five and eight minutes, sample the reaction products with a syringe and inject immediately into the Valco valve with a 50 µl sample loop. Measure the difference in peak height, and calculate the LD-1 activity from a calibration curve.

RESULTS AND DISCUSSION

Potential of Electrode

The determination of LD activity by an electrochemical method is effected by either direct measurement of NADH (Thomas and Christian, 1975) or by a coupled reaction with mediator (Smith and Olson, 1974). The NADH has half peak potentials of 0.65 V and 0.38 V vs. S.C.E. at a glass carbon electrode. When an applied potential is used at either of these potentials, the electrode exhibited a serious fouling problem, especially for high concentrations of NADH. To eliminate or minimize this problem, potassium ferricyanide can be used as an electron transfer mediator between NADH and the electrode (Thomas and Christian, 1976). Since the potential of ferricyanide is 0.15 V vs. S.C.E., it is possible to decrease the applied potential during analysis. Both Fe^{III} and Fe^{II} rapidly exchange electrons with the working electrode. The rate of current change by the concentration of Fe^{II} or Fe^{III} can be measured at two regions, either above or below 0.2 V. In the signal below 0.2 V, the cathodic current is decreased by the reduction of the Fe^{III} species. On the other hand, at above 0.2 V, the anionic current is increased by the oxidation of Fe^{II} species. Fig. 2 shows the relative activity of LD which is obtained at various applied potentials; a potential at 0.07 vs. Ag/AgCl was chosen because at this potential maximal LD activity resulted and the

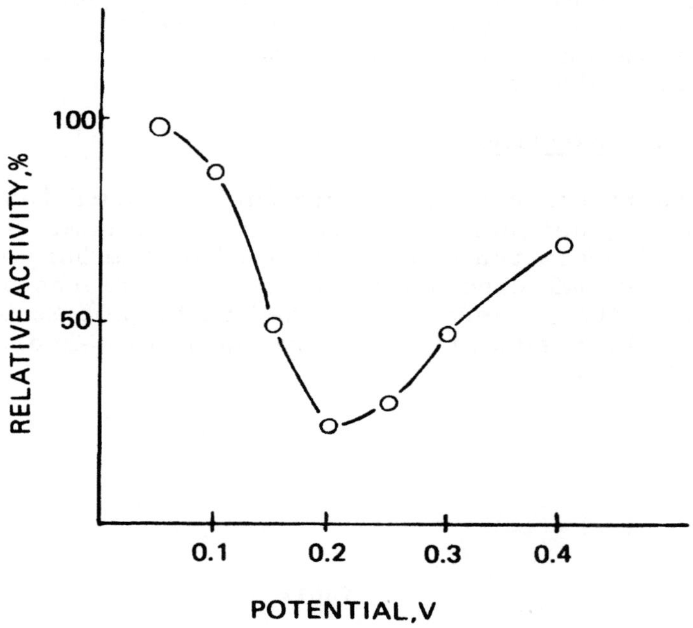

Fig. 2. Relative Activity of LD at Various Applied Potentials

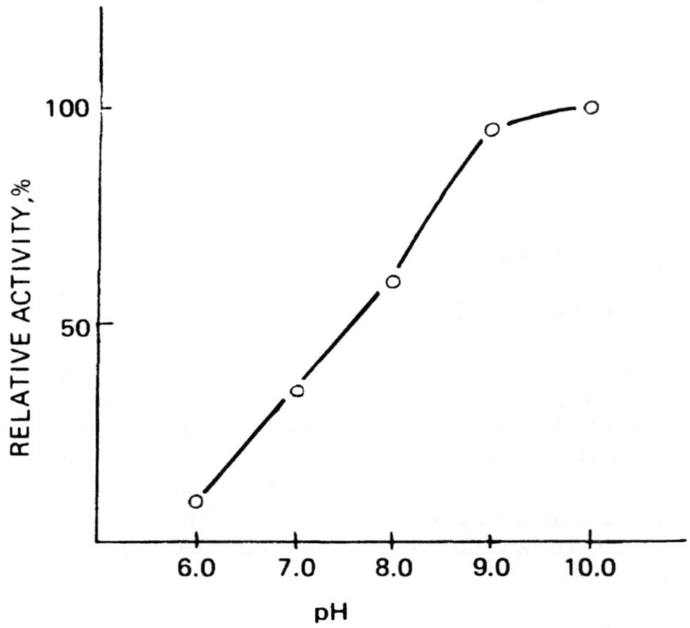

Fig. 3. Relative Activity of LD at Various pH's

interferences of protein or other biological constitutents in serum sample were minimized. A small amount of Fe^{III} (1.5 μM) was mixed into the flow solution to keep the surface of electrode clean in order to obtain reproducible results.

Optimal Reaction Conditions

The forward reaction of LD determination occurs best at high pH's (Fig. 3). However, neither the reagent nor the sample is stable at very high pH, and hence the experimental pH of the buffer solution was set at pH 9.0. Optimal concentrations for KCl, diaphorase, NAD^+, Li-lactate, and $K_3Fe(CN)_6$, were determined to be 0.05 M, 0.45 U, and 1.0 mg/ml, 0.05 mg/ml, respectively. The optimum temperature for the reaction was 37 ±1°C.

Table 1

Precision Study of LD-1 Assay

Run to Run	Normal Serum	Abnormal Serum
Ave. (6)	22.3 U/L	201.7 U/L
S.D	2.5	7.0
C.V	11.2%	3.5%
Day to Day (3 days)		
Ave.	22.2 U/L	202.4 U/L
S.D	1.9	6.5
C.V	8.7%	3.2%

Performance Characteristics

a. Calibration Curve of LD-1

Pooled serum (normal activity of LD-1) and LD control serum (abnormal activity of LD-1) were mixed to prepare the working standard samples. These working standard samples were first handled by the immunological separation as mentioned in the previous section. Then the activity of the separated LD-1 in the sample was measured by both the standard spectrophotometric method and by the new electrochemical method. Linear relationships were obtained up to 202 U/L of LD-1 and to 385 U/L of total LD.

b. Precision Study

The precision of this assay was evaluated by assaying twelve serum samples, six normal and 6 elevated activities of LD-1. The results of run-to-run and day-to-day (3 days) assays are summarized in Table 1, showing the standard deviations (S.D.) and coefficients of variation (C.V.)

Table 2

Comparison Study of LD-1 and LD-1/Total LD

Sample #	**Hospital** Electrophoresis -UV			**Immunoseparation -UV**			**Immunoseparation** - Electrochemical		
	Total	LD-1	LD-1/ Total	Total	LD-1	LD-1/ Total	Total	LD-1	LD-1/ Total
	U/L	U/L	%	U/L	U/L	%	U/L	U/L	%
1	386	108	28	477	189	40	373	87	26
2	216	54	25	240	81	34	216	46	24
3	175	39	22	206	57	28	193	28	16
4	166	35	21	187	49	26	198	46	26
5	130	36	28	126	35	28	193	41	24
6	125	36	29	157	59	38	138	37	30
7	559	229	41	975	558	57	570	304	60
8	380	61	16	392	61	16	221	37	19
9	262	71	27	391	117	30	212	46	24
10	235	52	22	333	88	26	170	46	30

c. **Comparison Study**

Ten fresh serum samples obtained from Touro Infirmary, New Orleans, LA were analyzed for both the total LD and LD-1 by the immunoseparation -UV method and the electrochemical method. The results obtained were compared to those obtained by the electrophoresis separation -UV method used at the Hospital (see Table 2). The summary of linear regression analysis is presented in Table 3. The test results showed some discrepancy between the immunoseparation-UV method and the hospital electrophoresis method. However, the agreement was better between the immunoseparation-electrochemical method proposed herein and the hospital method, thus indicating the potential usefulness of this method.

Table 3

Linear Regression Analysis of a Comparison Study

1. Activity of LD-1

$\bar{Y}_1 = 0.375\ \bar{X}_1 + 23.5$, $r_1 = 0.992$

$\bar{Y}_2 = 0.69\ \bar{X}_2 + 22.6$ $r_2 = 0.974$

2. Ratio, LD-1/Total LD

$\bar{Y}_1 = 0.57\ \bar{X}_1 + 7.5$ $r_1 = 0.947$

$\bar{Y}_2 = 0.47\ \bar{X}_2 + 12.8$ $r_2 = 0.846$

$\bar{Y}_1$, $\bar{Y}_2$: Hospital

$\bar{X}_1$: U.V. Method

$\bar{X}_2$: Electrochemical Method

DISCUSSION

The determination of isoenzymes has become a very important technique in clinical laboratory because it offers more specific and reliable information for diagnosis. Criss has reported that changes in LD isoenzymes are frequently observed in cancer patients. Dawson found significant increase in the M subunits of LD in human tumors. Increased levels of the M subunits also have been observed in breast tumor, uterine carcinoma, maligant colon, and other carcinomas. It seems clear in the future that greater effort will be made to determine the correlation between the activity of LD isoenzymes and, not only the myocardial infarction, but also the carcinomas, and that more screening tests for blood LD isoenzymes will be performed in the clinical laboratory. The development of a reliable automated procedure, which can handle large number of samples within a short time at minimal cost, is urgently needed. The method developed in this laboratory meets all the requirements just stated because the procedure is rapid, simple, accurate, and easy to automate and manipulate. Based on the same principle, a procedure for the measurement of other LD isoenzymes, eg. LD-5 (M_4) can be developed using anti-LD-H_4 antibody for the inhibition and separation of LD isoenzyme containing H subunits.

REFERENCES

Galen, R. S., Gambino, S. R., In beyond Normality: The predictive value and efficiency of medical diagnosis, John Wiley & Sons, Inc., N. Y. 1975, pp. 87.

Smith, M. D., and Olson, C. L., 1974, Differential amperometric determination of serum lactate dehydrogenase activity using Binschedler's Green, Anal. Chem., 46:1544-1548.

Thomas, L. C. and Christian, G. D., 1975, Voltammetric measurement of reduced nicotinamide-adenine nucleotides and application to amperometric measurement of enzyme reactions, Anal. Chim. Acta, 78:271-275.

Thomas, L. C. and Christian, G. D., 1976, Amperometric measurement of hexacyanoferrate(III) coupled dehydrogenase reactions, Anal. Chim. Acta, 82:265-270.

Toyoda, T., Kuan, S. S. and Guilbault, G. G., 1985, Determination of lactate dehydrogenase isoenzyme (LD-1) using flow injection analysis with electrochemical detection after immunochemical separation, Anal. Letters, 18:345-355.

Wilkinson, J.H., "Isoenzymes", J. P. Lippencott, Co., Philadelphia, PA., 1970, pp. 134.

RAPID SUBPICOMOLE ENZYME IMMUNOASSAY BY FLOW INJECTION WITH AMPEROMETRIC DETECTION

W. Uditha de Alwis and George S. Wilson

Department of Chemistry
University of Arizona
Tucson, Arizona U.S.A. 85721

INTRODUCTION

The immunoassay plays a particularly important role in the detection of trace levels of components in complex biological fluids. High sensitivity characteristic of such techniques as mass spectrometry is achievable at a much lower level of complexity and cost. We have concentrated our attention on the development of heterogeneous enzyme immunoassays where detection limits in the sub-nanogram per milliliter range are required. In addition, it is assumed that the total amount of sample available is limited to 5-50 microliters and that the total analysis time should be short (10-20 minutes). It was also desired to configure the immunoassay so that trace levels of analytes could be handled without contamination or sample loss and the overall methodology readily automated.

The flow injection analysis mode (FIA) with electrochemical detection using a thin-layer amperometric detector of the type commonly employed in HPLC has proven to effectively meet the above requirements. The heart of the system is the immunoreactor, a miniature packed column containing a support material to which an antibody has been covalently attached. The success of the assay depends critically upon the development of a stable, reproducible immunosorbent surface which is free from non-specific adsorption effects. The FIA mode significantly aids in the achievement of this goal by permitting highly reproducible conditioning of the immunosorbent surface both prior to and following the analysis step. Also critical to the assay is the availability of well-characterized conjugates (antibody-enzyme or antigen-enzyme) possessing both high enzymatic and immunological activity (Sittampalam, Wilson and Byers, 1982).

The FIA Immunoassay configuration is shown schematically in Figure 1. An HPLC pump is used to supply the mobile phase or assay buffer which is employed during the introduction of the analyte and any subsequent immunological or enzymatic reactions. An acidic or disruption buffer is used to dissociate any antibody-antigen complexes formed during the analysis step thus resulting in the regeneration of

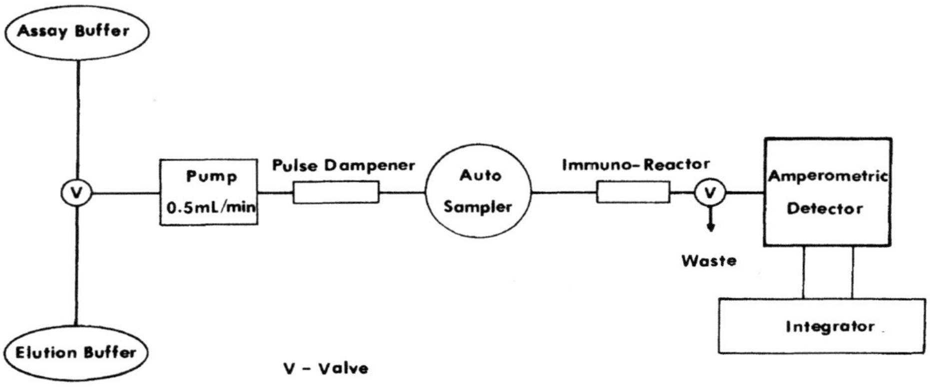

Figure 1 - Schematic of FIA Immunoassay Apparatus

the immunosorbent surface. Samples and appropriate reagents are introduced in a carefully timed sequence using an automatic sample injector. At the flow rates normally employed (0.5 mL/min) the residence time of the sample in the immunoreactor is very short (typically 6-8 sec.). We have demonstrated under FIA conditions that the antibody-antigen reaction in the immunoreactor is extremely fast so that even at very low analyte concentrations the reaction is more than 90% complete in this very short interaction time (Sportsman, Liddil and Wilson, 1983). Thus, contrary to the usual practice in heterogeneous immunoassays, long incubation times are not necessary and in some cases may be detrimental to both the precision and sensitivity of the assay. Residence time in the reactor can be easily defined by controlling the mobile phase flow rate so that complete reaction is not even required. In the assays to be described below, the final measurement step involves the determination of the activity of the enzyme label which is immobilized in the immunoreactor as a consequence of the assay sequence. At this point the analyte has been separated from the sample matrix by the immunosorbent so that the determination of the electroactive product of the enzymatic reaction can be made without interference. Until this point in the assay the mobile phase is diverted around the electrochemical detector thus avoiding possible contamination. As an additional precaution, the platinum electrode is covered with a polymeric film which protects the sensor from protein fouling (Sittampalam and Wilson, 1983).

It is possible to describe the behavior of the immunoreactor accroding to a very simple theoretical model (Sportsman, Liddil, and Wilson, 1983). Owing to the very large value of the equilibrium constant for the antibody-antigen reaction (log K ranging from 6 to 11), the number of theoretical plates in the microreactor column is very small. Thus the usual models for immunoaffinity chromatography do not apply and it is possible to consider the entire reactor as a "beaker" containing a certain concentration of binding sites. Using this model, one can evaluate the number of binding sites in the reactor and also obtain the value of the equilibrium constant for the immunological reaction. It is possible to demonstrate that the value of this constant is not altered as a result of the covalent attachment of the antibody to the support material. This information is very valuable in optimizing conditions within the reactor particularly for competitive binding assays.

We report here a sandwich ELISA (enzyme linked immunosorbent assay) for mouse anti-bovine IgG which serves as a model system to demonstrate the feasibility of the FIA mode for rapid immunoassays.

MATERIALS AND METHODS

Monoclonal mouse anti-bovine IgG and polyclonal goat anti-mouse IgG were generously donated by American Qualex International, La Mirada, CA. The monoclonal antibody was immunoaffinity purified before use while the goat antibody was used without purification. Crystalline bovine IgG (BIgG) and glucose oxidase Type X were obtained from Sigma Chemical Co., St. Louis, MO. Reactigel-6X was obtained from Pierce Chemical Co., Rockford, IL. All other chemicals were Reagent Grade unless otherwise specified. Buffers used for the mobile phase were prepared from doubly distilled water filtered through a 0.45 μm filter, and were then protected from dust.

The afinity purification of the monoclonal anti-bovine IgG, the preparation of the microreactor containing Reactigel-6X immobilized BIgG, the preparation of the goat anti-mouse IgG-glucose oxidase conjugates, and the experimental apparatus have been described previously (de Alwis and Wilson, 1985).

RESULTS AND DISCUSSION

The sequence of events for the sandwich assay is illustrated schematically in Figure 2. The analyte, anti-bovine IgG in this case, is introduced into the microreactor which contains an approximately 10,000-100,000 fold excess of binding sites. This creates a favorable situation for the essentially quantitative removal of the analyte from the sample matrix. We have demonstrated that high concentrations of protein (60 g/L) typical of undiluted serum do not significantly affect the efficiency of this process. The second step in the assay is the introduction of the goat anti-mouse-glucose oxidase conjugate. This second immunological reaction is slower than the first and injection of conjugate leads to binding on 60-75% of the sites on the analyte. A second injection increases this figure to 85-95%. By dividing the total conjugate introduced into two injections, greater reaction efficiency is achieved. This leads to an improved linear dynamic range for the assay particularly at the high end of the analyte concentration range.

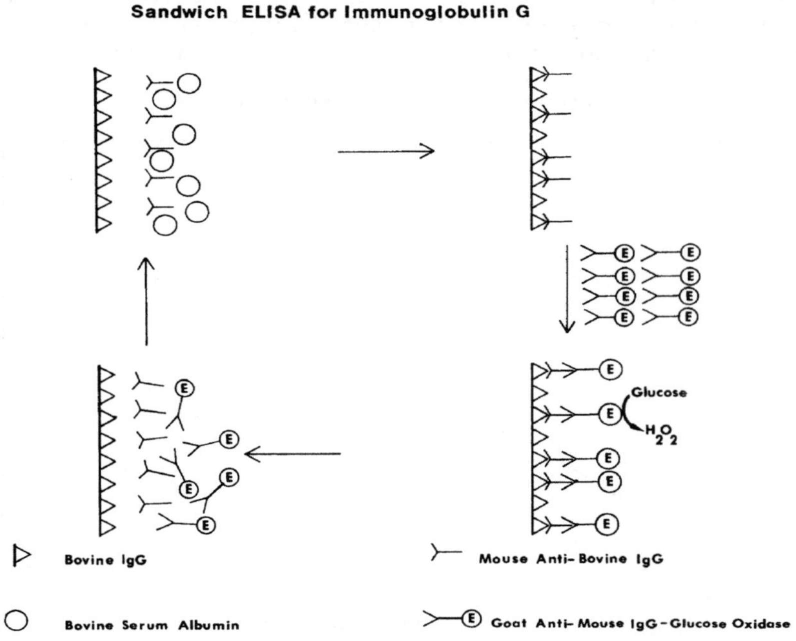

Figure 2 - Schematic of Sandwich ELISA for IgG

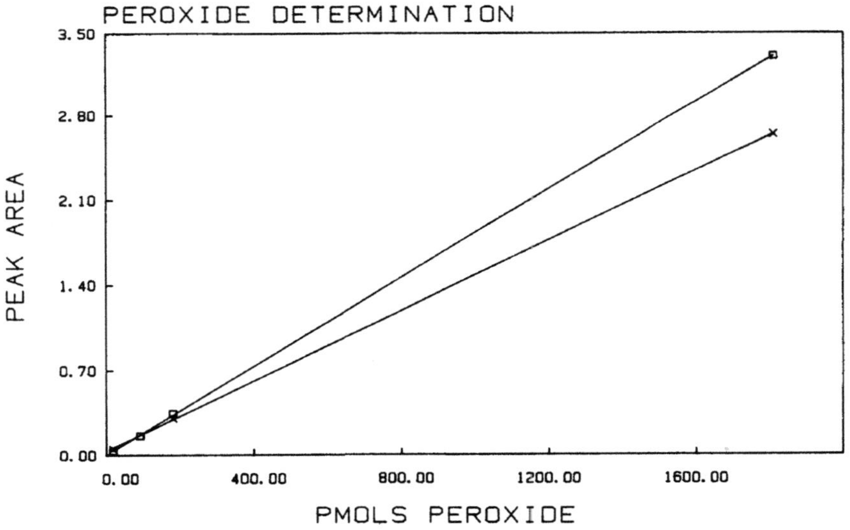

Figure 3 - Calibration Curve for Electrochemical Detection of Peroxide. Effect of Protective Membrane. Upper Curve - 50 μM cellulose Acetate Membrane. Lower Curve - "Bare" Pt Electrode.

The final step in the assay involves the introduction of glucose in an oxygen-containing buffer. The enzyme catalyzed reaction results in the production of hydrogen peroxide which is detected electrochemically. Because the concentrations of glucose injected are very high relative to oxygen, it is possible for the enzyme turnover to be limited by the oxygen supply. The use of oxygen saturated buffers improves both sensitivity and precision. The injection of glucose and the subsequent production of peroxide appears to release non-specifically bound conjugate. Subsequent injections of glucose are much more reproducible but lead to a smaller response. This observation has led to the use of at least two glucose injections.

Figure 3 shows the FIA response to the injection of known amounts of hydrogen peroxide. The upper curve corresponds to the response of the platinum indicator electrode which is "bare" versus the lower curve for the same electrode covered with a thin (approx. 50 μm cellulose acetate membrane). It should be noted that the response is measured

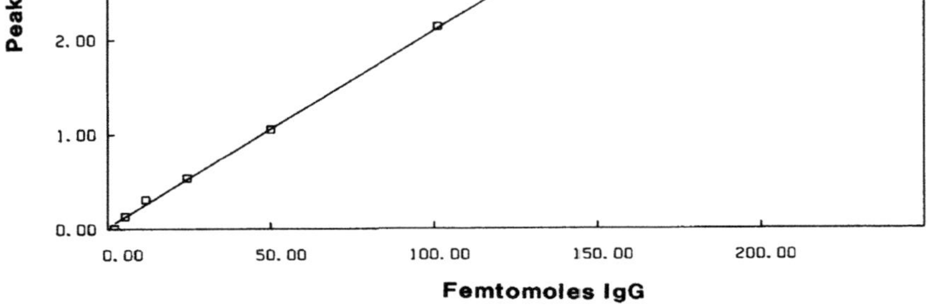

Figure 4 - Calibration Curve for Sandwich Assay

as the area under the amperometric detector peak. Under the conditions of our experiments the detection limit for peroxide is about 10 pmoles for a S/N ratio of 50 and this corresponds to a maximum peak current of about 10 nA. This detection limit could be lowered by nearly an order of magnitude by improving the pulse dampening characteristics of the pump.

Figure 4 shows a typical calibration curve for the sandwich assay. Over the range of about 2-200 fmol a correlation coefficient of greater than 0.99 is obtained. Both the precision and sensitivity is unaffected by the addition of high levels of protein as noted above. In the presence of control serum, the detection limit is elevated to 1 pmol. It seems likely that this is due to a cross reaction between the conjugate and some component which has been non-specifically adsorbed. The cross-reaction effects may be alleviated by screening the conjugate, which is polyclonal, to remove the cross reacting fraction or by diluting the serum with bovine serum albumin. At any rate, the precision of the measurements in the femtomole range is very good, with 2-4% relative standard deviation and accuracy of about 5% at the lower concentration levels and 1-2% at the upper end (de Alwis and Wilson, 1985). The very high precision of these measurements is specifically attributable to careful conditioning of the immunosorbent surface and to the very carefully timed exposure of reagents and sample under very favorable mass transfer conditions. After the disruption buffer (pH 2) is introduced as the mobile phase, the pH is subsequently raised to 6.8. Within about two minutes the pH of the effluent reflects this change, consistent with the flow rate and dead volume of the system. The poor precision of measurements made immediately following this step is indicative of an immunosorbent surface which has not been propertly conditioned. If one waits 10 minutes, a precision of 2-4% is obtained whereas shortening this time to 5 minutes yields 5-10% precision. It is not yet known whether this rather slow conditioning process is due to slow surface pH changes or to slow renaturation of the immobilized protein. This question is currently being examined in greater detail in our laboratories.

The significant advantage of the FIA configuration as we have employed it is the greatly improved stability and reproducibility obtained. This is attributable first to the stability of the immunosorbent prepared from carbonyldiimidazole-activated cross-linked agarose (Reactigel-6X). A stable carbamate bond is formed between the support and the immobilized protein without creating residual charge on the support surface. Repetitive cycling of the immunosorbent between the assay and disruption buffers serves to remove non-covalently immobilized protein. We have recently increased protein loading to 8-10 milligram/mL gel. The microreactor may be used for over 500 assays over a period of at least three months. The observed loss of activity is about 2-3% per 50 cycles. An important source of immunoreactor failure is bacterial contamination. Even if the bacteria do not destroy the immobilized protein, their cell surfaces provide sites for non-specific adsorption. To prevent such difficulties the buffer solutions are filtered and 0.01% sodium azide is added. This unfortunately causes an increase in the background current of the electrochemical detector but, because it is constant, can be easily subtracted out.

The quality of the goat anti-mouse-glucose oxidase conjugate is also extremely important. We have optimized the p-benzoquinone coupling procedure first described by Ternynck and Avrameas (1976), to yield a reproducible and stable product. Conjugates thus prepared can be stored at concentrations of 2-3 mg/mL for periods up to 6 months at

4°C in PBS pH 7.4 with 0.01% sodium azide added. No other stabilizers are necessary. Eventually precipitation occurs, however, the supernatant can still be used if the concentration of soluble conjugate is high enough.

As noted above, the reproducibility of reaction conditions make the FIA immunoassay far superior to techniques utilizing polyethylene tubes or other similar surfaces. These surfaces are extremely difficult to reproduce and are time consuming to prepare. In addition, the incubation and washing steps are not particularly easy to automate. However, if one has a large number of samples of analyze, some measure of efficiency is achieved because the incubation and washing steps can, in effect, be carried out in parallel. Thus in the case of the FIA immunoassay it is appropriate to ask what limitations are imposed on sample throughput. We have already been able to reduce the 30 minute analysis time originally reported (de Alwis and Wilson, 1985) to about 12 minutes. This has been accomplished using improved immobilization procedures and support materials (de Alwis and Wilson, 1986). In spite of the fact that the reaction rates are very high, the residence time of the reactants in the reactor is still very short. The apparent "slow" reaction of the conjugate with the immobilized analyte (anti-BIgG) results from the fact that the effective microreactor column length is not the geometric 4 cm. but rather only about 1 cm. When the analyte is introduced into the reactor it sees an overwhelming excess of immobilized reactant at the entrance to the reactor. Complete removal of the analyte occurs, confining it to this region. Thus the residence time for reaction of the conjugate is reduced to about 1-2 sec. This problem can be corrected simply by lowering the concentration of reactive sites in the reactor. This will spread the immobilized analyte out. Another alternative to increasing residence time is to lower the flow rate of the assay buffer. We have generally found this approach not to be especially effective. It increases both non-specific interactions and analysis time. Improved conjugate purity and biological activity has facilitated higher reaction rates.

There is no reason why the FIA immunoassay cannot be configured with parallel microreactor columns, thus increasing sample throughput. The reproducibility of these reactors from day to day is such that a one or perhaps two-point calibration curve would be sufficient. Immunosorbents prepared in a single batch and distributed in several columns could be expected to give similar behavior. Ultimately the detection limit for this enzyme immunoassay will be limited by the amount of peroxide produced during the residence time of the glucose substrate in the reactor. In a multiple column system with column switching there is no reason why flow could not be stopped to allow enzymatic reaction product to accumulate. It is important to have pulse-free pumping during the electrochemical detection step as the response is especially sensitive to changes in flow rate. Otherwise the pumping requirements are minimal as the pressure drop across the immunoreactor normally does not exceed 150 psi.

Finally it might be mentioned that the FIA immunoassay technique can be extended to competitive binding assays. In this case the exact concentration of binding sites in the column is understandably much more critical, but assays in the 10 fmol range have been developed (de Alwis and Wilson, 1986).

SUMMARY

The FIA immunoassay offers a convenient, rapid, and reproducible means for analyzing samples in the sub-picomole range. Many of the difficulties associated with the heterogeneous assay are avoided by careful control and timing of reaction conditions. Because the analyte has been physically separated from the sample matrix prior to the final measurement step, background interference in the electrochemical determination is minimal. If prior sample preparation is required, this can be easily integrated into the system at the sample injection step.

ACKNOWLEDGEMENTS

This work was supported in part by National Institutes of Health Grant AM30718. We thank American Qualex International for the generous donation of antibodies.

REFERENCES

de Alwis, W. U. and Wilson, G. S. (1985, Rapid Sub-Picomole Electrochemical Enzyme Immunoassay for Immunoglobulin G, Anal. Chem., 57, 2754-56.

de Alwis, W. U. and Wilson, G. S. (1986), Unpublished results.

Sittampalam, G., Wilson, G. S., and Byers, J. M. (1982), Characterization of Antigen-Enzyme Conjugates: Theoretical Considerations for Rate Nephelometric Assays of Immunological Reactivities, Anal.Biochem., 122, 372-378,

Sportsman, J. R., Liddil, J. D. and Wilson, G. S. (1983), Kinetic and Equilibrium Studies of Insulin Immunoaffinity Chromatography, Anal. Chem., 55, 771-775.

Ternynck, T. and Avrameas, S. (1976), A New Method Using p-Benzoquinone for Coupling Antigens and Antibodies to Marker Substances, Ann. Immunol. (Paris), 127C, 197-208.

CONTRIBUTORS

AIZAWA, MASUO
Department of Bioengineering, Faculty of Engineering, Tokyo Institute of Technology, Ookayama, Meguro-Ku, Tokyo 152, Japan

ALEXANDER, PETER W.
Department of Analytical Chemistry, University of New South Wales, P.O. Box 1, Kensington, N.S.W., Australia 2033

de ALWIS, W. UDITHA
Department of Chemistry, University of Arizona, Tucson, Arizona 85721, USA

BOITIEUX, JEAN-LOUIS
Laboratoire de Technologie Enzymatique, U.T.C. 60206 Compiegne, Cedex, France

BROYLES, C.A.
Department of Chemistry, University of Delaware, Newark Delaware 19716, USA

CONNELL, GEORGE R.
Department of Physiology, University of Nevada, School of Medicine, Reno, Nevada 89557, USA

DESMET, GERARD
Laboratoire d'Hormonologie, Hopital-Sud CHU d'AMIENS, 80036 AMIENS Cedex, France

D'ORAZIO, PAUL
Ciba-Corning Diagnostics Corp. 63 North Street, Medfield, MA 02052, USA

DOYLE, MATTHEW J.
The Procter and Gamble Company, Miami Valley Laboratories, P.O. Box 39175, Cincinnati, OH 45247, USA

GEBAUER, CARL R.
Technicon Instruments Corporation, 511 Benedict Avenue, Tarrytown, New York 10591, USA

GUILBAULT, GEORGE G.
Department of Chemistry, University of New Orleans, New Orleans, Louisiana 70148, USA

HAGA, MAKOTO
Faculty of Pharmaceutical Sciences, Science University of Tokyo 12, Ichigaya Funagawara-machi, Shinjuku-Ku, Tokyo 162, Japan

HALSALL, H. BRIAN
Department of Chemistry, University of Cincinnati, Cincinnati, OH 45221, USA

HEINEMAN, WILLIAM R.
Department of Chemistry, University of Cincinnati, Cincinnati, OH 45221, USA

HILL, H. ALLEN O.
Inorganic Chemistry Laboratory, University of Oxford, South Parks Road, OX1 3QR, England
KARUBE, ISAO
Research Laboratory of Resources Utilization, Tokyo Institute of Technology, Nagatsuta-cho, Midori-Ku, Yokohama, 227, Japan
KEATING, M.Y.
Central Research and Development Department, DuPont Company, Wilmington, DE 19898
KRULL, ULRICH J.
Department of Chemistry, University of Toronto, Toronto, Ontario, Canada M5S 1A1
KUAN, SHIA S.
Food & Drug Administration, 4298 Elysian Fields Avenue, New Orleans, LA 70122, USA
MATTIASSON, BO
Department of Biotechnology, Chemical Center, P.O.Box 124, S-221 00 Lund, Sweden
McCLINTOCK, SAM A.
McGill University, Department of Chemistry, 801 Sherbrooke St. W., Montreal, Quebec, Canada H3A 2K6
NGO, THAT T.
Department of Developmental and Cell Biology, University of California at Irvine, Irvine, California 92717, USA
PURDY, WILLIAM C.
McGill University, Department of Chemistry, 801 Sherbrooke St. W., Montreal, Quebec, Canada H3A 2K6
RECHNITZ, G.A.
Department of Chemistry, University of Delaware, Newark, Delaware 19716, USA
SANDERS, KENTON M.
Department of Physiology, University of Nevada, School of Medicine, Reno, Nevada 89557, USA
SUZUKI, SHUICHI
Department of Environmental Engineering, Saitama Institute of Technology, 1690 Fusaiji, Okabe, Saitama 369-02, Japan
TAMIYA, EIICHI
Research Laboratory of Resources Utilization, Tokyo Institute of Technology, Nagatsuta-cho, Midori-ku, Yokohama, 227, Japan
TAUSKELA, JOSEPH
Department of Chemistry, University of Toronto, Toronto, Ontario, Canada M5S 1A1
THOMAS, DANIEL
Laboratoire de Technologie Enzymatique, U.T.C. 60206 Compiegne, Cedex, France
THOMPSON, MICHAEL
Department of Chemistry, University of Toronto, Toronto, Ontario, Canada M5S 1A1
WALTON, NICHOLAS J.
Inorganic Chemistry Laboratory, University of Oxford, South Parks Road, Oxford, OX1 3QR, England
WEHMEYER, KENNETH R.
The Procter and Gamble Company, Miami Valley Laboratories, P.O. Box 39175, Cincinnati, Ohio 45247, USA

WILSON, GEORGE S.
Department of Chemistry, University of Arizona, Tucson, Arizona 85721, USA

WRIGHT, D. SCOTT
Warner Lambert, Pharmaceutical Research Division, 2800 Plymouth Road, Ann Arbor, Michigan 48105, USA

INDEX

GPSR Compliance

The European Union's (EU) General Product Safety Regulation (GPSR) is a set of rules that requires consumer products to be safe and our obligations to ensure this.

If you have any concerns about our products, you can contact us on ProductSafety@springernature.com

In case Publisher is established outside the EU, the EU authorized representative is:

Springer Nature Customer Service Center GmbH
Europaplatz 3
69115 Heidelberg, Germany

Batch number: 10762211

Printed by Printforce, the Netherlands